# GENETIC CONTROL OF AUTOIMMUNE DISEASE

We wish to express our grateful appreciation to the following for generous support:

Abbott Laboratories
Calbiochem-Behring Corporation
Difco Laboratories
Merck Sharp & Dohme
Ortho Diagnostics
Technicon Instruments Corporation
The Upjohn Company
National Institutes of Health:
    Fogarty International Center
    National Cancer Institute
    National Institute of Child Health and Human Development

*Workshop on Genetic Control of Autoimmune Disease, Bloomfield Hills, Mich.*

# GENETIC CONTROL OF AUTOIMMUNE DISEASE

Proceedings of the Workshop on Genetic Control of Autoimmune Disease held in Bloomfield Hills, Michigan, U.S.A. on July 10-12, 1978

*Edited by*

NOEL R. ROSE
*Department of Immunology and Microbiology*
*Wayne State University School of Medicine, Detroit, Michigan*

PIERLUIGI E. BIGAZZI
*Department of Pathology*
*University of Connecticut School of Medicine, Farmington, Connecticut*

NOEL L. WARNER,
*Departments of Pathology and Medicine*
*University of New Mexico School of Medicine, Albuquerque, New Mexico*

ELSEVIER/NORTH-HOLLAND
NEW YORK • AMSTERDAM • OXFORD

RC600
W67
1978

Published by:

Elsevier North Holland, Inc.
52 Vanderbilt Avenue, New York, New York 10017

Sole distributors outside U.S.A. and Canada:

Elsevier/North-Holland Biomedical Press
Jan Van Galenstraat 335, Box 211
Amsterdam, The Netherlands

Library of Congress Cataloging in Publication Data

Workshop on Genetic Control of Autoimmune Disease,
      Bloomfield Hills, Mich., 1978.
      Genetic control of autoimmune disease.

      Bibliography: p.
      Includes index.
      1. Autoimmune diseases — Genetic aspects — Congresses. 2. Autoimmune
      diseases — Animal models — Congresses. 3. Immunogenetics — Congresses.
      I. Rose, Noel R. II. Bigazzi, Pierluigi E. III. Warner, Noel Lawrence,
      1934–   IV. Title. [DNLM: 1. Autoimmune diseases — Familial and genetic —
      Congresses. WD305 W926g]
RC600.W67 1978      616.9'78      78-10135
ISBN 0-444-00297-9

Manufactured in the United States of America

# Contents

# Preface

In humans, evidence that genetic abnormalities play a role in the etiology of autoimmune diseases stems mainly from two kinds of studies. For quite some time, astute clinicians have noticed that certain autoimmune diseases clustered with unexpected frequency in some families. When objective markers of autoimmune disturbance became available in the form of tests for autoantibodies, a statistically significant increase in autoimmunity was documented even in clinically unaffected relatives of autoimmune patients. The other line of evidence has come from investigations relating autoimmune and other diseases with genetic cell-surface markers, especially HLA haplotypes. While rarely firm enough to be of practical diagnostic value, these associations hold great promise for defining disease-susceptibility genes and for distinguishing different varieties of the same disease.

The World Health Organization Collaborating Center for Autoimmune Disorders, located at Wayne State University, sponsored a consultation on genetic control of autoimmune disease in July, 1978, at the Cranbrook Educational Community, Bloomfield Hills, Michigan. The organizing committee consisted of Pierluigi E. Bigazzi, University of Connecticut; Noel L. Warner, University of New Mexico; and Noel R. Rose, Wayne State University. The purpose of the meeting was to see if recently generated information derived from human and animal investigations might suggest new approaches to the classification, diagnosis, and treatment of this important group of disorders. Clinicians interested in human disease were brought together with investigators studying spontaneous and induced models of autoimmunity in animals. Together, they reviewed critically the evidence of genetic influence on human autoimmune disease and examined the possible mechanisms of genetic control revealed through experimental studies. From their reports emerged some new concepts of the mechanisms by which the body normally discriminates between self and non-self. These concepts are an extension of the newer understanding of immune regulation generally. Recognition of self-antigens based on complementary receptors on lymphocytes is clearly demonstrable, and self-reactive lymphocytes can readily be induced to proliferate *in vitro* as well as *in vivo* under proper conditions. Two active controlling mechanisms normally seem to prevent proliferation of self-reactive clones of lymphocytes. The first is production of anti-receptor or anti-idiotype antibody; the second depends upon generation of regulatory T cells. The role of the major histocompatibility antigens in restricting cellular interactions was also discussed.

For those in attendance, the meeting was something of a scientific landmark. An opportunity now seems to be at hand not only to diagnose autoimmune disease in the human more accurately and to understand its pathogenetic mechanisms, but also to get at the etiological factors that trigger the disorders. Active intervention to interrupt specifically the cycle of autoimmunization and to arrest its progress has become a definite possibility for the first time, based on evidence that suppressor cells and factors can be generated *in vitro*. A new chapter in the application of immunological knowledge to the treatment of autoimmune disorders may be anticipated, growing in part out of the stimulating discussions held under the elms of Cranbrook.

To the many investigators from around the world who gave generously of their time and their knowledge, the WHO Collaborating Center expresses its gratitude. Thanks are also due to the members of the immunological community in Detroit who served as hosts for the meeting, and to Elsevier North-Holland for promptly publishing the proceedings.

Noel R. Rose  
Pierluigi E. Bigazzi  
Noel L. Warner

# List of Participants

| | |
|---|---|
| Abplanalp, H. | University of California, School of Medicine, Davis, California 95616 |
| Anders, R.F. | The Walter and Eliza Hall Institute of Medical Research, Royal Melbourne Hospital, Victoria 3050, Australia |
| Arnon, Ruth | The Weizmann Institute of Science, Rehovot 76100, Israel |
| Bacon, Larry D. | Regional Poultry Research Laboratory, U.S. Department of Agriculture, SEA, East Lansing, Michigan 48823 |
| Baloyannis, Stavros J. | University of Pennsylvania, School of Medicine, Philadelphia, Pennsylvania 19104 |
| Benacerraf, Baruj | Harvard Medical School, Boston Massachusetts 02115 |
| Benedict, A.A. | University of Hawaii, Honolulu, Hawaii 96822 |
| Bernoco, Domenico | University of California, School of Medicine, Los Angeles, California 90024 |
| Bigazzi, Pieluigi E. | University of Connecticut Health Center, Farmington, Connecticut 06032 |
| Binz, Hans | Zurich University, Zurich, Switzerland |
| Burakoff, Steven J. | Harvard Medical School, Boston, Massachusetts 02115 |
| Cantor, Harvey | Harvard Medical School, Boston, Massachusetts 02115 |
| Christadoss, Premkumar | Mayo Clinic, Medical School, Rochester, Minnesota 55901 |
| Christy, M. | Steno Memorial Hospital of Copenhagen, Denmark |
| Chused, Thomas M. | National Institutes of Health, Bethesda, Maryland 20014 |
| Coffey, M.F. | The Walter and Eliza Hall Institute of Medical Research, Royal Melbourne Hospital, Victoria 3050, Australia |
| Cohen, Irun R. | The Weizmann Institute of Science, Rehovot 76100, Israel |
| Cole, Randall K. | Cornell University, Ithaca, New York 14850 |
| Cooper, P.C. | The Walter and Eliza Hall Institute of Medical Research, Royal Melbourne Hospital, Victoria 3050, Australia |
| Datta, Syamal K. | Tufts University School of Medicine, Boston, Massachusetts 02111 |
| David, Chella S. | Mayo Clinic, Medical School, Rochester, Minnesota 55901 |
| Davidson, Wendy F. | National Institutes of Health, Bethesda, Maryland 20014 |
| Dupont, Bo | Sloan-Kettering Institute for Cancer Research, New York, New York 10021 |
| Eklund, J. | University of California, School of Medicine, Davis, California 95616 |
| Elrehewy, Mostafa | Assiut University, Assiut, Egypt |
| Engleman, Edgar G. | Stanford University, School of Medicine, Stanford, California 94305 |

| | |
|---|---|
| Erickson, K. | University of Hawaii, Honolulu, Hawaii 96822 |
| Flier, Jeffrey S. | National Institutes of Health, Bethesda, Maryland 20014 |
| Frischknecht, Hannes | Zurich University, Zurich, Switzerland |
| Gasser, David L. | University of Pennsylvania, School of Medicine, Philadelphia, Pennsylvania 19174 |
| Gershon, Richard K. | Yale University, School of Medicine, New Haven, Connecticut 06512 |
| Gershwin, M.E. | University of California, School of Medicine, Davis, California 95616 |
| Gibofsky, A. | The Rockefeller University, New York, New York 10021 |
| Giraldo, Alvaro A. | Wayne State University, School of Medicine, Detroit, Michigan 48201 |
| Gonatas, Nicholas K. | University of Pennsylvania, School of Medicine, Philadelphia, Pennsylvania 19104 |
| Hansen, Carl T. | National Institutes of Health, Bethesda, Maryland 20014 |
| Harrison, Len C. | National Institutes of Health, Bethesda, Maryland 20014 |
| Horowitz, Mark | Yale University, School of Medicine, New Haven, Connecticut 06512 |
| Hurtenbach, Ursula | Max-Planck-Institut fur Immunbiologie, D-78 Freiburg-Zahringen, West Germany |
| Ikeda, R.M. | University of California, School of Medicine, Davis, California 95616 |
| Indrieri, R.J. | University of California, School of Medicine, Davis, California 95616 |
| Irvine, W. James | Royal Infirmary of Edinburgh, Edinburgh EH3 9YW Scotland, United Kingdom |
| Jarrett, David B. | National Institutes of Health, Bethesda, Maryland 20014 |
| Jepsen, Peter | Washington University, School of Medicine, St. Louis, Missouri 63110 |
| Johnson, Armead H. | National Institutes of Health, Bethesda, Maryland 20014 |
| Kahn, C. Ronald | National Institutes of Health, Bethesda, Maryland, 20014 |
| Kemp, John D. | Yale University, School of Medicine, New Haven, Connecticut 06512 |
| Kincade, Paul W. | Sloan-Kettering Institute for Cancer Research, Rye, New York 10580 |
| Kohler, Heinz | La Rabida Children's Hospital and Research Center, University of Chicago, Chicago, Illinois 60649 |
| Kong, Yi-chi M. | Wayne State University, School of Medicine, Detroit, Michigan 48201 |
| Kunkel, H.G. | The Rockefeller Univerity, New York, New York 10021 |
| Kyriakos, Michael | Washington University, School of Medicine, St. Louis, Missouri 63110 |
| Lando, Zeev | The Weizmann Institute of Science, Rehovot 76100, Israel |

Lennon, Vanda      Mayo Clinic, Rochester, Minnesota 55901

Levine, Lenore S.      New York Hospital-Cornell Medical Center, New York, New York 10021

Mackay, Ian R.      The Walter and Eliza Hall Institute of Medical Research, Royal Melbourne Hospital, Victoria 3050, Australia

Mann, Dean L.      National Institutes of Health, Bethesda, Maryland 20014

Montero, J.      University of California, School of Medicine, Davis, California 95616

Morse, Herbert C.      National Institutes of Health, Bethesda, Maryland 20014

Moutsopoulos, H. M.      National Institutes of Health, Bethesda, Maryland 20014

Muggeo, Michele      National Institutes of Health, Bethesda, Maryland 20014

Mullen, Helen Braley      University of Missouri-Columbia, Columbia, Missouri 65201

Murphy, Donal B.      Yale University, School of Medicine, New Haven, CT 06512

Murphy, Edwin D.      The Jackson Laboratory, Bar Harbor, Maine 04609

Nerup, J.      Steno Memorial Hospital of Copenhagen, Denmark

Nun, Avraham Ben      The Weizmann Institute of Science, Rehovot, 76100 Israel

O'Neill, Geoffrey J.      Sloan-Kettering Institute for Cancer Research, New York, New York 10021

Palmer, Anthony C.      School of Veterinary Medicine, University of Cambridge, England

Patarroyo, M.E.      The Rockefeller University, New York, New York 10021

Peterson, Per      Uppsala University, Uppsala, Sweden

Pflugfelder, Christina      University of California, School of Medicine, Davis, California 95616

Platz, P.      Tissue-Typing Laboratory of the State University Hospital, Rigshospitalet, Copenhagen, Denmark

Pollack, Marilyn S.      New York Hospital-Cornell Medical Center, New York, New York 10021

Polley, Carl R.      Wayne State University, School of Medicine, Detroit, Michigan 48201

Porter, David D.      University of California, School of Medicine, Los Angeles, California 90024

Prester, Marlot      Max-Planck-Institut fur Immunbiologie, D-78 Freiburg-Zahringen, West Germany

Quimby, Fred      Tufts University School of Medicine, Boston, Massachusetts 02111

Roitt, Ivan M.      Middlesex Hospital, Medical School, London W1P 9PG, England

Rose, Noel R.      Wayne State University, School of Medicine, Detroit, Michigan 48201

| | |
|---|---|
| Roth, Jesse | National Institutes of Health, Bethesda, Maryland 20014 |
| Roths, John B. | The Jackson Laboratory, Bar Harbor, Maine 04609 |
| Roubinian, J.R. | University of California, Veterans Administration Hospital, San Francisco, California 94121 |
| Ryder, L.P. | Tissue-Typing Laboratory of the State University Hospital, Rigshospitalet, Copenhagen, Denmark |
| Schwartz, Janine Andre | Tufts University, School of Medicine, Boston, Massachusetts 02111 |
| Schwartz, Robert S. | Tufts University, School of Medicine, Boston, Massachusetts 02111 |
| Sege, Karin | Uppsala University, Uppsala, Sweden |
| Sharp, Gordon C. | University of Missouri-Columbia, Columbia, MO 65212 |
| Sharrow, Susan O. | National Institutes of Health, Bethesda, Maryland 20014 |
| Shultz, Leonard D. | The Jackson Laboratory, Bar Harbor, Maine 04609 |
| Svejgaard, A. | Tissue-Typing Laboratory of the State University Hospital, Rigshospitalet, Copenhagen, Denmark |
| Swanborg, Robert H. | Wayne State University, School of Medicine, Detroit, Michigan 48201 |
| Tait, Brian D. | The Walter and Eliza Hall Institute, Royal Melbourne Hospital, Victoria 3050, Australia |
| Talal, N. | University of California, Veterans Administration Hospital, San Francisco, California 94121 |
| Tam, L. | University of Hawaii, Honolulu, Hawaii 96822 |
| Teitelbaum, Dvora | The Weizmann Institute of Science, Rehovot 76100, Israel |
| Terasaki, Paul I. | University of California, School of Medicine, Los Angeles, California 90024 |
| Thomsen, M. | Tissue-Typing Laboratory of the State University Hospital, Rigshospitalet, Copenhagen, Denmark |
| Volpe, Robert | University of Toronto, The Wellesley Hospital, Toronto, Ontario M4Y 1J3 |
| Warner, Noel L. | University of New Mexico, School of Medicine, Albuquerque, New Mexico 87131 |
| Wekerle, Hartmut | Max-Planck-Institut fur Immunbiologie, D-78 Freiburg-Zahringen, West Germany of Germany |
| Welch, Andrew M. | Wayne State University, School of Medicine, Detroit, Michigan 48201 |
| Wigzell, Hans | University of Uppsala, Uppsala, Sweden |
| Winchester, R.J. | The Rockefeller University, New York, New York 10021 |
| Yang, Soo Young | Sloan-Kettering Institute for Cancer Research, New York, New York 10021 |
| Zabriskie, J. | The Rockefeller University, New York, New York 10021 |
| Zurier, Robert B. | University of Connecticut Health Center, Farmington, Connecticut 06032 |

# PART I
# HUMAN DISEASE
Pierluigi E. Bigazzi *and* Ivan M. Roitt, *Co-Chairmen*

# HLA-D AND DISEASE

DOMENICO BERNOCO AND PAUL I. TERASAKI
Department of Surgery, UCLA School of Medicine, University of California, Los
Angeles, California

ABSTRACT

   The associations that have been reported for the HLA-D and -DR loci are summarized. From the overall data accumulated to date, it appears that the D locus is
indeed closer to the postulated disease susceptibility locus in diseases such
as in the DW2-associated multiple sclerosis (MS) and the DW3-associated autoimmune
diseases previously linked with HLA-B8. DW4 is associated with rheumatoid arthritis
which had not previously been associated with HLA-ABC.

   The HLA-D locus, however, is probably not unique in being associated with
diseases since the B27-ankylosing spondylitis association and A3-idiopathic
hemochromatosis and CW5-psoriasis association is higher than those noted for the
HLA-D locus. Thus, if disease susceptibility genes exist, they are probably
spread along the HLA complex.

   The HLA-D association with disease also suggests that we are dealing with
disease susceptibility genes that are in linkage disequilibrium with different HLA
specificities. This means that in different populations the linkage will be
different and that the evolutionary history of disease susceptibility genes might
be traced through the disequilibria in different populations.

INTRODUCTION

   There has been great interest in the possibility that HLA-D may be highly
linked to disease susceptibility genes because of the analogy of the HLA-D locus
with the mouse Ia locus. Perhaps the greatest deterrent to this work has been the
technical difficulty in performing the tests because the earlier work on HLA-D was
done with MLC typing using homozygous cells. The difficulties in classifying the
cells by this reaction are well known. The more recent method in the last few years
has been by serologic typing of B lymphocytes. Because the technique and reagents
are just being established, sharp definition of each specificity has been difficult.

   We have attempted to list most of the current HLA-D studies with disease associations that were readily available to us. Without going into detail, the data of
percent positive for the patients as compared with the controls are given in the
tables. For purposes of this discussion, the D and DR specificities will be
considered to be operationally equivalent.

### DW2

Perhaps the most thoroughly studied disease for the D locus is multiple sclerosis (MS). Over 1,000 patients have been tested for these antigens, and there now seems to be no doubt that DRW2 and DW2 are of statistically higher frequency in MS patients (Table 1). It has, however, been somewhat disappointing that the relative risk connected with this specificity has proved to be in the range of about 3. Some of the earlier studies had indicated a higher relative risk. but this figure has been lower in more recent studies. The relative risk, however, is higher than the previously found relative risks for HLA-A3 and B7.

It is interesting that DW2 is not significantly associated with MS in the Shetland and Orkney Islands, where the incidence of MS is the highest in the world. It should be noted, however, that the frequency of DW2 is also extremely high in the control population in these islands. Conceivably, the disease susceptibility gene is of older origin in this population and multiple crossovers may have destroyed the original linkage disequilibrium. On this basis one would expect the linkage disequilibrium to be the weakest in the population where the mutation first occurred and to be stronger when there had been a recent mixture into a new population that did not have the susceptibility genes.

In this connection it is interesting that, in Israel, an association of HLA-A3, B7, or DW2 with MS could not be found.[27] Thus, the association seems to be present in most Caucasians and those who had migrated to the United States, Australia, and Canada. Among Japanese, there is also no association with B7 or DW2.[28]

In Negroes an association of MS and optic neuritis has been noted.[7]

There is a weak association of DW2 with idiopathic hemochromatosis but it should be noted that in this instance a higher association with HLA-A3 and B7 is found. Thus, in this particular disease, the disease susceptibility gene is in fact not as closely linked to the D locus as it is to the A locus.

### DW3, DRW3

As shown in Table 2, the diseases that are associated with these antigens are those that have also previously been associated with HLA-B8. It is interesting that the relative risk for each of these diseases with DW3 is somewhat higher than it is with HLA-B8, indicating that the disease susceptibility gene must be closer to the D locus. In some diseases as many as 90% of the patients had the DRW3 antigens. If these figures could be confirmed, the DW3 test could become useful in diagnosis.

As we suggested earlier, this group of diseases provides the best evidence that an autoimmunity gene must be genetically linked to the B8-DW3 haplotype. Either a second gene exists that determines which target organs are attacked or environmental exposure determines which of the organs would be the target of autoimmune disturbance. It will be important to determine whether pedigrees would have multiple cases of the different diseases within this grouping of autoimmune diseases.

The diseases associated with DW3 also point to the important role of sex in determining disease susceptibility. Although the DRW3 antigen occurs in equal frequencies in males and females, there is a marked preponderance of females in these diseases. Whether this susceptibility is at the genetic or endocrine levels remains to be determined.

## DW4

The association of DW4 with rheumatoid arthritis found by Stastny[21] is unique as of this date in that an HLA-D association is found despite the absence of any association with HLA-A, B, and C loci antigens. The confirmation at the Oxford workshop and other laboratories seems to establish that this association is valid (Table 3). It will now be important to determine whether juvenile rheumatoid arthritis falls in this category.or is associated with another antigen TMo.[19]

Juvenile onset diabetes seems to have a higher incidence of DW4 with the relative risk of 3 to 4. This disease also has an association with DW3 as shown above.

## Other DRW associations

Other associations with DRW antigens are given in Table 4. Many of them remain to be confirmed by other studies. As techniques for determination of the D specificities improve and as larger numbers of specificities are added, new associations will undoubtedly emerge. The field is currently engaged in the refinement of the specificities.

## Genetic linkage

Relatively few family studies have been done as of this date. From studies of juvenile onset diabetes families, Rubenstein and Suciu-Foca have concluded that a recessive gene with 50% penetrance linked to HLA is involved.[29] These studies, as well as haplotype linkge analysis,[30] should aid in the localization of the postulated disease susceptibility genes and their rate of expression. It seems likely that the linkage disequilibrium of susceptibility genes to HLA-D and other HLA antigens would depend on the time since the original mutation[31] and the distance of the gene from the HLA loci.

TABLE 1
DW2 (DRW2)[a]

| Disease | Group Studied | Patients Total | % | Controls Total | % | Relative Risk | Probability | Reference |
|---|---|---|---|---|---|---|---|---|
| Multiple Sclerosis | Cauc* | 56 | 84 | 72 | 33 | 10.4 | $5\times10^{-5}$ | Terasaki, et al. (1976)[1] |
| Multiple Sclerosis | Cauc | 59 | 83 | 30 | 33 | 9.8 | $5\times10^{-4}$ | Compston, et al. (1976)[2] |
| Multiple Sclerosis | Cauc | 734 | 47-70 | 1095 | 15-31 | 4.3 | $1\times10^{-10}$ | Svejgaard and Ryder (1977)[3] |
| Multiple Sclerosis | Cauc | 121 | 47 | 98 | 20 | 3.5 | $1\times10^{-5}$ | Paty, et al.(1977)[4] |
| Multiple Sclerosis | Cauc | 139 | 53 | 156 | 21 | 4.2 | $1\times10^{-6}$ | Oxford (1977)[5] |
| Multiple Sclerosis | Cauc (N.Eur.) | 134 | 41 | 332 | 22 | 2.5 | $1\times10^{-5}$ | Oxford (1977)[5] |
| Multiple Sclerosis | Cauc*(Aus.) | 28 | 52 | 332 | 22 | 3.8 | $1\times10^{-4}$ | Oxford (1977)[5] |
| Multiple Sclerosis | Cauc*(Can.) | 22 | 50 | 294 | 28 | 2.6 | .01 | Oxford (1977)[5] |
| Multiple Sclerosis | Cauc(Orkney) | 41 | 44 | 78 | 40 | 1.2 | .40 | Poskanzer, et al.(in press)[6] |
| | Total Cauc. | 1334 | 41-83 | 2487 | 15-40 | | | |
| Multiple Sclerosis | Black | 31 | 35 | 34 | 0 | Indef. | $1\times10^{-4}$ | Dupont, et al.(1976)[7] |
| Optic Neuritis | Cauc | 84 | 40-50 | 232 | 18-30 | 2.9 | $7\times10^{-5}$ | Svejgaard and Ryder (1977)[3] |
| Idiopathic Hemochromatosis | Cauc | 75 | 45 | 156 | 21 | 3.1 | $1\times10^{-4}$ | Oxford (1977)[5] |
| Idiopathic Hemochromatosis | Cauc* | 132 | 32 | 332 | 22 | 1.7 | .03 | Oxford (1977)[5] |

[a] DRW2 studies are indicated by * in "Group" column.

TABLE 2
DW3 (DRW3)[a]

| Disease | Race | Patients Total | % | Controls Total | % | Relative Risk | Probability | Reference |
|---|---|---|---|---|---|---|---|---|
| Juvenile Onset Diabetes | Cauc | 42 | 50 | 157 | 21 | 3.8 | $3 \times 10^{-4}$ | Thomsen, et al. (1975)[8] |
| Insulin Dependent Diab. | Cauc* | 293 | 27 | 384 | 17 | 1.8 | $2 \times 10^{-3}$ | Oxford (1977)[5] |
| Insulin Dependent Diab. | Cauc* | 123 | 59 | 176 | 14 | 9.16 | $2 \times 10^{-4}$ | de Moerloose, et al.(1978)[9] |
| Insulin Dependent Diab. | Japn | 34 | 47 | 123 | 12 | 6.4 | $1 \times 10^{-5}$ | Oxford (1977)[5] |
| Myasthenia Gravis | Cauc | 110 | 25-36 | 164 | 14-19 | 2.3 | $6 \times 10^{-3}$ | Svejgaard and Ryder (1977)[3] |
| Myasthenia Gravis | Cauc* | 106 | 30 | 340 | 17 | 2.1 | $5 \times 10^{-3}$ | Oxford (1977)[5] |
| Chronic Hepatitis | | 38 | 68 | 91 | 24 | 6.8 | $3 \times 10^{-6}$ | Opelz, et al.(1976)[10] |
| Chronic Hepatitis | Cauc* | 46 | 43 | 332 | 17 | 3.8 | $1 \times 10^{-5}$ | Oxford (1977)[5] |
| Dermatitis Herpetiformis | | 66 | 62-93 | 293 | 19-20 | 13.5 | $1 \times 10^{-10}$ | Svejgaard and Ryder (1977)[3] |
| Dermatitis Herpetiformis | * | 29 | 97 | 120 | 25 | 84 | $5 \times 10^{-3}$ | Solheim, et al.(1977)[11] |
| Thyrotoxicosis | Cauc | 112 | 52-54 | 202 | 16-21 | 4.4 | $5 \times 10^{-9}$ | Svejgaard and Ryder (1977)[3] |
| Idiopathic Addison's Disease | | 30 | 70 | 157 | 21 | 8.8 | $3 \times 10^{-7}$ | Thomsen, et al. (1975)[8] |
| Coeliac Disease | | 28 | 96 | 100 | 27 | 73 | $1 \times 10^{-10}$ | Keuning, et al.(1976)[12] |
| Sicca Syndrome | | 29 | 69 | 58 | 10 | 19 | $4 \times 10^{-8}$ | Ivanyi, et al.(1975)[13] |
| Juvenile RA | Cauc* | 21 | 95 | 99 | 49 | 20 | $5 \times 10^{-3}$ | Gershwin, et al.(1975)[14] |
| CLL | * | 476 | 20 | 1180 | 13 | 1.67 | .05 | Oxford (1977)[5] |
| Sjögren's Disease | | 19 | 74 | 96 | 24 | 8.9 | $1 \times 10^{-5}$ | Fye, et al.(1978)[15] |
| Sjögren's Disease | | 25 | 84 | 91 | 24 | 6.7 | $1 \times 10^{-5}$ | Chused, et al.(1977)[16] |
| Sjögren's Disease | | 29 | 69 | 58 | 10 | 18.9 | $1 \times 10^{-5}$ | Hinzova, et al.(1977)[17] |

[a] DRW3 studies are indicated by * in "Group" column.

TABLE 3
DW4 (DRW4)[a]

| Disease | Studied | Patients Total | % | Controls Total | % | Relative Risk | Probability | Reference |
|---|---|---|---|---|---|---|---|---|
| Rheumatoid Arthritis | Cauc | 38 | 45 | 157 | 17 | 3.9 | $6\times10^{-4}$ | Thomsen, et al.(1976)[18] |
| Rheumatoid Arthritis | Cauc(F) | 98 | 60 | 384 | 13 | 10 | $1\times10^{-6}$ | Oxford (1977)[5] |
| Rheumatoid Arthritis | Cauc(F) | 98 | 62 | 384 | 24 | 5.2 | $1\times10^{-6}$ | Oxford (1977)[5] |
| Rheumatoid Arthritis | Cauc(M) | 24 | 100 | 384 | 13 | Indef. | $1\times10^{-6}$ | Oxford (1977)[5] |
| Rheumatoid Arthritis | Cauc(M) | 24 | 50 | 384 | 24 | 3.17 | $9\times10^{-3}$ | Oxford (1977)[5] |
| Rheumatoid Arthritis | Cauc* | (?) | 59 | (?) | 16 | 7.5 | $1\times10^{-3}$ | Stastny and Fink (1977)[19] |
| Rheumatoid Arthritis | Mixed | 43 | 72 | 45 | 13 | 16.8 | $2\times10^{-8}$ | Stastny (1976)[20] |
| Rheumatoid Arthritis | Cauc | 80 | 54 | 69 | 16 | 6.1 | $1\times10^{-5}$ | Stastny (1978)[21] |
| Rheumatoid Arthritis | Cauc* | 53 | 70 | 68 | 28 | 6.0 | $1\times10^{-5}$ | Stastny (1978)[21] |
| Juvenile Onset Diabetes | Cauc | 79 | 42 | 157 | 19 | 3.5 | $6\times10^{-5}$ | Thomsen et al.(1975)[8] |
| Juvenile Onset Diabetes | Cauc | 293 | 38 | 384 | 13 | 4.07 | $1\times10^{-6}$ | Oxford (1977)[5] |
| Multiple Sclerosis | Cauc*(Ital) | 29 | 28 | 54 | 4 | 9.9 | $1\times10^{-3}$ | Oxford (1977)[5] |
| Chronic Active Hepatitis | Cauc* | 46 | 41 | 332 | 24 | 2.2 | $5\times10^{-3}$ | Oxford (1977)[5] |
| CLL | * | 476 | 31 | 1180 | 16 | 2.34 | $1\times10^{-6}$ | Oxford (1977)[5] |
| Myasthenia Gravis | * | 31 | 32 | 47 | 13 | 3.25 | .02 | Oxford (1977)[5] |

[a] DRW4 studies are indicated by * in "Group" column.

TABLE 4

MISCELLANEOUS SPECIFICITIES

| Disease | Specificity | Patients Total | Patients % | Controls Total | Controls % | Relative Risk | Probability | Reference |
|---|---|---|---|---|---|---|---|---|
| Subacute Thyroiditis | DW1 | 24 | 33 | 157 | 19 | 2.1 | .05 | Bech, et al. (1976)[22] |
| CLL | DRW6 | 476 | 26 | 1180 | 13 | 2.36 | $1\times10^{-5}$ | Oxford (1977)[5] |
| Myasthenia Gravis | M13 | 31 | 25 | 47 | 8 | 3.73 | .04 | Oxford (1977)[5] |
| Myasthenia Gravis | M15 | 31 | 16 | 47 | 2 | 8.8 | .02 | Oxford (1977)[5] |
| Juvenile RA | LDTMO | 100 | 25 | 60 | 0 | Indef. | $1\times10^{-3}$ | Stastny and Fink (1977)[19] |
| Chronic Merangiocapillary Glomerulonephritis | MCG3 | 13 | 77 | 24 | 17 | 16.6 | $5\times10^{-4}$ | Friend, et al. (1977)[23] |
| Psoriasis | DMA | 39 | 42 | 142 | 6 | 10.4 | $1\times10^{-6}$ | McMichael, et al.(1978)[24] |
| Glutensensitive Enteropathy | B1 | 16 | 81 | 37 | 0 | Indef. | $1\times10^{-6}$ | Mann, et al. (1976)[25] |
| Glutensensitive Enteropathy | W1 | 16 | 94 | 37 | 0 | Indef. | $1\times10^{-6}$ | Mann, et al.(1976)[25] |
| Dermatitis Herpetiformis | B1 | 19 | 79 | 37 | 0 | Indef. | $1\times10^{-6}$ | Mann, et al.(1976)[25] |
| Dermatitis Herpetiformis | W1 | 15 | 100 | 37 | 0 | Indef. | $1\times10^{-6}$ | Mann, et al.(1976)[25] |
| Rheumatoid Arthritis | M58 | 43 | 74 | 37 | 27 | 7.8 | $5\times10^{-4}$ | Panayi and Wooley (1977)[26] |
| Multiple Sclerosis (Ca. Italy) | DRW5 | 29 | 31 | 54 | 12 | 3.6 | .02 | Oxford (1977)[5] |
| Multiple Sclerosis (Japn) | DRW5 | 11 | 55 | 123 | 18 | 5.5 | $4\times10^{-3}$ | Oxford (1977)[5] |
| CLL | DRW5 | 476 | 21 | 1180 | 14 | 1.63 | .02 | Oxford (1977)[5] |

REFERENCES

1. Terasaki, P.I., Park, M.S., Opelz, G. and Ting, A. (1976) Science, 193, 1245.

2. Compston, D.A.S., Batchelor, J.R. and McDonald, W.I. (1976) Lancet, ii, 7998.

3. Svejgaard, A. and Ryder, L.P. (1977) in HLA and Disease, Dausset, J. and Svejgaard, A. eds., Williams and Wilkins, Baltimore, p. 46.

4. Paty, D.W., Cousin, H.K., Stiller, C.R., Boucher, D.W., Furesz, J., Warren, K.G., Marchuk, L., Dossetor, J.B. (1977) Transplant. Proc. 9, 1845.

5. Joint Report of the Seventh International Histocompatibility Workshop, in Histocompatibility Testing 1977, Bodmer, W., ed., Munksgaard, Copenhagen (in press).

6. Poskanzer, D.C., Terasaki, P.I., Park, M.S., Prenney, L.B. and Sheridan, J.L. (submitted for publication).

7. Dupont, B., Lisak, R.P., Jersilk, C., Hansen, J.A., Silberberg, D.H., Whitsett, C., Zwieman, B. and Ciongoli, K. Submitted to Transplant. Proc.

8. Thomsen, M., Platz, P.,, Ortved Anderson, O., Christy, M., Lyngsoe, J., Nerup, J., Rasmussen, K., Ryder, L.P., Nielsen, L.S. and Svejgaard, A. (1975) Transplant. Rev. 22, 125.

9. de Moerloose, P.H., Jeannet, M., Bally, C., Raffoux, C., Pointel, J.P. and Sizonenko, P. (1978) Brit. Med. J. April, 823.

10. Opelz, G., Terasaki, P., Myers, L., Ellison, G., Ebers, G., Zabriskie, J., Weiner, H., Kempe, H. and Sibley, W. (1977) Tissue Antigens, 9,54.

11. Solheim, B.G., Albrechtsen, D., Thorsby, E. and Thune, P. (1977) Tissue Antigens, 10, 114.

12. Keuning, J.J., Pena, A.S., Van Hooff, J.P., Van Leeuwen, A. and van Rood, J.J. (1976) Lancet, i, 506.

13. Ivanyi, D., Drizhal, I., Erbenova, E., Horej, J., Salavec, M., Macurova, H., Dostal, C., Balik, J. and Juran, J. (1975) Tissue Antigens, 7, 45.

14. Gershwin, E., Terasaki, P.I. and Castles, J.J. (1978) Tissue Antigens, 11, 71.

15. Fye, K.H., Terasaki, P.I., Michalski, J.P., Daniels, T.E., Opelz, G. and Talal, N. (1978) Arth. Rheum. 21, 337.

16. Chused, T.M., Kassan, S.S., Opelz, G., Moutsopoulos, H.M. and Terasaki, P.I. (1977) N. Engl. J. Med. 296, 895.

17. Hinzova, E., Ivanyi, D., Sula, K., Horejs, J., Dostal, C. and Drizhal, I. (1977) Tissue Antigens, 9, 8.

18. Thomsen, M., Sorensen, S.F., Platz, P., Ryder, L.P., and Svejgaard, A. (1976) (Copenhagen/Denmark).

19. Stastny, P. and Fink, C.W., (1977) Transplant. Proc. 9, 1863.

20. Stastny, P. (1976) J. Clin. Invest. 57, 1148.

21. Stastny, P., (1978) N. Engl. J. Med. 298, 869.

22. Bech, K., Nerup, J., Thomsen, M., Platz, P., Ryder, L.P., Svejgaard, A., Siersbaek-Nielsen, K. and Hansen, J. (1976) Molholm (Copenhagen/Denmark).

23. Friend, P.S., Noreen, H.J., Yunis, E.J. and Michael, A.F. (1977) Lancet, i, 562.

24. McMichael, A.J., Morhenn, V., Payne, R., Sasazuki, T. and Farber, E.M. (1978) Brit. J. Dermatol. 98, 287.

25. Mann, D.L., Nelson, D.L., Katz, S.I., Abelson, L.D., Strober, W. (1976) Lancet, i, 110.

26. Panayi, G.S. and Wooley, P.H. (1977) Ann. Rheum. Dis. 36, 365.

27. Brautbar, C., Cohen, I., Kahana, E., Alter, M., Jorgensen, F. and Lamin, L. (1977) Tissue Antigens, 10, 291.

28. Naito, S., Kuroiwa, K., Itoyama, T., Tsubaki, T., Horikawa, A., Sasazuki, T., Noguchi, S., Ohtsuki, S., Tukuami, H., Miyatake, T. and Kawanami, S. (1977) Tissue Antigens, 10, 191.

29. Rubinstein, P., Suciu-Foca, N. and Nicholson, J.F. (1977) N. Engl. J. Med. 297, 1036.

30. Terasaki, P.I. and Mickey, M.R. (1975) Transplant. Rev. 22, 105.

31. Terasaki, P.I. and Mickey, M.R. (1976) Neurology, 26, 56.

DISCUSSION

WARNER: Have you attempted to follow the inheritance pattern of the non-HLA B cell specific antigens to determine whether they are coded for a single gene?

TERASAKI: It must be remembered that these antibodies often react with cells from siblings who are HLA-identical to the antibody producer and even react to autologous cells. There is no clear inheritance pattern in families. Some children are positive with certain sera when both parents do not react with the serum.

TALAL: Many laboratories have found mostly antibodies to T cells in SLE and have suggested that these antibodies are selectively destructive for suppressor T cells. Have you studied the functional properties of antibodies to T or B cells?

TERASAKI: Yes, as we also have shown, SLE patients do have antibodies that react against T lymphocytes. In fact, it is unique among the diseases in having strong non-HLA antibodies to T lymphocytes. However, in common with the other diseases, the sera from SLE patients react more strongly against B lymphocytes than T lymphocytes.
To date, we have had variable evidence of suppression of MLC with these sera. Many sera block MLC, but not all. The specificity of blocking did not parallel cytotoxicity.

ROITT: Have you carried out studies at body temperature with lymphocytes plus these antibodies and complement to see whether actions other than cytolysis, e.g., phagocytosis by macrophages, may lead to elimination of cells in vivo?

TERASAKI: There does not seem to be a direct relationship between the number of circulating B lymphocytes and the titer of cytotoxic cold antibodies. However, Winfield and Kunkel have shown that there is an association between lymphopenia and cold IgM cytotoxins in SLE. We have not tested for opsinization or other antibody effects.

CANTOR: Can you separate SLE autoantibodies against lymphocytes into anti-T cell and anti-B cell fractions by appropriate absorptions with T and B cell lines? If so, is there a correlation between αT cell autoantibody titer and the clinical course of the disease?

TERASAKI: We have not done the specific absorptions you suggest, but part of the activity can be removed by absorption with even platelets or red cells. Any weakening of the antisera by such absorption or dilution results in antibodies reacting with B lymphocytes rather than T lymphocytes.
Other workers have noted an association of non-HLA antibodies in SLE with severity of disease.

MORSE: Has any work been done to determine the biochemical nature of the antigen(s) with which the "anti-B cell" sera are reactive?

TERASAKI: No, I am not aware of any work on the biochemical nature of the autoantigen. We are working on the hypothesis that the antigen might be some part of the immunoglobulin molecule which is attached to the B cell membrane.

SCHWARTZ: Is there any correlation between antibodies to B cells and presence of rheumatoid factor in the serum?

TERASAKI: Yes. Parallel tests for rheumatoid factor by latex agglutination and cold B lymphocyte cytotoxicity have shown a statistically significant positive correlation. However, it appears that the cold cytotoxin is not exactly rheumatoid factor since some activity remains after absorption with IgG. The distribution of the cytotoxin in different diseases is somewhat similar to rheumatoid factor in being most prevalent in autoimmune diseases and in females. It is only discrepant in being more frequent in SLE than in RA. The cold antibody is present at higher levels in aged persons as is rheumatoid factor.

SCHWARTZ: What hypothesis do you have about linkages between HLA and diseases like hemochromatosis and psoriases, in which there is no known immunological pathogenetic factor?

TERASAKI: As immunologists we tend to propose an immunological explanation to phenomena that may have nothing to do with immunology. Just as a gene for tail anomaly could be close to the H-2 genes in the mouse, a 'tail' anomaly gene for ankylosing spondylitis in man could be in close proximity to the HLA locus. Since there is room for hundreds of genes in the HLA region, it should not be surprising if some genes that ultimately result in susceptibility might be mixed into the HLA region.

WINCHESTER: Paul, in this context I would like to mention the earlier studies by Winfield and Kunkel that employed absorption with B or T cell lines. The preponderance of autoreactive antibodies in patients with SLE reacted with both B and T cells as well as other non-lymphoid cells. These absorption experiments, however, also provided clear evidence for antibodies with separate specificities for either T or B cells.

TERASAKI: Thank you for reminding me of these results. Our data indicate that T cell autoantibodies are extremely rare in diseases other than SLE, with most of the autoantibodies being directed at B lymphocytes. We also wish to stress that autoantibodies to the B lymphocytes are distinct from the HLA-D antibodies to B cells. They are IgM, react better in the cold, do not have clear specificities and appear after a wide variety of antigenic challenges.

MACKAY: You ascribed certain autoimmune diseases to a mutation within the major histocompatibility complex, but for maintenance of such a mutation some compensating survival advantage can be assumed. Would you comment on this?

TERASAKI: In many of the autoimmune diseases, natural selection may not play a role since their effects are often manifested after the reproductive period. In fact most of the diseases thus far linked to HLA are evolutionarily irrelevant. Life expectancy in primitive times was in the range of 30 years; consequently, some diseases are hidden genetic defects revealed only by artificial prolongation of life.

MACKAY: It is often assumed that diseases occurring after the reproductive period of life are irrelevant in terms of survival advantage. However, biological advantage could be assumed for the child of healthy surviving parents.

TERASAKI: That is a good concept, which I had not thought of. Of course, these diseases are sufficiently rare that it would be difficult to find instances of both parents dying of the disease. Perhaps even in early times a child could survive if one parent were left to help with his survival needs.

ROSE: There may be some analogy between the cold-reactive B cell antibodies you have described and cold-reactive blood-group-specific alloantibodies. Have you studied lymphocytes of individual donors to see if some rare individuals may lack a putative B cell alloantigen?

TERASAKI: We have applied perhaps the most critical test of whether the cold B cell antibodies might have the HLA-D specificity, that is, to see whether the specificities "fit" at alleles with the HLA-D locus. They do not have correlation with the known DRw1-8 specificities nor do they react with the "blank" cells as you suggest. Moreover, we had initially grouped the cold B cell cytotoxins into a group called "group 2", outside the main D locus. This group of sera have a positive association with each other and vary in percentage of reactivity from 5% to 100%.

SVEJGAARD: Returning to the question of natural selection, I agree with Paul Terasaki that none of the diseases so far found associated with HLA have had large impacts on our present genetic constitution. Large infectious diseases, often great epidemics, have been much more important in this respect. It may be worth noting that Van Rood and co-workers have recently found evidence that HLA is involved in the development of leprosy. Concerning the autoimmune diseases, we have not encountered the Dw2 antigen in a sample of 118 unrelated patients with insulin-dependent diabetes, a disease which previously certainly was characterized by low fitness (fertility). This very same antigen is associated with susceptibility to multiple sclerosis, i.e., a gene may protect against one disease and confer susceptibility to another.

TERASAKI: Yes. This phenomenon may be related to that of immunologic perspective. If the Dwz-linked Ir gene product is composed of amino acids XYZ, a microorganism ABC can be readily reacted against, whereas one with YXZ could not be recognized. In this view, no one can be immune to everything and everyone will have his vulnerable blind spots. The evolution of the HLA complex may have been an attempt to minimize the chances of mimicry by pathogens.

GERSHWIN: Aging has been attributed to immunologic phenomena. Are there any alterations of HLA allele frequencies with age and is there any evidence that such influences may suggest a critical relationship between the HLA system and evolution?

TERASAKI: Again, aging as we understand it today, may be an artifact of modern civilization. There has probably not been any natural selection for longevity in excess of the period of reproduction and child care. Yet it is tempting to think that if all those with disease susceptibility genes die off, we would be left with only the fit survivors if those who are 80 years old or more are tested. Our group and Yunis et al. tested aged persons for their HLA frequencies; in both series, no single HLA type was found to be significantly increased or decreased as compared to healthy normal persons 20-30 years of age.

MORSE: In studies of xenotropic murine leukemia viruses in mice, we have shown that the gp70 antigen is found in higher levels on B cells than T cells; is variably expressed by various inbred strains; and some strains spontaneously produce antibodies against this determinant whereas others do not.

TERASAKI: It would be extremely interesting if an analogous antigen were present on human B lymphocytes. Thus far we have not noted a difference in reactivity of lymphocytes from patients with SLE, R.A., normal or cold blood. We should note that leukemic lymphocytes tend not to react with the cold B cell antibody (Billings et al., Lancet).

GENETIC LINKAGE OF DISEASE-GENES TO HLA

BO DUPONT, GEOFFREY J. O'NEILL, SOO YOUNG YANG,
MARILYN S. POLLACK, AND LENORE S. LEVINE

Sloan-Kettering Institute for Cancer Research,
and New York Hospital-Cornell Medical Center, New York, New York  10021

INTRODUCTION

It is now well-established that a broad variety of diseases of multifactorial
nature are associated with certain HLA determinants (for review see ref.[1]).
These studies of HLA-disease associations have provided new insights into genetic
host factors involved in the etiology and pathogenesis of disease.  It is impressive to see that so many diseases of very different nature are associated with
HLA.  However, no common denominator can be identified at present which would
account for these HLA disease associations.  In fact, the apparent diversity
among HLA associated diseases indicates that no single mechanism is likely to
be ·found which will provide a fundamental biological explanation.  One conclusion
can, however, be made, namely that a number of different HLA linked genes are
involved in resistance or susceptibility of the host to disease development.
Moreover it is now possible to identify diseases  in which at least one of the
important genetic factors is HLA linked.  Diseases such as Multiple Sclerosis,
Juvenile Diabetes Mellitus, Graves Disease, Psoriasis and Coeliac Disease,for
example,have been known to have a major genetic component and it is now
recognized that one of the genetic factors in each of these diseases is HLA-
linked.  This has provided new approaches for the continued study of pathogenic
and etiological factors in these diseases.

The H-2 complex of the mouse provides an excellent model for the analysis of
the HLA complex in man.  The classical observations of Lilly et al.[2] that
susceptibility to murine leukemogenesis is dependent on determinants within the
H-2 complex and the reports by McDevitt[3] and Benacerraf[4] that the H-2 complex
contains immune response (Ir) genes, provided the basic concept that HLA-disease
associations could be explained by assuming the presence or absence of particular
Ir genes in patients with some HLA associated diseases.  Studies by Doherty and
Zinkernagel[5,6] demonstrating an H-2 restriction of cytotoxic T-lymphocytes in
response to virus infection and other reports that T-helper cell functions are
associated with H-2 linked genes suggest that cell mediated immunity in man is
also strongly associated with HLA linked genes.  However,the wide variety of
HLA associated diseases suggests that immunological factors alone cannot account
for all known HLA-disease associations.

The observation that the H-2 system influences total complement levels by Hinzová et al.[7] in 1972 and subsequently, that the S-region contains the gene coding for C4 (Démant et al. 1973[8], Lachmann et al. 1975[9], Curman et al. 1975[10], Meo et al. 1975[11]) prompted similar studies of human complement components and HLA. It is now established, that the genes coding for C2, C4 and Bf are closely linked to HLA. This has provided a new dimension for the possible explanation of HLA-disease associations and demonstrates that genes different from those controlling T- and B-lymphocyte function are present within the HLA complex which may account for the deviation in homeostatic function leading to disease.

In this paper we summarize these and other recent findings which indicate that the HLA complex contains certain genes which are not of primary importance in immunological mechanisms but which may, however, exert important regulatory functions in the maintenance of homeostasis.

## The H-2 Complex and Physiological and Pathological Traits

Several physiological parameters and quantitative traits of non-immunological nature have recently been found to be influenced by the H-2 complex. Iványi & Michová (1971)[12] have demonstrated, for example, that H-2 haplotypes influence vesicular gland weight and that also testis and thymus weight are H-2 dependent (Iványi et al. 1972a)[13]. These investigators demonstrated subsequently that serum testosterone levels and levels of testosterone binding globulin are H-2 dependent (Iványi et al. 1972b[14], 1976[15]). H-2 haplotypes were found to also influence target organ sensitivity to testosterone (Michová & Iványi, 1975)[16].

Meruelo & Edidin (1975)[17] demonstrated that cAMP levels in liver cells are H-2 dependent. Since steroid hormones that are incorporated into cells can induce profound effects on cell function, it is possible that the H-2 system exerts an important non-immunological regulatory function on target cells via testosterone and cAMP.

Recently, Goldmann et al. (1977)[18] found that the level of a cytosol cortisol-binding protein was regulated by the H-2 system. This finding again implicates the H-2 system in the control of steroid-hormone function.

In addition, several H-2 linked genetic traits have been found which influence embryonal development. These include spermatogenesis, expression of Brachyury (T) genes and development of the mandible bone (reviewed by Iványi 1978)[64].

These and other studies indicate that the H-2 complex contains genes or is closely linked to genes which exert a broad variety of important regulatory function. At present, therefore we can identify a number of different classes of H-2 region genes which are involved in six different functions as listed in the following table (Table 1):

Table 1.                    FUNCTIONS OF H2 REGION GENES

    1.  Production of humoral antibodies;

    2.  Production of killer T-cells and helper T-cells;

    3.  Regulation of immune response;

    4.  Biosynthesis of some complement components;

    5.  Regulation of embryogenesis;

    6.  Regulation of some steroid-hormones.

Evidence for the occurrence of most of these classes of genes in the human HLA
region has also been accumulating, as discussed below.


The HLA-Linkage Group

   The HLA-complex is defined today as a genetic system of five closely linked
genes: HLA-A; HLA-C; HLA-B; HLA-D and HLA-DR[19]. The sequence of these genes
on the short arm of chromosome 6 is given below together with an estimate of
the recombinant fractions between the different genes[20]. (fig. I).

   Additional genes are known to be linked to HLA. The first to be identified
was the gene coding for the electrophoretic polymorphism of phosphoglucomutase-3
($PGM_3$)[21,22]. Subsequently it was found that the structural gene for factor B
(Bf) of the alternative complement pathway was located close to HLA[23] and that
the genes for complement C2 deficiency[24,25] and C4 deficiency[26,27] were also
HLA linked. The combined data for mapping of C2 deficiency and Bf demonstrate
that both genes are close to the HLA-B locus and most likely are located between
HLA-B and HLA-D.

   It is interesting to point out that C2 and Bf have similar functional
properties in that they both form C3 converting enzymes. C2 with C4 forms
the C3 convertase of the classical pathway whereas the C3 convertase of the
alternative pathway is formed from C3, factor B (Bf) and factor D. C2 and Bf
are both single polypeptide chains of similar molecular size and it is tempting
to postulate that these proteins have arisen as a result of gene duplication as
has been suggested recently for C6 and C7 (Lachmann et al. 1978)[28]. Recent
studies have shown that the gene responsible for the polymorphic variants of
C2 also is HLA linked (Meo et al.)[29].

   The structural gene for Glyoxalase-I (GLO) is also linked to HLA[30] and is
located approximately 5 cm outside HLA-B[20].

   The genes coding for the two red cell antigens Chido (Ch) and Rodgers (Rg)
are also known to be linked to HLA and these genes are located close to HLA-B
[31,32] as described in detail below.


C4, Chido and Rodgers Red Cell Antigens

   The two erythrocyte blood group substances Chido ($Ch^a$) and Rodgers ($Rg^a$) are
found on the red cells and in the plasma of approximately 97% of the population

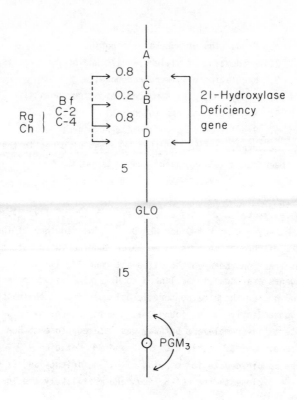

Figure 1.   HLA-Linkage group.

(Middleton & Crookston, 1972[33]; Longster & Giles, 1976[34]).  The genes for $Ch^a$ and $Rg^a$ are both HLA-linked and lack of $Ch^a$ is associated with B12 and Bw35 while lack of $Rg^a$ is associated with B8 (reviewed in O'Neill et al. 1978). Genetic analysis of family data and population studies have shown, that $Ch^a$ and $Rg^a$ are not alleles.  The frequency of Ch(a+) Rg(a+) is too high compared with the frequencies of Ch(a+)Rg(a-) and Ch(a-)Rg(a+).  Furthermore, until recently no Ch(a-)Rg(a-) individuals have been identified.  (Giles et al. 1976)[32].

We have recently shown that $Ch^a$ and $Rg^a$ are antigenic components of complement C4  (O'Neill et al. 1978b)[36].  An electrophoretic method was developed in which patterns of bands of C4 could be identified in EDTA plasma.  Three different C4 patterns were found.  The most common pattern of C4 (the FS form) consists of 8 bands and occurs in 93% of the population. The C4S pattern with only the 4 slow moving cathodal bands is found in 1% and the C4F pattern with only the 4 fast moving anodal bands is found in 6% of the population. These findings of the relationship between C4, Chido and Rodgers are summarized in Table II. (O'Neill et al. 1978 a,b,c).

Table II

Relationship Between C4, Chido and Rodgers

1. C4F individuals were all Ch(a-),Rg(a+)
2. C4S individuals were all Rg(a-),Ch(a+)
3. C4FS individuals were all Ch(a+),Rg(a+)
4. 2 individuals who were C4 homozygous deficient were Ch(a-),Rg(a-)
5. Purified C4FS inhibits both anti $Ch^a$ and anti $Rg^a$
6. Purified C4F inhibits anti $Rg^a$ but not anti $Ch^a$
7. Purified C4S inhibits anti $Ch^a$ but not anti $Rg^a$

Genetic analysis of 79 families studied for HLA, C4, $Ch^a$ and $Rg^a$ demonstrated that the electrophoretic patterns of bands of C4 (F,S and FS) could not be explained by assuming a single genetic locus with two codominantly expressed alleles F and S. Another genetic model for the inheritance of the C4 patterns was found to be in agreement with the experimental data. In this model, C4 was assumed to be coded for by two genes both of which are linked to HLA. One locus (C4S locus) has the two alleles S and $s^o$ with $s^o$ recessive to S. This locus codes for the 4 slow moving cathodal bands of C4. The second locus (C4F) has the two alleles F and $f^o$ with $f^o$ recessive to F. This locus codes for the 4 fast moving anodal bands of C4. The C4S locus should code for the Chido antigen and the C4F locus should code for the Rodgers antigen. The results of segregation analysis of the alleles F and $f^o$ at one locus and S and $s^o$ at the other locus are in agreement with this hypothesis, and linkage analysis of the C4F locus and the C4S locus to HLA demonstrate that both loci are linked to HLA. The peak Lod scores were obtained for the recombinant fraction $\Theta \simeq 0.00$ for C4S to HLA-B and C4F to HLA-B[37].

These data demonstrate that two HLA linked genes C4F and C4S are of importance in the expression of C4 and that individuals who are homozygous for the $s^o$ allele and the $f^o$ allele are Ch(a-)Rg(a-) and homozygous deficient for C4.

Since the C4 $f^o$ allele (Rg[a-]) is in positive genetic linkage disequilibrium with B8 and the C4 $s^o$ allele (Ch[a-]) is in positive linkage disequilibrium with B12, Bw35 and B5[36], this provides a new dimension for the study of the B8 associated diseases, many of which are in the group of organ specific autoimmune diseases. Such studies may imply that the complement system via C4 is involved as a genetic host factor in the pathogenesis of these diseases.

From studies in the mouse it is known that the S-region codes not only for the Ss-protein which is C4 but also for the Slp protein (for review see Shreffler & David, 1975)[38]. Recent evidence has suggested that the Ss and Slp proteins are distinct antigenic subclasses of C4 (Roos et al. 1978)[39] with possibly distinct functional properties (Ferreira et al. 1977)[40]. The H-2G region adjacent to the S-region codes for erythrocyte antigens which can also be detected in plasma (David et al. 1975)[41]. It remains to be seen if these red cell antigens in the mouse, as has been shown for $Ch^a$ and $Rg^a$ antigens in humans,[36] are complement components bound to the cell surface.

Studies of complement components and HLA now indicate that four of these complement genes are located within the HLA complex and that they all most likely are located between HLA-B and HLA-D. Two of the genes are the C4S and C4F genes and the other two are the Bf gene and the C2 gene. The present concept regarding the evolution of HLA and H-2, which is derived from amino acid sequence analysis of HLA-A and HLA-B, and H-2K and H-2D antigens, respectively indicates that the Major Histocompatibility Complex in each case is a result of multiple gene duplications. The studies of C4, C2 and Bf support this concept. A hypothetical genetic model for these complement components is shown below:

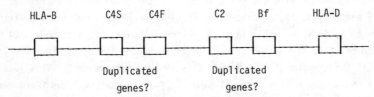

Genetic linkage disequilibrium has been found for some of the C4 alleles with HLA-B and also for C2, and Bf with HLA (O'Neill et al. 1978a, Dausset et al. 1978[42], Meo et al. 1976[43]). The importance of such HLA-B/C2/HLA-D haplotypes for disease susceptibility has been demonstrated for the B18/C2 deficiency/Dw2-haplotype which is associated with certain diseases such as systemic lupus erythematosis[24,25].

## HLA-Linked Diseases

In addition to studies of HLA disease associations a number of genetic disorders have been studied for linkage to HLA. Cystic Fibrosis has been studied by Kaiser et al. 1976[44], Polymenidis et al. 1973[45] and Lamm et al. 1975[46] and found not to be HLA linked.

von Willebrand's disease has been studied by Goudemand et al. (1976)[47] and Muller et al. (1976)[48]. No linkage to HLA was found.

Negative studies have also been performed in acute prophyria, familial mediterranean fever, Hurler's disease and G-6-PD deficiency.

Fotino et al.[49] have recently shown that Paget's disease of the bone probably is HLA linked but the Lod scores have not yet passed 3. A study by Nyulassy et al.[51] of families with pyrophosphate chondrocalcinosis indicates that this disease also may be HLA-linked.

One large family with Osler-Weber-Rendu disease (hereditary haemorrhagic telangiectasia) has been studied, and linkage to HLA is possible (Kissel et al. 1976)[50].

Familial spinocerebellar ataxi is shown in one large pedigree to be HLA linked with a peak Lod score of 3.5 for $\Theta \simeq 0.12$[52].

Two other diseases have been studied, where the linkage to HLA is convincingly documented. These diseases are Idiopathic Hemochromatosis and Congenital Adrenal Hyperplasia of the 21-hydroxylase deficiency type.

Idiopathic Hemachromatosis has been assumed to be an autosomal dominant inherited disease with incomplete penetrance in females due to loss of blood in menstruation and pregnancy. It has also been suggested that an autosomal recessive form of the disease exists (McKusick, 1975)[53]. Linkage studies to HLA in families with this disease indicate that probably two HLA linked genes are involved in the disease development or that the disease is simple autosomal recessive in character with the disease gene located 10-15 cm outside HLA (Bomfeld et al. 1977[54,56] Simon et al. 1976[55], 1977[57]). In either case, the studies demonstrate that gene(s) of importance in iron metabolism are located close to the HLA complex. In this disease it has also been shown that the disease gene is in genetic linkage disequilibrium with the HLA determinants A3 and B14[55,56,57].

The most recent finding of HLA linked disease genes is the mapping of the 21-hydroxylase deficiency gene close to the HLA-B locus. The initial study of six families of patients with Congenital Adrenal Hyperplasia of the 21-OH-def. type, placed the 21-OH-def. gene close to the HLA-B locus with a peak Lod score of 3.4 for $\Theta \simeq 0.00$ (Dupont et al. 1977)[58]. Three additional families were reported by Weitkamp et al. 1978[59] and six more families were reported by Price et al. 1978[60]. We have recently completed studies of 34 Caucasian families with a total of 48 patients, 48 healthy siblings and their parents. (Levine et al. 1978[61], Yang et al. 1978[62]). Lod score analysis for HLA-B:21-OH-def. now shows a peak at $\Theta \simeq 0.00$ of 9.5.

The gene for 21-hydroxylase deficiency was separated by genetic recombination from the HLA-A locus and from the locus for glyoxalase I (GLO). No HLA-A, B or C antigen was selectively increased among the 34 unrelated patients (Levine et al. 1978[61]).

These studies demonstrate that at least one structural gene coding for a critical enzyme in the biosynthesis of steroid hormones can be mapped within the HLA complex.

Studies of the 11-β-hydroxylase deficiency which is another autosomal recessive inherited disease leading to CAH have shown that the 11-β-OH-def. gene is not HLA linked (Brautbar et al. 1978[63]).

## SUMMARY

In addition to genes affecting immunological function, the HLA region contains a number of genes which appear to influence disease etiology in other ways. Although only a limited number of inborn errors of metabolism and genetically inherited diseases have been studied so far for linkage to HLA, such studies have already disclosed that genes coding for C2, C4 and Bf of the complement system and gene(s) involved in iron metabolism and in the biosynthesis of steroid hormones are closely linked to HLA.

The HLA linked complement components are obvious candidates for genetic host factors involved in resistance and susceptibility to autoimmune diseases. It now seems possible that genes involved in the biosynthesis of steroid hormones may well also play a role in making the host susceptible to environmental factors leading to the development of autoimmune diseases.

It is tempting to postulate that the HLA-complex plays a central role in the regulation of a broad range of functions of which some are immunological and some are metabolic while other HLA-linked genes play roles in embryogenesis and cell differentiation.

The continued study of HLA-linkage to inherited diseases will surely provide information on the genetic composition of the HLA complex. As these genes are further identified, it may be possible to analyze the significance of this genetic framework and explain the presently confusing conglomeration of diseases which are HLA associated.

## ACKNOWLEDGEMENTS

Original work quoted in this paper was supported in part by grants from USPHS, NIH, 1 PO1 CA 22507, CA 08748, CA 19267, CA 17404, RO1 EY 01616, NIH-5-SO-7-RR 05534 and HD 00072 and also from a grant (RR47) from the General Clinical Research Centers Program of the Division of Research Resources, NIH.

## REFERENCES

1. Svejgaard, A. and Ryder,L .P. (1977) in HLA and Disease. Dausset, J. and Svejgaard, A. eds. Munksgaard, Copenhagen, pp.46-71.

2. Lilly, F., Boyse, E.A. and Old, L.J. (1964) Lancet, 2, 1207.

3. McDevitt, H.O., Dead, B.D., Shreffler, D.C., Klein, J., Stimpfling, J.H. and Snell, G.D. (1972) J. exp. Med. 135, 1259.

4. Debre, P., Kapp, J.A., Dorf, M.E. and Benacerraf, B. (1975) J. exp. Med. 142, 1447.

5. Doherty, P.C. and Zinkernagel, R.M. (1975) Nature (Lond.) 256, 50.

6. Zinkernagel, R.M. and Doherty, P.C. (1974) Nature (Lond.) 251, 547.

7. Hinzová, E., Démant, P. and Ivány, P. (1972) Folia biol. (Praha) 18, 237.

8. Démant, P., Capková, J., Hinzová, E. and Voracová, B. (1973) Proc. natn. Acad. Sci (U.S.A.) 70, 863.

9. Lachmann, P.J., Grennan, D., Martin, A. and Démant, P. (1975) Nature (Lond.) 258, 242.

10. Curman, B., Ostberg, L., Sandberg, L., Malmheden-Eriksson, I., Stalenheim, G. Rask, L. and Peterson, P.A. (1975) Nature (Lond.) 258, 243.

11. Meo, T., Krasteff, T. and Shreffler, D.C. (1975) Proc. natn. Acad. Sci. 72, 4536.

12. Ivány, P. and Mickova, M. (1971) in Immunogenetics of the H-2 system. Lengerova, A. and Vojtiskova, M. eds. Karger, Basel, pp. 104-119.

13. Ivány, P., Gregorová, S. and Mickova, M. (1972a) Folia biol. (Praha) 18, 81.

14. Ivány, P., Hampl, R., Starka, L. and Mickova, M. (1972b) Nature New Biol. 238, 280.

15. Ivány, P., Hampl, R., Mickova, M. and Starka, L. (1976) Folia biol. (Praha) 22, 42.

16. Mickova, M. and Ivány, P. (1975) Folia biol. (Praha) 21, 435.

17. Meruelo, D. and Edidin, M. (1975) Proc. natn. Acad. Sci. (U.S.A.) 72, 2644.

18. Goldmann, A.S., Katsumata, M., Yaffe, S.J. and Gasser, D.L. (1977) Nature (Lond.) 265, 643.

19. Nomenclature Committee on the HLA-System, 1977 (1978) Tissue Antigens, 11, 81.

20. Report of the committee on the genetic constitution of chromosome 6 in Human Gene Mapping 4. (Winnipeg Conference 1977) in press.

21. Lamm, L.U., Kissmeyer-Nielsen, F., and Henningsen, K. (1970) Human Heredity 20, 305.

22. Lamm, L.U., Kissmeyer-Nielsen, F., Svejgaard, A., Petersen, B.G., Thorsby, E., Mayr, W. and Högman, C. (1972) Tissue Antigens 2, 205.

23. Allen, F.H. (1974) Vox. Sang. 27, 382.

24. Fu, S.M., Kunkel, H.G., Brusman, H.P., Allen, F.H. and Fotino, M. (1974) J. exp. Med. 140, 1108.

25. Fu, S.M., Stern, R., Kunkel, H.G., Dupont, B., Hansen, J.A., Day, N.K., Good, R.A., Jersild, C. and Fotino, M. (1975) J. exp. Med. 142, 495.

26. Rittner, C., Hauptmann, G., Grosse-Wilde, H., Tongio, M.M. and Mayr, S. (1975) in Histocompatibility Testing 1975. Kissmeyer-Nielsen, F. ed. Munksgaard, Copenhagen, p. 945.

27. Ochs, H.D., Rosenfeld, S.I., Thomas, E.D., Giblett, E.R., Alper, C.A., Dupont, B., Scholler, J.G., Gilliland, B.C., Hansen, J.A. and Wedgwood, R.J. (1977) New Engl. J. Med. 296, 470.

28. Lachmann, P.J., Hobart, M.J. and Woo, P. (1978) Clin. exptl. Immunol. in press.

24

29. Meo, T., Alkinson, J., Bernoco, M., Bernoco, D. and Ceppellini, R. (1977) Proc. atn. Acad. Sci (U.S.A.) 74, 1672.

30. Weitkamp, L.R. and Guttormsen, S.A. (1976) in Human Gene Mapping 3. p. 364.

31. Middleton, J., Crookston, M.C., Falk, J.A., Robson, E.B., Cook, P.J.L., Batchelor, J.R., Bodmer, J., Ferrara, G.B., Festenstein, H., Harris, R., Kissmeyer-Nielsen, F., Lawler, S.D., Sachs, J.A. and Wolf, E. (1974) Tissue Antigens 4, 366.

32. Giles, C.M., Gedde-Dahl, Jr. T., Robson, E.B., Thorsby, E., Olaisen, B., Arnason, A., Kissmeyer-Nielsen, F. and Schreuder, I. (1976) Tissue Antigens 8, 143.

33. Middleton, J. and Crookston, M.C. (1972) Vox Sang. 23, 256.

34. Longster, G. and Giles, C.M. (1976) Vox Sang. 30, 175.

35. O'Neill, G.J., Yang, S.Y. and Dupont, B. (1978a) Transpl. Proc. in press.

36. O'Neill, G.J., Yang, S.Y., Tegoli, J., Berger, R. and Dupont, B. (1978b) Nature (Lond.) 273, 668.

37. O'Neill, G.J., Yang, S.Y. and Dupont, B. (1978c) Proc. natn. Acad. Sci. (U.S.A.) in press.

38. Shreffler, D.C. and David, C.S. (1975) Adv. Immunol. 20, 125.

39. Roos, M.H., Atkinson, J.P. and Shreffler, D.C. (1978) J. Immunol. 120, 1794.

40. Ferreira, A., Takahashi, M. and Nussenzweig, V. (1977) J. exp. Med. 146, 1001.

41. David, C.S., Stimpfling, J.H. and Shreffler, D.C. (1975) Immunogenetics 2, 131.

42. Dausset, J., Legrand, L., Lepage, V., Contu, L., Marcelli-Barge, A., Wildloecher, L., Benajam, A., Meo, T. and Degos, L. (1978) Tissue Antigens in press.

43. Meo, T., Atkinson, J., Bernocio, M., Bernocio, D. and Ceppellini, R. (1976) 6, 916.

44. Kaiser, G., Lazlo, A., Gyurkovits, K. and Gyodi, E. (1976) Proc. HLA and Diseases, INSERM, Paris.

45. Polymenidis, Z., Ludwig, H. and Gotz, M. (1973) Lancet 2, 1452.

46. Lamm, L.U., Thorsen, I.L., Petersen, G.B., Jorgensen, J., Henningsen, B., Beck, B. and Kissmeyer-Nielsen, F. (1975) Ann. Hum. Genetics 38, 383.

47. Goudemand, J., Mazurier, C. and Parquet-Gernez, A. (1976) Proc. HLA and Disease Symposium, INSERM, Paris.

48. Muller, N., Budde, U. and Etzel, F, (1976) Proc. HLA and Disease Symposium, INSERM, Paris.

49. Fotino, M., Haymovits, A. and Falk, C.T. (1977) Transpl. Proc. 9, 1867.

50. Kissel, P., Raffoux, C., Faure, G., Andre, J.M., Netter, P. and Streiff, F.C. (1976) Proc. HLA and Disease Symp. INSERM, Paris.

51. Nyulassy, S., Stefanovic, J. and Sitaj, S. (1975) in Histocompatibility 1975. Kissmeyer-Nielsen, F. ed. Munksgaard, Copenhagen, p.805.

52. Jackson, J.F., Currier, R.D., Terasaki, P.I. and Morton, N.E. (1977) New Engl. J. Med. 296, 1138.

53. McKusick, V. (1975) Mendelian Inheritance in Man. The Johns Hopkins University Press, Baltimore, p.124 and p.446.

54. Bomford, A., Eddleston, A.L.W.F., Kennedy, L.A., Batchelor, J.R., and Williams, R. (1977) Lancet 1, 327.

55. Fauchet, R., Simon, M., Bourel, M., Genetet, B., Genetet, N., Alexandre, J.L. (1976) Proc. HLA and Disease Symp. INSERM, Paris.

56. Bomford, A., Eddleston, A.L.W.F., Williams, R., Kennedy, L. and Batchelor, J.R. (1976) Proc. HLA and Disease Symp. INSERM, Paris.

57. Simon, M., Bourel, M., Fauchet, R., Genetet, B. (1976) Gut, 17, 332.

58. Dupont, B., Oberfield, S.E., Smithwick, E.M., Lee, T.D., and Levine, L.S. (1977) Lancet 2, 1309.

59. Weitkamp, L.R., Bryson, M., Bacon, G.E. (1978) Lancet 1, 931.

60. Price, D.A., Klouda, P.T., Harris, R. (1978) Lancet 1, 930.

61. Levine, L.S., Zachmann, M., New, N.I., Prader, A., Pollack, M.S., O'Neill, G.J., Yang, S.Y., Oberfield, S.E. and Dupont, B. (1978) submitted.

62. Yang, S.Y., Levine, L.S., Zachmann, M., New, N.I., Prader, A., Oberfield, S.E., O'Neill, G.J., Pollack, M.S. and Dupont, B. (1978) Transpl. Proc. in press.

63. Brautbar, C., Rosler, A., Landau, H., Cohen, I., Cohen, T., Levine, C., Sack, J., Benderli, A., Moses, S., Lieberman, E., Dupont, B., Levine, L.S., and New, N.I. (1978) submitted.

64. Ivanyi, P. (1978) Proc. R. Soc. (London) in press.

DISCUSSION

TALAL: In the mouse, several complement components including C4 are induced by androgen. Their serum concentrations are higher in males than in females. Do you have any evidence that the Chido or Rodgers antigens are quantitatively different in males and females?

DUPONT: The levels of C4 have been determined immunochemically and functionally by us in families where one child is Ch(a-) or Rg(a-). The Ch(a-) Rg(a+) and Ch(a+) Rg(a-) have half levels of C4. We do not yet have enough data on C4 levels in males and females to determine a sex difference.

WARNER: Could you clarify for me the relationship of the Ch and Rg blood group antigens to C4? Is C4 passively adsorbed to the red cell, or has the red cell precursor synthesized this cell membrane component?

DUPONT: It has been claimed that Ch(a) and Rg(a) is passively taken up from plasma by erythrocytes. This may be correct. It is of more interest to **see if Ch(a) and Rg(a) are synthesized by** nucleated cells. We have previously found that the anti-Ch antibody will precipitate an antigen from the lymphocyte membrane, which has tyrosine residues similar to the HLA A, B, and C antigens and which occurs together with $\beta_2$-microglobulin (C. Cunningham-Rundles et al., Transpl. Proc., 1976). We think that the Ch(a) and Rg(a) component of C4 very well may be a normal constituent of the cell membrane and that these antigens could be the insertion site in the cell membrane for the complement-induced cell lysis.

VOLPE: I was interested in your studies of families of 21-hydroxylase deficiency. Did you study families of other enzyme deficiencies?

DUPONT: We have studied the 11-β-hydroxylase deficiency together with Dr. Brautbar and colleagues in Israel, and this gene is not HLA linked. The testicular feminization syndrome (androgen insensitivity) has also been studied by us in a few families. This is in search for the HLA-linked genes controlling serum testosterone, which we would predict from the H-2 data to be within HLA. So far the data are not sufficient for conclusive statements.

ROITT: Does the heterozygote for the steroid synthesizing enzyme deficiency show defects in steroid production and have you established whether there is any undue prevalence of this defect in autoimmune disorders?

DUPONT: Heterozygous deficiency for the 21-hydroxylase deficiency gene cannot be determined with certainty by biochemical testings. No data are available regarding autoimmune diseases in heterozygous carriers of this disease gene. However, the close linkage to HLA for this gene now allows the identification of heterozygous carriers for this disease in high risk families and studies as the one you request can now be performed.

SVEJGAARD: As Dr. Dupont pointed out, it is extremely important from a genetic point of view that we here have a disorder (21-hydroxylase deficiency), which is very closely linked to the HLA-B locus without showing association. However, it could be noted on your slide that the B8 antigen had a very low frequency in your patients, and I would just like to point out that you only need another group of the same size to show if this negative association is true.

DUPONT: It is correct that there is an under-representation of B8 in our material and I agree that new and larger materials of unrelated patients should be collected to answer this question.

HLA ASSOCIATION WITH CHRONIC ACTIVE HEPATITIS

IAN R. MACKAY AND BRIAN D. TAIT

Clinical Research Unit, The Walter and Eliza Hall Institute of
Medical Research, and Tissue Typing Laboratory, Royal Melbourne
Hospital, Post Office, Royal Melbourne Hospital, Victoria, 3050,
Australia

Chronic active hepatitis (CAH) is one of a number of immune-
mediated diseases which are associated with antigens coded for by
genes of the major histocompatibility complex (HLA).  Such diseases
fall into four groups (Table 1):

TABLE 1
ASSOCIATIONS OF HLA AND DISEASE

| NON-HLA ASSOCIATED | HLA ASSOCIATED | | |
| AUTOIMMUNE | AUTOIMMUNE | IMMUNOPATHIC | NON-IMMUNOPATHIC |
|---|---|---|---|
| Hashimoto<br>Gastritis, PA type | B8 Thyrotoxicosis<br>B8 Adrenalitis<br>B8 Diabetes<br>(insulin<br>dependent) | B27 Ankylosing<br>spondylitis<br>B8 Celiac disease<br>(gluten<br>enteropathy | A3 )<br>B14) Hemochromatosis<br>CW6 Psoriasis |
| Myasthenia gravis,<br>with thymoma | B8 Myasthenia<br>gravis,<br>juvenile | B7 Multiple sclerosis | |
| Autoimmune diseases<br>of skin, blood,<br>kidney, muscle,<br>heart, nerve, etc. | B8 Sjögren's<br>disease<br>B8 Chronic active<br>hepatitis<br>B8 SLE | D4 Rheumatoid<br>arthritis | |

Chronic active hepatitis began to be recognised as a disease
entity from 1950[1], but there have long been differences of opinion
on definition, nomenclature and criteria for diagnosis.  As defined
by the Fogarty-IASL Conference Committee[2] it results from a continuing

inflammatory lesion of the liver with a potential to progress to more
severe disease, including cirrhosis, to continue unchanged or to sub-
side spontaneously or with treatment. Other features often included
in definitions of CAH are histological changes of "bridging" necrosis
and destruction of the limiting plate of the liver lobule together
with prominent lymphocyte-plasma cell infiltration into the liver
parenchyma, amelioration of the disease by corticosteriod drugs, and
autoimmune serological reactions.

The following subtypes may be recognized[3]: A, characterized by
hypergammaglobulinemia (>20g/l) and autoimmune serological markers,
particularly antinuclear antibody (ANA), and also anti-smooth
muscle antibody (ASMAb) or, in some cases, anti-mitochondrial antibody (AMA) or anti-microsomal antibody; B, characterized by presence
of hepatitis B surface antigen (HBsAg) in blood and persistence of
the hepatitis B virus genome in the liver; C, cryptogenic with no
specific markers; and D, initiation by drug sensitivity. CAH-A is
the predominant type in Australia and in countries with a predominantly Northern Caucasian origin of the population, and there are two
peaks of high age incidence, one among adolescent females and a
second among older females (Fig.1) among whom there appear to be
more extrahepatic autoimmune manifestations such as thyroiditis and
Sjögren's disease. CAH-B in contrast to CAH-A occurs mostly among
males and differs also in the lower frequency of immunoserologic
abnormalities[3].

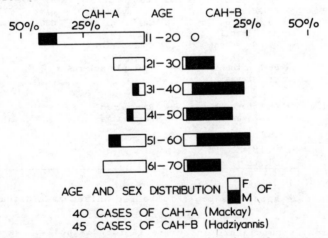

Fig. 1. Contrasting age and sex distribution of CAH-A and CAH-B

From Mackay, I.R., Genetic studies in chronic active hepatitis, in Proceedings
of International Symposium on Immune Reactions in Liver Disease: Pathogenetic
Role and Developments in Immunotherapy, (in press, 1978).

METHODS

## Histocompatibility typing

Typing for HLA A, B antigens was performed by a standard micro-
lymphocytotoxic assay on lymphocytes separated from peripheral
blood by an isopaque-ficoll density gradient method[4]; the HLA
specificities detected were as previously described [5]. Typing for
D locus related antigens on B cells (DRW antigens) was performed
by a lymphocytotoxicity assay on B lymphocytes from peripheral
blood prepared as follows.

Separation of B and T lymphocytes was performed by a modification
of the method described by Nelson et al.[6] and Grier et al.[7] in which
there is selection for B cells by use of a goat anti-human F (ab')$_2$
serum. In our procedure, a rabbit anti-human IgG F(ab')$_2$ serum
(Behringwerke AG, Marburg) was purified by affinity chromatography
by passage through cyanogen bromide Sephadex 4b conjugated with
human immunoglobulin. A titer of rabbit anti-human IgG F(ab')$_2$ was
chosen which resulted in high enrichment for B cells, identified
by the presence of surface immunoglobulin. The antiserum was
stored in aliquots of 2 mg protein/ml at -70°C, and diluted in
phosphate buffered saline (PBS) preparatory to use.

A standard 30 ml tissue culture flask (Costar, Cambridge, Mass.,
USA) was coated with 1-2 ml of rabbit anti-human IgG F(ab')$_2$
appropriately diluted in PBS and held with the coated side down
for 30 minutes at 37°C. The flask was then washed with PBS and
medium 199 containing 10% fetal calf serum (FCS). To this flask
were added peripheral blood lymphocytes prepared by an isopaque-
ficoll density gradient method and washed three times in medium
199 containing 10% FCS, and adjusted to 5-10 million cells per ml
in medium 199. Three ml of cell suspension was added to each
flask so that the cells were in contact with the antiserum-coated
surface for 30 minutes at 37°C. The supernatant containing the
non-adherent fraction (T cells) was removed and the flask was
washed four times with medium 199 containing 10% FCS. Ten ml of
an elution medium consisting of 4 ml of pooled human serum, 1 ml
of 5% EDTA and 5 ml of medium 199 was added to the flask and held
for 30 minutes at 37°C. The flask was vigorously shaken and the
supernatant containing B cells was removed. The flask was then
washed four times with medium 199 to collect remaining B cells.

Seven DRW antigens were determined using 40 antisera, the specificities of which were designated on the basis of their reactivity

with cells which had been typed for DRW antigens as part of the
7th International Histocompatibility Workshop. The standard prolonged microlymphocytotoxicity dye exclusion assay recommended at
the 7th International Workshop was used for typing cells for DRW
antigens in the present studies; 1 µl of antiserum was used and the
incubation periods were 60 minutes for cells and antiserum, and 120
minutes with complement.

Mixed lymphocyte reactions were performed as previously described[8].

## Serological methods

Tests for ANA, ASMAb and other non-organ-specific autoantibodies
were performed as previously described[5]. Antibodies to hepatitis A
virus were determined by a solid phase radioimmunoassay[9] (Dr. I.
Gust, Fairfield Infectious Diseases Hospital, Melbourne) and for the
core antigen of hepatitis B virus by radioimmunoassay[9] (Dr. G. Vyas,
Department of Laboratory Medicine, University of California, San
Francisco).

## Diseases studied

The cases of chronic hepatitis (CAH-A) considered in this study
fulfilled the definitions of the Fogarty-IASL Conference Committee[2]
and were all seropositive for ANA and ASMAb and seronegative for
HBsAg. Forty-eight have been typed for HLA A and B locus antigens,
and 17 for "Ia-like" D-related (DRW) antigens on B cells. Six
families of patients with CAH-A were typed for A, B and DRW antigens
to examine haplotypes. Other types of liver disease studied included
primary biliary cirrhosis, cryptogenic cirrhosis and alcoholic
cirrhosis; there were not sufficient cases of CAH-B available in
Melbourne for meaningful analysis.

## RESULTS

### CAH and HLA associations - A and B locus antigens

The results of typing for HLA A and HLA B locus antigens in cases
of liver disease studied in this 'Unit were published by Morris et al.[5]
and may be summarized as follows. In CAH-A there were highly significant associations with HLA A1 (65%, $2P = 1.5 \times 10^{-5}$) and HLA B8
(69%, $2P = 8.0 \times 10^{-12}$). With the exception of hemochromatosis, for
which there is a known association with A3 and B14[11], there were no
significant HLA associations with other types of chronic parenchymal
liver disease. There were minor increases in alcoholic cirrhosis of
B5 (15%) and B8 (34%), and minor differences in primary biliary
cirrhosis in A1 and B8 (low) and B5 (raised).

HLA B phenotypes were not examined in detail, although note can
be made of the higher than expected association in CAH-A of B8-B27
(3 of 48 cases, 6%), and the same held for systemic lupus erythematosus (5 of 46 cases, 10%); the HLA B8-B27 phenotype existed in
0.7% of 404 Melbourne controls[12].

CAH and HLA associations - DRW antigens

   In Table 2 are presented results for 44 cases of CAH-A derived
from this Unit and from King's College Hospital, London, selected
according to the same criteria, and Table 3 shows for 35 of the 44
cases correlations between HLA B8, HLA DRW3 and the occurrence of
ANA and ASMAb: these data were as presented at the 7th International
Histocompatibility Workshop[13]. There is a significantly increased
frequency for A1, B8, and DRW3, and a decreased frequency of HLA
BW35. DRW3 was significantly associated with ANA and ASMAb, and B8
only with  ASMAb. Since the original typing for the Histocompatibility
Workshop, there have been 17 cases of CAH-A typed or retyped in
Melbourne for DRW3, and in this experience, the 13 cases positive
for DRW3 (76%) reflect a higher association, probably attributable
both to case selection and improved B cell techniques. The known
linkage disequilibrium between HLA B8 and DRW3 appeared stronger for
our cases of CAH-A than for our control population.

TABLE 2
HLA ANTIGENS IN CAH-A*

| | | HLA ANTIGEN FREQUENCIES % | | | |
|---|---|---|---|---|---|
| | | A1 | B8 | B35 | DRW3 |
| PATIENTS | (44) | 64 | 64 | 4 | 49 |
| CONTROLS | (52) | 46 | 23 | 17 | 25 |
| $\chi^2$ | | | 2.9 | 16.1 | 3.8 | 5.8 |

   * 7th International Workshop (1977) Data on London
     and Melbourne Cases.

TABLE 3

CORRELATIONS BETWEEN DRW3 AND AUTOANTIBODIES ASSOCIATED WITH CHRONIC
ACTIVE HEPATITIS*

|  |  | ANA | | |
|---|---|---|---|---|
|  |  | strong | absent or weak |  |
| DRW3 | + | 13 | 2 | 15 |
|  | - | 8 | 9 | 17 |
|  |  | 21 | 11 | 32 |

$\chi^2 = 5.3$   $p < 0.05$

|  |  | SMAb | | |
|---|---|---|---|---|
|  |  | strong | absent or weak |  |
| DRW3 | + | 14 | 5 | 19 |
|  | - | 4 | 11 | 15 |
|  |  | 18 | 16 | 34 |

$\chi^2 = 7.4$   $p < 0.01$

|  |  | strong | absent or weak |  |
|---|---|---|---|---|
| B8 | + | 15 | 5 | 20 |
|  | - | 7 | 6 | 13 |
|  |  | 22 | 11 | 33 |

$\chi^2 = 1.6$   N.S.

|  |  | strong | absent or weak |  |
|---|---|---|---|---|
| B8 | + | 16 | 7 | 23 |
|  | - | 3 | 9 | 12 |
|  |  | 19 | 16 | 35 |

$\chi^2 = 6.3$   $p < 0.05$

* From 7th International Histocompatibility Workshop, Oxford, 1977.

Haplotypes in CAH-A

Typing for A, B and DRW antigens has been performed in six families
(Figs 2a and 2b); the propositus in each family was a young female
with classic CAH-A.  In all families the propositus carried A1-B8-DRW3
on the one chromosome and in two cases there was homozygosity for
DRW3; the A1-B8-DRW3 haplotype was inherited from the mother in 3
and the father in 3 families.  In two families, B8-DRW3 was present
in family members with no overt disease.  In one family, No. 4,
there were two siblings with CAH-A; the disease was of protracted
and relapsing type in a female sibling who carried B8-DRW3, but
milder and remitting on treatment in an older brother in whom neither
B8 nor DRW3 was represented.  The female propositus in a Greek
family, No. 5, had severe CAH and was homozygous for DRW3; the
maternal haplotype A1-B8-DRW5 may have been the result of a BD
recombination because of the known linkage disequilibrium between
A1, B8 and DRW3.

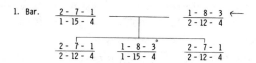

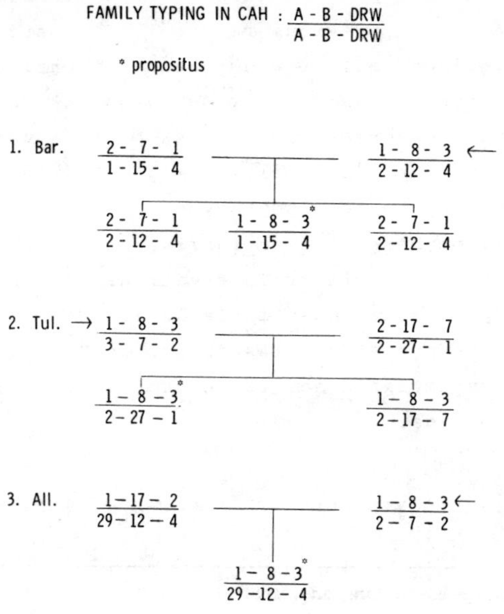

FAMILY TYPING IN CAH : $\dfrac{A - B - DRW}{A - B - DRW}$

° propositus

1. Bar. $\dfrac{2 - 7 - 1}{1 - 15 - 4}$      $\dfrac{1 - 8 - 3}{2 - 12 - 4}$ ←

$\dfrac{2 - 7 - 1}{2 - 12 - 4}$   $\dfrac{1 - 8 - 3°}{1 - 15 - 4}$   $\dfrac{2 - 7 - 1}{2 - 12 - 4}$

2. Tul. → $\dfrac{1 - 8 - 3}{3 - 7 - 2}$      $\dfrac{2 - 17 - 7}{2 - 27 - 1}$

$\dfrac{1 - 8 - 3°}{2 - 27 - 1}$     $\dfrac{1 - 8 - 3}{2 - 17 - 7}$

3. All. $\dfrac{1 - 17 - 2}{29 - 12 - 4}$      $\dfrac{1 - 8 - 3}{2 - 7 - 2}$ ←

$\dfrac{1 - 8 - 3°}{29 - 12 - 4}$

Fig. 2a

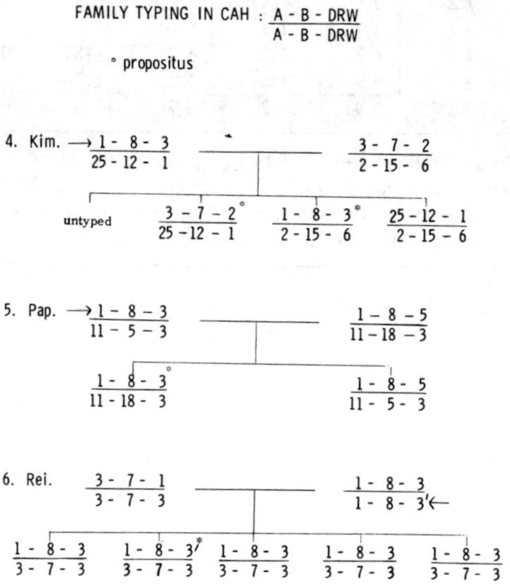

FAMILY TYPING IN CAH : $\dfrac{A - B - DRW}{A - B - DRW}$

° propositus

4. Kim. → $\dfrac{1 - 8 - 3}{25 - 12 - 1}$      $\dfrac{3 - 7 - 2}{2 - 15 - 6}$

untyped   $\dfrac{3 - 7 - 2°}{25 - 12 - 1}$   $\dfrac{1 - 8 - 3°}{2 - 15 - 6}$   $\dfrac{25 - 12 - 1}{2 - 15 - 6}$

5. Pap. → $\dfrac{1 - 8 - 3}{11 - 5 - 3}$      $\dfrac{1 - 8 - 5}{11 - 18 - 3}$

$\dfrac{1 - 8 - 3°}{11 - 18 - 3}$     $\dfrac{1 - 8 - 5}{11 - 5 - 3}$

6. Rei. $\dfrac{3 - 7 - 1}{3 - 7 - 3}$      $\dfrac{1 - 8 - 3}{1 - 8 - 3'}$ ←

$\dfrac{1 - 8 - 3}{3 - 7 - 3}$   $\dfrac{1 - 8 - 3°}{3 - 7 - 3}$   $\dfrac{1 - 8 - 3}{3 - 7 - 3}$   $\dfrac{1 - 8 - 3}{3 - 7 - 3}$   $\dfrac{1 - 8 - 3}{3 - 7 - 3}$

Fig. 2b.

Family No. 6 showed a typing pattern of considerable interest and has been the subject of a previous report[8]. The outcome of this mating was five siblings, all of whom gave an identical pattern of histocompatibility typing, namely A1, B8, DRW3, A3, B7, DRW3. The female propositus had severe relapsing CAH-A of 13 years' duration, whilst the other four HLA identical siblings, 2 males and 2 females, were completely healthy; mixed lymphocyte reactions (MLR) were performed among all members of this family, using blood lymphocytes both as stimulators (after mitomycin treatment) and responders, and the pattern of responses is shown in Table 4. Whereas the four healthy siblings were mutually unresponsive in MLR, the cells of propositus DR both stimulated and responded to cells of all of the histocompatible siblings.

TABLE 4

RESULTS OF MIXED LYMPHOCYTE CULTURE OF CAH-A PATIENT (DR), PARENTS AND SIBLINGS

| Responding cells | Stimulating cells (mitomycin-treated) | | | | | | | |
|---|---|---|---|---|---|---|---|---|
| | F | M | S1 (DR) | S2 | S3 | S4 | S5 | C |
| F | (1,240) | 9.8 | 3.8 | 4.9 | 4.4 | 5.2 | 4.1 | 20.1 |
| M | 16.1 | (509) | 6.2 | 5.8 | 7.2 | 6.4 | 6.8 | 14.3 |
| S1 (DR) | 7.4 | 9.3 | (390) | 4.1 | 3.6 | 4.8 | 5.2 | 15.4 |
| S2 | 6.9 | 7.2 | 5.1 | (414) | 1.1 | 0.8 | 1.2 | 20.9 |
| S3 | 5.8 | 10.2 | 3.7 | 1.3 | (470) | 1.9 | 2.0 | 17.2 |
| S4 | 7.1 | 7.8 | 4.3 | 1.0 | 1.7 | (548) | 0.9 | 16.6 |
| S5 | 5.7 | 8.7 | 4.9 | 1.6 | 1.2 | 1.7 | (212) | 15.9 |
| C | 18.6 | 20.2 | 15.9 | 19.2 | 18.7 | 17.6 | 20.8 | (809) |

F = Father; M = mother; S1 (DR) = patient; S2–S5 = patient's siblings; C = unrelated control. Results are presented as cpm in autologous combination and as SI in homologous situation: patient's results outlined in solid line, siblings in dotted line: SI = cpm against mitomycin-treated stimulating cells/cpm with no stimulating cells.

(From Dumble and Mackay[8], reproduced with permission of the Editor of Digestion and S. Karger AG, Basel).

Virus antibodies.

Because of the possibility that infection with hepatitis virus A(HAV) or B(HBV) may contribute to the pathogenesis of CAH-A, antibody to hepatitis A virus and to the core antigen of hepatitis B virus (HBc) was determined in 46 cases, and results were correlated with typing for HLA B8 (Table 5). The frequency of anti-HAV (50%) did not differ from that of the Melbourne population, and in 15 cases there was no serological evidence of past infection with HAV

or, as judged by anti-HBc, with HBV.  In those cases with reactivity
to HBc, there had been no previous evidence of an episode of HBV
infection.  The frequency of HLA B8 did not differ according to the
presence or absence of antibodies to hepatitis virus antigens in
these 46 cases.

TABLE 5

HEPATITIS VIRUS ANTIBODIES IN CAH-A[*]

| ANTIBODY TO | | CASES | | HLA B8 | |
|---|---|---|---|---|---|
| HAV | HBc | NO. | | | |
| + | - | 23 | (50%) | 12/17 | 71% |
| - | - | 15 | (33%) | 9/12 | 75% |
| - | + | 6 | (13%) | 2/3 ) | |
| + | + | 2 | (4%) | 1/1 ) | 75% |
| All | cases | 46 | | 24/33 | 72% |

* From Mackay, I.R., Genetic studies in chronic active hepatitis,
  in Proceedings of International Symposium on Immune Reactions
  in Liver Disease: Pathogenetic Role and Developments in
  Immunotherapy, (in press, 1978).

Family studies on autoantibodies

There has been no study for frequency of autoantibodies among
families of Melbourne cases of CAH.  Hence a review was made of two
published studies on the frequency of autoantibodies in blood
relatives of patients with CAH, from Galbraith et al.[14] in England
and Salaspuro et al.[15] in Finland, a and b respectively in Table 6.

TABLE 6

ANTIBODIES IN RELATIVES OF CASES CAH

| SUBJECTS | | NO. NO. | POSITIVE RESULTS % | | | | |
|---|---|---|---|---|---|---|---|
| | | | AMA | SMA | ANA | THY | GPC |
| CAH relatives | a. | 165 | 3.6[*] | 8.5[*] | 7.9 | 27 | 6.1 |
| | b. | 58 | 0 | 1.7 | 19[*] | 23[*] | 10[*] |
| Controls | a. | 260 | 0.4 | 1.9 | 4.6 | 18 | 4.6 |
| | b. | 504 | 0.6 | 1.4 | 3.0 | 14 | 2.2 |

* $p < 0.05$

Overall, there was an increase in frequency of all types of autoantibodies in the relatives, but to differing degrees in the two
series.

DISCUSSION

The association of HLA B8 with the autoimmune type of CAH has been reported from several centers[3]. Cases of CAH-B, associated with seropositivity for HBsAg, were not included in the present study; HLA B8 is reported as not increased in CAH-B, whilst the specificities A3[16] and Bw35[17,18] may be increased. Another autoimmune disease of the liver, primary biliary cirrhosis (PBC), is due to destruction of cholangioles rather than hepatocytes, and there is known to be over-lapping clinical and/or autoimmune serologic features of CAH and PBC in the one patient, or among related family members[14]; on the other hand, HLA B8 is not increased in PBC, indicating that the disease - association between CAH and PBC must depend on genetic factors other than those associated with HLA.

HLA DW3, which exists in linkage disequilibrium with B8, was first reported by Opelz et al.[19] to be increased in CAH, 68% versus 24% in controls, and to have a closer disease-association than did HLA B8. This did not hold for DRW3, according to the 7th Histocompatibility Workshop data[13], but with technical improvements in DRW typing in our laboratory, the frequency of DRW3 in CAH (76%) is similar to that of B8. In one particular family the HLA specificities of five siblings were identical on serologic typing, but lymphocytes of one sibling with CAH-A both stimulated and responded to the lymphocytes of all other siblings in mixed lymphocyte culture, suggestive of a "split" DW3 detected by MLC, but not serologically; the alternate gene product DRW3' (Figure 2b) could represent a genetic determinant of CAH-A.

There is no accepted explanation for the association of B and D locus specificities with immunopathic diseases including CAH and, were Ir gene effects to be operative, the antigens to which these are directed are unknown. In cases of CAH-A, HLA B8 has been correlated with hyperresponsiveness to particular viruses, measles and rubella[20], and DRW3 has been correlated with high titers of autoantibodies, ANA and ASMA. A gene associated with the B-D subregion of the histo-compatibility complex could determine responsiveness to an autoantigen of liver, possibly the liver-specific protein claimed to represent the autoantigen relevant to CAH-A[21].

Applications of knowledge of HLA associations with CAH include the facilitation of assigning cases of CAH to subgroups which, for example, may differ in clinical features, response to treatment[5,19] or outcome of disease. Indeed CAH is but one of several diseases which can be subgrouped according to HLA frequencies, others including myasthenia gravis (B8 increased in young females with "thymitis"), diabetes mellitus (B8 and Bwl5 increased in insulin-dependent type), thyro-gastric autoimmune disease (B8 increased in thyrotoxicosis but not in other components of this group). Moreover, the splitting of CAH into types associated with autoimmunity and hepatitis B virus has interesting geo-ethnic connotations[22]; notably, the former appears to predominate in populations derived from countries in which the prevalence of HLA-B8 is high, Great Britain and Scandinavia[23], whereas the latter appears to predominate among populations derived from countries in which the prevalence of HLA B8 is low, namely Mediterranean, African and Asian countries.

Genetic determinants for CAH other than those associated with HLA B8 might include the following. Firstly an immunoaugmentative effect related to the X-chromosome is suggested because CAH-A predominates among females. Secondly a genetic defect in immuno-regulation or maintenance of tolerance is suggested because among families of patients with CAH there is a higher than expected frequency of various types of autoantibodies. Thirdly a determinant not often considered in human autoimmune disease is inherited vulnerability of the target organ, postulated as a factor in auto-immune thyroiditis of the obese chicken[24], and in autoimmune gastritis in man[25]. The importance of this determinant in human liver disease may not be great, as judged by the scanty literature on familial pre-disposition to cirrhosis[26], although mouse strains are known to differ widely in susceptibility to hepatotoxins, notably carbon tetra-chloride[27].

CONCLUSIONS

The genetic contribution to the cause of chronic active hepatitis could have four components, as follows.

Firstly, the evidence from associations with HLA points to an Ir gene-related effect, similar in nature to the determinant in other HLA B8 associated organ-specific autoimmune disease.

Secondly, the predominance in females of this and other immunopathic diseases, together with evidence from human and animal studies of higher immune responsiveness in females, points to an immunopotent-iating effect associated with the X-chromosome.

Thirdly, the coexistence within some families of CAH-A and other autoimmune conditions, and the increased prevalence of various auto-antibodies in relatives, points to a genetic defect in immunoregu-lation or suppressor T cell function.

Fourthly, there may be an inherited vulnerability of the target organ itself, the liver, to any type of damaging influence.

ACKNOWLEDGEMENTS

We thank Miss Carole Wilson, SRN, for arranging attendance of patients and families, and Mrs Susan Mansfield, BSc, for her technical expertise in HLA typing procedures. Dr Mackay is in receipt of a grant from the National Health and Medical Research Council of Australia.

REFERENCES

1. Mackay, I.R. (1975) in Frontiers of Gastrointestinal Research, vol. 1. Van der Reis, L. ed., Karger, Basel, pp. 142-187.

2. Fogarty International Center and International Association for for Study of the Liver Standardization Conference (1976) Fogarty International Center Proceedings No. 22, DHEW Publication No. (NIH) 76-725, Bethesda, pp. 1-212.

3. Mackay, I.R. (1977) in HLA and Disease, Dausset, J. and Svejgaard, A. eds., Munksgaard, Copenhagen, pp. 186-195.

4. Böyum, A. (1968) Scand. J. Clin. Lab. Invest. 21, Suppl. 97, 31-50.

5. Morris, P.J., Vaughan, H., Tait, B.D. and Mackay, I.R. (1977) Aust. N.Z. J. Med. 7, 616-624.

6. Nelson, D.L., Strober, W., Abelson, L.D., Bundy, B.M. and Mann, D.L. (1977) J. Immunol. 118, 943-946.

7. Grier, J.O., Abelson, L.A., Mann, D.L., Amos, D.B. and Johnson, A.H. (1977) Tissue Antigens, 10, 236.

8. Dumble, L.J. and Mackay, I.R. (1977) Digestion, 15, 254-259.

9. Lehmann, N.I. and Gust, I.D. (1977) Med. J. Aust. 2, 731-732.

10. Vyas, G.N. and Roberts, I.M. (1977) Vox Sang. 33, 369-372.

11. Simon, M., Bourel, M., Fauchet, R. and Genetet, B. (1976) Gut, 17, 332-334.

12. Mathews, J.D., Mathieson, I.D. and Tait, B.D. (1978) in Proc. 1st International Symposium on Immunological Factors in Atherosclerosis, Florence, Academic Press, New York, London (in press).

13. Histocompatibility Testing, 1977, in Proc. 7th International Histocompatibility Workshop, Oxford (in press, 1978).

14. Galbraith, R.M., Smith, M., Mackenzie, R.M., Tee, D.E., Doniach, D. and Williams, R. (1974) New Engl. J. Med. 290, 63-69.

15. Salaspuro, M.P., Laitinen, O.I., Lehtola, J., Makkonen, H., Rasanen, J.A. and Sipponen, P. (1976) Scand. J. Gastroent, 11, 313-320.

16. Mazzilli, M.C., Trabace, S., Di Raimondo, F., Visco, G. and Gandini, E. (1977) Digestion, 15, 278-285.

17. Hillis, W.D., Hillis, A., Bias, W.B. and Walker, W.G. (1977) New Engl. J. Med. 296, 1310-1314.

18. Penner, E., Grabner, G., Dittrich, H. and Mayr, W.R. (1977)
    Tissue Antigens, 10, 63-64.

19. Opelz, G., Vogten, A.J.M., Summerskill, W.H.J., Schalm, S.W.
    and Terasaki, P.I. (1977) Tissue Antigens, 9, 36-40.

20. Galbraith, R.M., Eddleston, A.L.W.F., Williams, R., Webster,A.D.B.,
    Pattison, J., Doniach, D., Kennedy, L.A. and Batchelor, J.R.
    (1976) Lancet, 1, 930-934.

21. Hopf, U., Meyer zum Büschenfelde, K.-H. and Arnold, W. (1976)
    New Engl. J. Med. 294, 578-582.

22. Whittingham, S., Mackay, I.R., Thanabalasundrum, R.S.,
    Chuttani, H.K., Manjuran, R., Seah, C.S., Yu, M. and
    Viranuvatti, V. (1973) Brit. Med. J. 4, 517-519.

23. Ryder, L.P., Andersen, E. and Svejgaard, A. (1978) Human Hered.
    28, 171-200.

24. Rose, N.R., Bacon, L.D. and Sundick, R.S. (1976) Transplant.
    Rev. 31, 264-285.

25. Ungar, B., Francis, C.M. and Cowling, D.C. (1976) Med. J. Aust.
    2, 900-902.

26. Maddrey, W.C. and Iber, F.L. (1964) Ann. Intern. Med. 61, 667-679.

27. Hill, R.N., Clemens, T.L., Liu, D.-K., Vesell, E.S. and
    Johnson, W.D. (1975) Science, 190, 159-160.

DISCUSSION

GASSER: If the 1-8-3 haplotype, or DRw3 allele in particular, predisposes an individual toward chronic active hepatitis, wouldn't you expect homozygotes to be even more susceptible than heterozygotes?

MACKAY: Possibly so. One study (Page et al., J. Clin. Invest., 56:530, 1975) found increased homozygosity for HLA-B8 in CAH, but another study (Freudenberg et al., Digestion, 15:260, 1977) gave no support for this.

VOLPE: I am interested in the vulnerability of the target organ to immune attack. Is there any evidence in patients who will develop chronic active hepatitis <u>later</u>, that their livers are more vulnerable to injury?

MACKAY: I have no information on this. Vulnerability in relation to the target organ in human autoimmune disease is still a rather vague concept and would be hard to assess. There are various liver diseases in which such vulnerability may be one component, notably alcoholic cirrhosis which develops in only a relatively small proportion of heavy drinkers of alcohol.

WARNER: Have there been any studies on T cell responsiveness to either hepatitis A or B antigens? In view of the animal studies that indicate genetic control at the T cell level, this may be of relevance to possible response to viral antigens in chronic active hepatitis.

MACKAY: There are no studies on T cell responsiveness to hepatitis A virus. There do exist studies on T cell responses to hepatitis B surface antigen, by leucocyte migration inhibition and by lymphoblast transformation, and these show that hyporesponsiveness exists in HBsAg positive CAH (CAH-B). However, in such studies, HLA or other genetic markers have not been included.

VOLPE: I would like to respond to the point made by Dr. Roitt indicating that since relatives of patients with autoimmune thyroid disease have a high incidence of thyroid autoantibodies and thyroid abnormalities, this might mean that this target organ might be more vulnerable. However, these data could well be explained by an abnormality of the immune system, to which the target organ is captive, without the target tissue necessarily being more susceptible in any way.

ROITT: That is correct. The interpretation of the data, however, would say that there are (presumably) genetic factors which affect the production of an autoimmune response to <u>either</u> the thyroid <u>or</u> the stomach in these families.

ENGLEMAN: It is disappointing that there are so little data to support the concept of an antigen-specific abnormality of T proliferation in these HLA-B8, Dw3 associated, organ-specific autoimmune disorders. Is it possible that the "important" antigens have yet to be discovered? Dr. Dupont has presented some interesting findings on this point elsewhere, and I wonder if he would share these findings with us here.

DUPONT: Dr. S. Cunningham-Rundles in my laboratory has demonstrated that normal blood donors who are HLA-B8 are high responders in <u>in vitro</u> lymphocyte transformation to wheat gluten antigen (Frazer's fraction III). There was no general increase in lymphocyte

responses among the B8 positive donors as compared to the B8 negative
donors when tested with mitogens, Candida albicans antigen, and allogeneic cells. These data are consistent with the possibility that
response to wheat gluten is regulated by a gene in linkage disequilibrium with B8. This may provide a new tool for evaluation of Ir-genes
in patients with B8 associated autoimmune diseases.

MACKAY: I agree with Dr. Dupont's interpretation and will set about
establishing this by typing with known Dw3 cells in MLC. The question
is whether the alternate Al B8 DRw3 haplotype has a particular relationship to CAH in the patient.

IRVINE: While individuals may vary in the susceptibility of
particular tissues to damage, it should also be stressed that they
may vary in their ability to regenerate damaged cells in these organs.
This, for example, may make the difference between Hashimoto goiter
and primary atrophic hypothyroidism, and why some persons with organ-
specific autoantibodies in the serum fail to progress to clinical
disease of that organ.

MACKAY: Perhaps particular types of autoantibodies may specifically
interfere with regeneration.

IRVINE: Consideration of the relationship between HLA type and Ir
gene should take into account not only the presence of the antibody
but also the persistence of the antibody in the serum. This is seen
to be particularly relevant in Type I (insulin-dependent) diabetes,
and probably also in thyrotoxicosis.

MACKAY: Mostly autoantibodies tend to persist. The pancreatic
islet cell antibody is exceptional in this regard. This may be
related to the relatively early occurrence of "antigen-exhaustion",
as islets are destroyed.

TALAL: You postulate an X-linked immune response gene as a
causative factor in autoimmune liver disease. Don't you think it
equally likely that the sex factor in autoimmunity is related to
sex hormones, with estrogen predisposing to autoimmunity?

MACKAY: Not equally likely. The second peak of incidence of CAH
occurs in older women, past 60 years of age, in whom hormonal
influences should not be operative.

TALAL: The development of autoimmunity in the aged, or indeed
in males, does not argue against the role of estrogen. Autoimmune
diseases are multifactorial and sex hormones are not the only factor
involved. There are immunoregulatory defects associated with aging
which themselves predispose to autoimmunity.

MACKAY: While recognizing your evidence for regulatory effects of
estrogenic hormones in NZ mice, there is evidence for an X chromosomal
influence on immune responsiveness, particularly as it has been shown
for various species, including man, that females as compared to males
show greater responsiveness to antigenic stimulation. Besides there
is an excess female predisposition to autoimmune states before puberty,
and after menopause.

HARRISON: In relation to tissue susceptibility to autoimmunity, is
there any evidence that organ-specific autoantibodies are more reactive
with tissues from affected patients and their relatives compared to
normals?

MACKAY: Such studies requiring biopsy material from patients with organ-specific autoimmune diseases have not been done. Moreover, the conventional technique of immunofluorescence might not be sufficiently sensitive to detect such differences. Some 20 years ago in my laboratory, it was shown for non-organ-specific autoimmune diseases that autologous tissues were less reactive with certain autoantibodies than homologous tissues.

BIGAZZI: You have mentioned that in disorders like Hashimoto and pernicious anemia one may postulate a genetic disposition to damage of the target organs. However, there does not seem to be an association between HLA (serologically defined) and production of thyroglobulin autoantibodies, microsomal autoantibodies and parietal cell autoantibodies. How do you explain this discrepancy versus other autoimmune diseases?

MACKAY: Some studies have shown an association between HLA specificities and presence and/or titer of autoantibody, but most show association of HLA specificities with disease per se rather than with marker autoantibodies. Dr. Irvine has drawn attention to associations between HLA B8 and persistence of certain autoantibodies.

ROITT: We are still awaiting results of DRw typing in Hashimoto's disease. Supposing, as is likely, that there is some gene in the 'B-D' region predisposing to the development of thyroid autoimmunity, it surely is not mandatory for there to be linkage disequilibrium with the existing B and especially the less well-defined D specificities?

SVEJGAARD: Empirically we know that all HLA-factors known today show linkage-disequilibrium with one or more other HLA factors from other segregant series. However, this does not exclude that some yet unknown HLA factors may not show this phenomenon even when large numbers of individuals are studied.

ROSE: In connection with your observation of heightened stimulation indices of the chronic active hepatitis patients in mixed lymphocyte cultures (MLC) with siblings, do you have information in other MLC responses or T cell mitogenic responses (PHA, Con A) of CAH patients?

MACKAY: In general patients with CAH tend to be hyporesponsive in assays for cell-mediated immune functions, including PHA responses, but MLC responsiveness has not been studied.

DUPONT: With regard to the question from Dr. Rose, I would like to clarify, that the particular family MLC study presented by Dr. Mackay demonstrated that the mother, who is homozygous for the A1, B8, DRw3 haplotype is HLA-D heterozygous. This was shown in the MLC by the fact that her cells stimulated the cells of all the children. This also demonstrates that the affected child is the only one who has inherited one of the A1,B8,DRw3 haplotype while all other children inherited the other maternal A1, B8, DRw3 haplotype. This would then fully explain, why the patient's cells stimulated and responded to all other siblings.

# THE GENETICS AND IMMUNOLOGY OF GRAVES' AND HASHIMOTO'S DISEASES*

ROBERT VOLPÉ

Physician-in-Chief, The Wellesley Hospital; Professor, Department of
Medicine, University of Toronto, Toronto, Ontario, Canada

It is clear that both Graves' and Hashimoto's diseases aggregate in
specific families, and thus appear to be genetically induced[1].  Indeed
these two disorders tend to occur in the same families, and may even
coexist within the same thyroid gland[1,2].  Moreover, homozygous twins
have been reported where one twin has Graves' disease and the other has
Hashimoto's thyroiditis[3,4].  The autoimmune nature of these two
closely-related disorders is beyond dispute and is the subject of
recent reviews[5,6,7].

The increased incidence of other autoimmune diseases in patients with
Graves' or Hashimoto's disease (as well as in their families) is now
well-known;  these include diabetes mellitus[1], pernicious anaemia[1],
myaesthenia gravis[1], rheumatoid arthritis[8], idiopathic thrombocytopenic
purpura[9], Addison's disease[1], vitiligo[1] and chronic active hepatitis[1].
In addition, functional thyroid disturbances and thyroid autoantibodies
occur in about half of the first-order relatives of patients with
autoimmune thyroid disease[10].

## Clinical observations in twins

The occurrence of Graves' disease in both siblings of dizygotic
twins is reported to be about 3-9% and of monozygotic twins to be
30-60%[3,11,12,13,14,15].  The fact that monozygotic twins have a
higher concordance rate is strong evidence for a genetic basis for
Graves' disease, but genetic factors alone do not explain why some
develop the disease, while others do not, i.e., the lack of concordance
in 40-70%.  It is of further interest that in even highly selective
case reports of twin studies of Graves' disease, the ages of occurrence
vary greatly, even as much as 10 years in siblings under 18 years of
age.  Thus it would appear that influences other than purely genetic
are necessary before the disease is expressed[14,15].

---

*Work cited from this laboratory was supported by a grant from the
Medical Research Council of Canada (MT859).

## Age-specific incidence rates in Graves' disease

In 1963, Burch and Rowell[16] showed mathematical evidence to support
the notion that Hashimoto's thyroiditis occurs randomly in a
genetically-predisposed population.  From the statistics obtained by
Burch and Rowell, it was concluded that there was no reason to suppose
that Hashimoto's thyroiditis was primarily a "disturbed antigen"
disease, as had been generally assumed;  rather it appeared to be a
condition due to disturbed tolerance, initiated by somatic mutations
of "forbidden clones" of lymphocytes.  Their data also were not in
keeping with the probability of infection or any other thyroid gland
injury.

Following their mathematical analysis, there was some criticism of
their mathematics, with respect to the proportion of the population at
risk, resulting in a lively exchange of letters[17-20].  However, even
the critics conceded that the original data did indicate a random
appearance of the disorder in a genetically-predisposed group, as
proposed by Burch[21,22].

Since Hashimoto's thyroiditis had a relatively high incidence in
females, Burch and Rowell[16] felt that there would be a central role for
the X chromosome in antibody synthesis.  Since the age-specific
incidence rates suggested a high penetrance (almost all of the population at risk develop the disease) within the normal lifespan, it
seemed probable that this subpopulation is characterized, at least in
part, by functionally dominant genes on the X chromosome.  On the basis
of the statistics cited by these authors, the most obvious interpretation was that individuals in this subpopulation inherit dominant
mutant genes on the X chromosomes.  Thus females had at least twice the
chance of having the genetic defect.  If several genetic defects are
necessary for the production of the disease, the ratio can be even
higher.

We have reviewed data for the age-specific incidence rates and sex-
incidence of Graves' disease in relation to the proportion of the
general population in each decade of life[14,15].  The relative age-
specific incidence rate in Graves' disease is shown on Figure 1.
This represents the age-specific incidence rates of 873 cases of
Graves' disease (184 males, 689 females) obtained at a University of
Toronto clinic over the interval 1959-69.

In order to arrive at an age-specific incidence rate for each decade
interval, it was necessary to relate the number of cases of the disease
appearing in each decade of life to the total population of the same

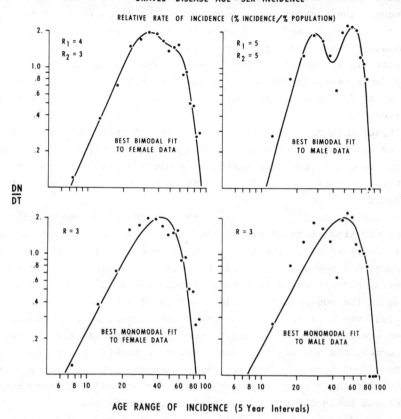

GRAVES' DISEASE AGE - SEX INCIDENCE

RELATIVE RATE OF INCIDENCE (% INCIDENCE/% POPULATION)

$R_1 = 4$
$R_2 = 3$

BEST BIMODAL FIT
TO FEMALE DATA

$R_1 = 5$
$R_2 = 5$

BEST BIMODAL FIT
TO MALE DATA

$\dfrac{DN}{DT}$

$R = 3$

BEST MONOMODAL FIT
TO FEMALE DATA

$R = 3$

BEST MONOMODAL FIT
TO MALE DATA

AGE RANGE OF INCIDENCE (5 Year Intervals)

Fig. 1. Reprinted from Volpé[15] with permission.

age group. The latter was arbitrarily chosen as the population of
Ontario taken from the 1966 census. We are permitted the assumption
that the age proportions of the population in our referral area are the
same as the age proportions generally in the province. One thus
divides the proportion of cases of Graves' disease in that decade by
the proportion of people in the province in that age decade. Separate
age-specific incidence rates are expressed for each sex.

It may be noted that there appears to be a surprising drop in the
age-specific incidence rate in the male group (and to a lesser degree
female group) in the fifth decade. These rather unexpected reductions
might represent either some manner of bias unconsciously introduced
into the selection of cases, or that the curves are bimodal rather than
monomodal. A bimodal curve would indicate that patients with Graves'

disease arise from two genetic populations.  For a variety of reasons, however, the bimodal variations between male and female groups seem most improbable, and it is unlikely that Graves' represents two populations.  For the present, therefore, we will take the view that the curves are monomodal, at least until larger series of age-specific incidence rates for Graves' disease from other centres become available. We have also obtained similar data for Hashimoto's thyroiditis in Toronto[23].

In non-mathematical terms, the meaning of these data is as follows: if at birth a subpopulation is at special risk through inheritance, with respect to the development of a given disease, then as this subpopulation ages, more and more members of it will succumb to the disease.

If the probability of onset of the disease increases rapidly with age, then the proportion of predisposed members who have not succumbed to it will decrease rapidly with age.  When this proportion becomes small enough, the age-specific incidence rate, expressed with respect to the general population, will fail to increase with age.  Consequently, a peak in the age-specific incidence rate will be attained, and this will be followed by a falling incidence with increasing age.  Theoretically, when every member of the predisposed population has finally been affected, the age-specific incidence rate should fall to zero.

Burch and Rowell[16] have suggested that the aetiology of conditions which fit the type of curves obtained for Graves' and Hashimoto's diseases would conform to the following conditions:

(i)   the disease is restricted to a (genetic) carrier subpopulation;

(ii)  onset of the disease requires the accumulation in a carrier individual of at least two discrete changes;

(iii) these discrete changes or events are random in nature and their average probability of occurrence is constant with respect to time (somatic mutation is a perfect example of such events);

(iv)  the penetrance of the inherited tendency to the disease approaches unity within the normal lifespan;

(v)   the average age-specific mortality rates are the same in the general population and in the carrier subpopulation before the onset of the autoimmune disease.

Thyroid autoimmunity and chromosomal abnormalities

Graves' disease, other clinical thyroid disorders and thyroid auto-antibodies have all been observed more often than expected amongst patients with Down's syndrome (mongolism) and their maternal relatives[1]. Some patients with gonadal dysgenesis and perhaps their mothers also have increased frequency of thyroid autoimmunity[1]. The biological implications of these associations are unclear, but it is possible that common pathogenic mechanisms are involved in the development of chromosomal aberrations and thyroid autoimmunity.

## HLA associations with Graves' and Hashimoto's diseases

HLA is a region on chromosome 6 which includes several genetic loci. The effects of most of these genes are expressed on cell surfaces and involve cell-cell interactions. HLA is structurally and functionally homologous to the major histocompatibility region in mice, rats, dogs, Rhesus monkeys and other higher vertebrates. Many biologically important functions, including qualitative and quantitative control of the immune responses to certain antigens, killing of viral-infected cells, and synthesis of several complement components have been associated with the major histocompatibility complexes of several species. However, these complicated effects are not directly assessed in testing for disease associations in man. For such purposes, the HLA cell surface "antigens" are used, because they are relatively easy to identify and because many frequent alternative genes are known for each locus. At least four loci have been clearly demonstrated within the HLA region. The A and B loci determine the classical transplantation antigens identified by serological methods. The C locus determines another series of serologically detected antigens, but their functions are unknown. D locus antigens cannot be detected by conventional serological testing, and are detected by mixed leukocyte cultures[1].

In Graves' disease, there have been several studies of HLA association. In Caucasians, an increased frequency of the B-locus antigen HLA-B8 has been found[1,24]. The relative risk of Graves' disease in persons with HLA-B8 compared to persons lacking this antigen is 2.4[1]. In addition, HLA-Dw3 has been found increased in incidence in Graves' disease in Caucasians; the relative risk for Graves' disease is 5.2 in persons with this HLA antigen[1,25,26].

It is of interest that Irvine et al.[27] have shown that patients with Graves' disease who go into remission following antithyroid drugs are much more unlikely to have HLA-B8 than those who do not go into remission. That is, the presence of HLA-B8 appears to be associated with an increased incidence of persistence of the disease without remissions.

In Japanese, HLA-B8 is only rarely found. None of a series of Japanese patients with Graves' disease or their controls had HLA-B8, but

HLA-Bw35 was significantly more often observed amongst the patients than amongst their controls[28]. The relative risk for Graves' disease in Japanese persons with HLA-Bw35 was 5.1[1]. There is also an increased incidence of HLA-DHO in Graves' disease in Japanese[29,30].

In Chinese persons, at least one study has shown an increase in HLA-B46[31]. If confirmed, this will indicate a difference in the genetic make-up of Chinese and Japanese persons. However, many patients with Graves' disease in these various populations do not have the "appropriate" HLA antigen. Furthermore, their first-order relatives, despite having functional disturbances of the thyroid gland, have no increased incidence of HLA-B8 (in Caucasians)[10]. Moreover, in Hashimoto's thyroiditis, only one study has suggested a relationship of HLA-B8 to the disorder, whereas three other studies have failed to show any such relationship[7].

It therefore seems more probable that linkage of defined HLA loci to disease predisposing genes is responsible for the HLA associations observed with Graves' disease. It might be that disease susceptibility is due to human homologues of the murine Ir (immune response) or Is (immune suppression) genes, which modulate the strength and characteristics of the immune response to certain specific antigens. Ir genes appear to be important in experimental autoimmune thyroiditis in mice, but as yet neither Ir nor Is genes have been unequivocally demonstrated in man[1].

Immune nature of Graves' and Hashimoto's diseases

Space will not permit more than a brief outline of the immune mechanisms that might be involved in autoimmune thyroid disease. However, it does seem increasingly evident that Graves' disease is closely related to Hashimoto's thyroiditis, both genetically and pathogenetically, and that each is caused by closely related immunological disorders[5-7]. Despite the relationship, there are at least a few aspects of each condition which continue to separate the two maladies, so that they cannot be considered merely different expressions of a spectrum of a single entity[7].

While the pathogenesis of these two conditions is not yet clarified, a working hypothesis may be proposed which can act as a framework on which further studies can be based. It seems probable that both disorders are due to specific genetic defects in immune surveillance or control; abnormalities in specific populations of suppressor T lymphocytes may be the basis of such defects[6,7]. With such a defect, a normally randomly mutating, thyroid-directed, self-reactive "forbidden clone" of thymus-dependent (T) helper lymphocytes is permitted to survive. It may well be that all normal persons have the capacity to produce "forbidden clones" of self-reactive lymphocytes (by a process of random mutation) directed towards normal body constituents. In normal individuals, however, suppressor T lymphocytes will exercise normal immunological surveillance or control, thus suppressing these "forbidden clones" of lymphocytes and preventing them from interacting with their complementary antigen.

In persons genetically predisposed to either Graves' or Hashimoto's disease, it may only take the random appearance of the appropriate thyroid-directed "forbidden clone" of helper T lymphocytes, which then escape normal control because of the presumed genetic defect. While there have been suggestions that it is necessary to have some thyroidal antigenic alteration (possibly induced by viral interaction), it is at least equally possible that no such antigenic alteration is necessary. In any event, once the appropriate "forbidden clone" of T lymphocytes has survived, it would then interact with its complementary antigen, presumably on the thyroid cell membrane, setting up a localized cell-mediated immune (CMI) response. Subsequently, the same helper T lymphocytes cooperate with and direct groups of already-present appropriate bursa-equivalent (B) lymphocytes, which, in consequence, produce the polyclonal thyroid autoantibodies. In the case of Graves' disease, these take the form of thyroid-stimulating immunoglobulins (TSI), which appear to be antibodies directed against the thyrotrophin (TSH) receptor. The TSI, after its interaction with the TSH receptor on the thyroid cell membrane, then appears to stimulate the thyroid follicular cells in a manner indistinguishable from TSH[7]. The other thyroid auto-antibodies may exert deleterious effect on the thyroid cells, either alone, or as antigen-antibody complexes, or in combination with lympho-cytes or "killer" cells[6,7] (Figure 2).

Space will also not permit a description of the evidence on which the above hypothesis is based; similarly, no further discussion of the role of thyroid-stimulating immunoglobulins or other thyroid antibodies is possible. For an examination of these topics, as well as speculations on the role of stress on the precipitation of hyperthyroidism, the nature of remissions in Graves' disease, and the aetiology of exophthal-mos, the reader is referred to reviews published elsewhere[5,6,7]. However, it may be of interest to discuss results of lymphocyte culture experiments in Graves' disease and preliminary attempts to produce an experimental model of this disorder.

50

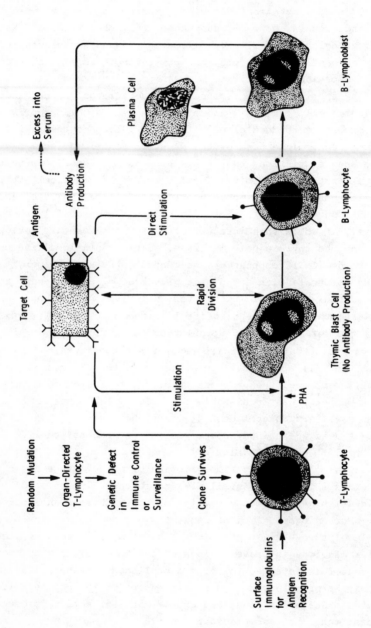

Fig. 2. Suggested schema for the initiation of Graves' disease.
Reprinted with permission from Volpé6.

## Results of lymphocyte culture

Lymphocytes when cultured with phytohaemagglutinin will undergo blast transformation;  this has been detected morphologically by the uptake of tritiated thymidine by these lymphocytes[7].  PHA has been shown to stimulate T lymphocytes only (and not B lymphocytes)[32] Unfortunately, this has not proved to be a useful method of detecting cell-mediated immune responses in autoimmune thyroid diseases.  Conflicting reports of evidence for blast transformation when Graves' or Hashimoto's lympho-cytes are cultured with PHA, or indeed various thyroid antigens, allow no firm conclusions to be drawn[7].  However, when the end-point is the production of thyroid-stimulating immunoglobulin (TSI), interesting results with lymphocyte cultures can be demonstrated.  If lymphocyte cultures from Graves' disease are cultured with either PHA[33], or crude normal thyroid antigen[34], or a membrane-rich human thyroid fraction[35], the lymphocytes will produce thyroid-stimulating immunoglobulin into the medium.  This can be detected by incubating the medium with human thyroid cell slices, which consequently show evidence of increased cyclic AMP production.  Since PHA stimulates only T lymphocytes, the former study implies that PHA stimulated thyroid-directed helper T lymphocytes, which then interacted with thyroid-directed B lymphocytes; the B lymphocytes then responded by releasing TSI into the culture medium.  Lymphocytes from patients with non-immune thyroid disease, or healthy controls, generally do not produce TSI into the medium under these conditions.  The fact that normal human thyroid fractions can stimulate previously sensitized Graves' lymphocytes suggests that no antigen alteration is necessary for lymphocyte interaction[34,35].

## Experimental production of Graves' disease

While experimental autoimmune thyroiditis has been convincingly demonstrated, an experimental model for Graves' disease has not yet been produced.  Attempts have been made to immunize animals with thyroid antigens, so as to produce TSI, but the demonstration of this immunoglobulin has not been convincing, and the animals have not been hyperthyroid.  Recently, Ong et al.[36] have been able to raise an anti-body in rabbits against bovine thyroid, which exhibited thyroid-stimulating properties in vitro.

However, Silverberg et al.[37] have recently shown that the infusion of lymphocytes from patients with Graves' disease into "nude" (athymic) mice who have been previously prepared with cyclophosphamide has result-ed in a transient rise in the serum thyroxine, reaching a peak at about two weeks, and falling thereafter to normal.  This does not occur when mice are infused with normal lymphocytes, nor will Graves' lymphocytes cause such changes in normal mice. This finding cannot be explained by the immunoglobulin trapped within the lymphocytes at the time of the injection, since such an effect (if it existed at all) would be short-lived. Since the athymic mice have very little immune surveillance function, it seems likely that the lymphocytes survived for at least several days under these circumstances, interacted with the mouse thyroid, and stimulated the thyroid (possibly by virtue of the production of TSI) for a matter of several days. These findings are still preliminary, and require confirmation and exploitation. They appear to be at least consistent with the view that the basic defect in Graves' disease is one of immune control.

Summary

It is the author's view that Graves' disease is a genetically-based disorder, where the genetic abnormality results in a specific defect in the immune control mechanisms, possibly a suppressor T lymphocyte defect. This permits a randomly-appearing thyroid-directed "forbidden clone" of helper T lymphocytes to survive, interact with previously normal thyroid cell membranes, and initiate the disease by cooperating with groups of already-present appropriate B lymphocytes. The latter in consequence produce thyroid-stimulating immunoglobulin to complete the pathophysiological expression of Graves' disease.

REFERENCES

1. Friedman, J.M. and Fialkow, P.J. (1978) Clin. Endocr. Metab., 7, 47-65.

2. Fatourechi, V., McConahey, W.M. and Woolner, L.B. (1971) Mayo Clin. Proc., 46, 682-689.

3. Jayson, M.I.V., Doniach, D., Benhamou-Glynn, N., Roitt, I.M. and El Kabir, D.J. (1967) Lancet, 2, 15-18.

4. Chertouv, B.S., Fidler, W.J. and Fariss, B.L. (1973) Acta Endocrinologica, 72, 18-24.

5. Volpé, R., Farid, N.R., von Westarp, C. and Row, V.V. (1974) Clin. Endocr., 3, 239-261.

6. Volpé, R. (1977) Ann. Int. Med., 87, 86-99.

7. Volpé, R. (1978) Clin. Endocr. Metab., 7, 3-29.

8. Monroe, R.T. (1935) N. Engl. J. Med., 212, 1074-1077.

9. Dunlap, D.B., McFarland, K.F. and Lutcher, C.L. (1974) Am. J. Med. Sci., 268, 107-111.

10. Chopra, I.J., Solomon, D.H., Chopra, U., Yoshihara, E., Terasaki, P.I. and Smith, F. (1977) J. Clin. Endocr. Metab., 45, 45-54.

11. Lehman, W. (1964) in Humangenetek, Ein kurzes Handbuch in 5 Banden, Becker, P.E. ed., Georg Thieme Verlag, Stuttgart, V.3, part 1, pp. 182-223.

12. Vogel, F. (1959) Ergebn. inn. Med. Kinderheilk, 12, 52-125.

13. Bartels, E.D. (1941) Heredity in Graves' Disease, Munksgaard, Copenhagen, pp. 34-36.

14. Volpé, R., Edmonds, M.W., Clarke, P.V. and Row, V.V. (1970) Acta Endocr. Panamericana, 1, 155-170.

15. Volpé, R., Edmonds, M.W., Lamki, L., Clarke, P.V. and Row, V.V. (1972) Mayo Clin. Proc., 47, 824-834.

16. Burch, P.R.J. and Rowell, N.R. (1963) Lancet, 2, 507-513.

17. Burch, P.R.J. (1963) Lancet, 2, 636-637.

18. Burch, P.R.J. (1963) Lancet, 2, 836-837.

19. Maynard-Smith, S. and Maynard-Smith, J. (1963) Lancet, 2, 357-359.

20. Maynard-Smith, S. and Maynard-Smith, J. (1963) Lancet, 2, 738.

21. Burch, P.R.J. (1966) J. Theor. Biol., 12, 397-409.

22. Burch, P.R.J. (1966) in Radiation and Ageing, Lindop, P.J. and Sacher, G.A. eds., Taylor and Francis Ltd., pp. 117-155.

23. Volpé, R., Clarke, P.V. and Row, V.V. (1973) Canad. M.A.J., 109, 898-901.

24. Grumet, F.C., Konishi, J., Payne, R.O. and Kriss, J.P. (1974) J. Clin. Endocr. Metab., 39, 1115-1119.

25. Thorsby, E., Segaard, E., Solem, J.H. and Kornstad, L. (1975) Tissue Antigens, 6, 54-64.

26. McMichael, A., Sasazuki, R., Payne, R.O., Grumet, F.C., McDevitt, H. and Kriss, J.P. (1975) in Histocompatibility Testing, Munksgaard, Copenhagen, pp. 769-772.

27. Irvine, W.J., Gray, R.S., Morris, P.J. and Ting, A. (1977) Lancet, 2, 898-900.

28. Grumet, F.C., Konishi, J., Payne, R.O. and Kriss, J.P. (1975) Tissue Antigens, 6, 347-352.

29. Kawa, A., Nakamura, S., Nakazawa, M., et al. (1978) Proc. 6th Asia and Oceania Congress of Endocrinology, Singapore, Jan. 22-27, Abstract 142.

30. Sasazuki, T., Kohns, Y., Iwamoto, I., Kosaka, K., Okimoto, K., Maruyama, H., Ishiba, S., Konishi, J., Takeda, Y., and Naito, S. (1978) New Engl. J. Med., 298, 630-631.

31. Chan, S.H., Yeo, P.P.B., Lui, K.F., et al. (1978) Proc. 6th Asia and Oceania Congress of Endocrinology, Singapore, Jan. 22-27, Abstract 141.

32. Greaves, M., Janoussy, G. and Doenhoff, M. (1974) J. Exp. Med., 140, 1-18.

33. Knox, A.J.S., von Westarp, C., Row, V.V. and Volpé, R. (1976) Metabolism, 25, 1217-1223.

34. Knox, A.J.S., von Westarp, C., Row, V.V. and Volpé, R. (1976) J. Clin. Endocr. Metab., 43, 330-337.

35. Sugenoya, A., Silverberg, J., Kidd, A., Row, V.V. and Volpé, R. (1978) J. Invest. Endocr., in press.

36. Ong, M., Malkin, D.G., Tay, S.K. and Malkin, A. (1976) Endocrinology, 98, 880-885.

37. Silverberg, J., Kidd, A., Sugenoya, A., Row, V.V. and Volpé, R. (1978) Proc. European Thyroid Assoc., Berlin, Sept. 4-7.

DISCUSSION

MACKAY: Since there can be overall few reports on thyroid auto-immune disease in monozygous twins, could not a precise figure rather than a range (30-60%) be cited for concordance?

VOLPE: The studies that I have cited from the literature have used differing sources for their original data (see references from my presentation). Thus, without a common data base, one could hardly anticipate a precise figure. While, therefore, I have quoted a range, the point is still valid: the concordance rate is very much higher in monozygotic twins as opposed to dizygotic twins. Further-more, even if one accepts the highest reported figure for concor-dance in monozygotic twins, i.e., 60%, there still is a lack of concordance of 40%.

MACKAY: Vitiligo was listed among autoimmune diseases associated with thyroid autoimmunity. I do not regard the evidence for vitiligo as an autoimmune disease as sufficient.

VOLPE: From recent immunofluorescent studies, there is evidence that antibodies against melanocytes can be demonstrated in vitiligo. This would appear to mark it as at least a putative autoimmune disease. There is, of course, the additional circumstantial evi-dence of "guilt by association" since vitiligo is so commonly associated with organ-specific autoimmune disorders.

MACKAY: You referred to subcellular components of thyroid cells as reactants in thyroid autoimmune disease. If these are relevant to initiation of disease, how do lymphocytes gain access to these?

VOLPE: Perhaps you misunderstood me. The data that I presented indicated that the only thyroid fraction which stimulated the Graves' lymphocytes to produce thyroid-stimulating immunoglobulin was a fraction rich in cell membranes. It is our prejudice that the truly appropriate antigen in Graves' disease is a cell membrane component, presumably the TSH receptor. I believe our data are in accord with that view.
I agree with you that lymphocytes should have no access to antigens that remain within the cell and am sorry that you had that impression.

MACKAY: Regarding the interesting data on the nude mice, mortality should not be high under good housing conditions. Also, have you studied the survival time of human lymphocytes xenografted into nude mice?

VOLPE: We have not had the opportunity as yet to study the survival time of human lymphocytes xenografted into the nude mice. Judging from the return to the baseline values of the blood thyroxines by day 21, it may be assumed that the human lymphocytes have been destroyed by then. Moreover, the histology of the mouse thyroid was normal at sacrifice at 21 days, without lymphocytic infiltration. We intend to follow the fate of injected lymphocytes in future studies and to sacrifice the mice at the time at which the blood thyroxine values are at their peak, i.e., between 10-14 days, so as to deter-mine any morphological correlates.

MACKAY: Estroff et al. (Eur. J. Immunol. 6:683, 1976) have indi-cated that human monocytes and lymphocytes are capable of restoring the in vitro response of the "nude" mouse to sheep red cells. This suggests that human lymphocytes (and monocytes) may confer a new ability to the "nude" mice to ultimately reject those self-same human leukocytes. However, this may take at least several days. We will attempt to obtain data on this point.

IRVINE: Some TSH receptor antibodies have the ability to stimulate the synthesis of thyroid hormones while other such antibodies may actually block the stimulation of the thyroid cell by endogenous TSH. Long-term follow up of thyrotoxic patients may show a persistent tendency to relapse in some but a tendency to become hypothyroid in others. Have you seen in your follow-up studies any change in the nature of your TSH receptor antibodies, e.g. from stimulating to blocking? My own impression is that one cannot predict except in the short term what the subsequent course is going to be from a measurement of TSH receptor antibodies.

VOLPE: There are indeed changes with time in Graves' disease in respect to the presence of the antibody to the TSH receptor. Firstly, this antibody when determined by a radioligand technique measures binding only, and thus it should be termed "thyrotropin displacing activity (TDA)"; it does not always equate with thyroid stimulating activity. It certainly equates quite well in active, untreated hyperthyroid Graves' disease, but does not equate well in other conditions in which the TDA is positive (such as in some cases of Hashimoto's thyroiditis, subacute thyroiditis, and thyroid carcinoma); in many of these instances, despite a positive TDA, there is no thyroid stimulation, and in some, the antibody is a blocking antibody. Thus there can be antibodies to the TSH receptor that may be blocking antibodies, antibodies that bind but neither block nor stimulate, and antibodies that bind and then stimulate (thyroid-stimulating immunoglobulin, TSI).
TSI does change during the course of Graves' disease. Generally, it either persists or disappears (immunological remissions), but occasionally it may remain positive in the TDA assay while becoming negative in the TSI assay--and even may change to a blocking antibody in time. (It may be pointed out that stimulating qualities and TSH-blocking qualities can coexist in the same IgG.)
TDA and TSI assays can be used as predictors of remission, although they are not ideal. Positive assays can occasionally be found when patients remain in remission long after antithyroid drug therapy, possibly due to morphological damage due to thyroiditis preventing the thyroid from responding, or due to the change of TSI to a blocking antibody. Moreover, even if the assays are negative, and the patients are in immunological remission, one cannot determine how long that remission may last. Indeed, it may be quite short-lived. Thus we do not have an assay that predicts long-term remissions!

DUPONT: A report in the New England Journal of Medicine from the group at Duke University, North Carolina, described that familiar occurrence of polyendocrine gland disease showed linkage to HLA. Do you have similar findings?

VOLPE: We have observed large numbers of patients who have two or more autoimmune organ-specific endocrinopathies, and a small number of patients who combine several such defects (e.g., autoimmune thyroid disease, hypoparathyroidism, adrenalitis, oophoritis). However, we have not done HLA typing on them. I believe Dr. Irvine has some data of this nature, and will be reporting it this afternoon.

LENNON: You mentioned finding no evidence of lymphoid cell infiltration in the thyroids of nude mice 21 days after injection of peripheral blood lymphocytes from patients with Graves' disease. Were thyroids examined at earlier time points? Your in vitro experiments suggested that lymphocytes from Graves' patients needed antigen contact to release thyroid stimulating Ig (T.S. Ig), but PHA could substitute for thyroid membrane antigen. If Graves' lymphocytes were transferred to thyroidectomized nude mice, it would be possible to directly test in vivo whether contact with autoantigen is necessary for the release of the autoantibody. I suspect that the foreign environment of the nude mouse would provide a sufficiently strong polyclonal stimulant for release of T.S. Ig from sensitized B lymphocytes.

VOLPE: The studies on the nude mice that I have presented are preliminary, and we sacrificed the animals only at 21 days, at a time when the blood thyroxine levels had already returned to normal. Since we did not receive the thyroxine determinations quickly, we did not realize that by day 21, the transient rise in blood thyroxine values was already over. Thus, we can now accept that it should be no surprise that lymphocytes were not seen within the thyroid.
We are, of course, going to sacrifice further groups of mice between 10-14 days after infusion of the lymphocytes, at a time when the thyroxine levels should be at their peak, in the hope that we will then observe lymphocytic infiltration, and perhaps morphological evidence of thyroid cell stimulation, within the mouse thyroid.
We also plan to repeat these studies after removal of the thyroid, with a view to determining whether the lymphocytes will produce stimulating immunoglobulin under those circumstances. The limiting factor here is the volume of mouse blood, but this can be overcome by utilizing large numbers of mice.

HARRISON: (a) Is there any difference in the response of Graves' lymphocytes to your 50,000 x g fraction from Graves' thyroid versus normal thyroid?
(b) Do lymphocytes have TSH receptors; if so, what might be their significance?
(c) What is the effect of propylthiouracil on immune function?

VOLPE: The 50,000 g fraction of thyroid tissue that stimulated the Graves' lymphocytes was equally stimulatory whether it came from a normal or Graves' thyroid gland, as far as we were able to tell.
There are TSH receptors in leukocytes, but these appear to be confined to monocytes. TSH receptors on lymphocytes have not been detected (as yet).
There are minor immunosuppressive effects of propylthiouracil that can be demonstrated in vivo. However, these cannot account for remissions, since the immunosuppressive effects of propylthiouracil are quite short-lived. Furthermore, significant remission rates can be achieved by nonspecific measures, such as **propranolol** or sedative therapy, or rest alone. (See my review in Clinics in Endocrinology and Metabolism 7:3-29, (1978, for a discussion of the nature of remissions in Graves' disease.)

# Ia-ANTIGENS IN SJOGREN'S SYNDROME

THOMAS M. CHUSED, HARALAMPOS M. MOUTSOPOULOS, ARMEAD H. JOHNSON AND
DEAN L. MANN
National Institutes of health, Bethesda, Maryland 20014 (USA)

ABSTRACT

All individuals with Sjögren's syndrome (SS) which we have tested
carry two genetically and immunologically unrelated B lymphocyte antigens coded by genes of the major histocompatability complex. These two
antigens are present in 37 and 24 percent of normal controls. This
suggests that two distinct immune response genes may be required for
the development of SS.

INTRODUCTION

Our group has examined patients with Sjögren's syndrome (SS) for B
cell Ia-like antigen specificities[1]. SS is known to be associated with
HLA-B8 (50-55%)[2-4] and HLA-DW3 (69%)[5,6]. We utilized a panel of 60
antisera from multiparous females absorbed with pooled platelets to
remove HLA-A, -B and -C specificities. B lymphocytes were isolated and
tested by conventional cytotoxicity assay. Of the twenty-four patients
tested, three had rheumatoid arthritis and three had systemic lupus
erythematosus in addition to SS.

RESULTS AND DISCUSSION

The results obtained with antisera that detected significant differences between the SS and control groups are shown in Table 1. Two
antisera, 172 and AGS, reacted with the B lymphocytes from all the SS
patients, compared to 37 and 24 percent of the control population,
respectively. Four additional antisera (35, 350, 590, and 715) reacted
more frequently (67, 63, 58, and 54 percent) with the patients' B cells
than with those of the controls (17, 21, 24, and 14 percent).

In order to determine whether antisera 172 and AGS recognized the
same or different antigens, their reactions with the lymphocytes from
184 normal individuals were compared by the $\chi^2$ test. All the correlations significant at $p<.05$ are shown in Table 2. For example, Ia-172
correlated with both Ia-35 and Ia-350 but not with Ia-AGS or Ia-715;
Ia-AGS correlated only with Ia-715. Table 2 also shows that Ia-35 is in
linkage disequilibrium with HLA-B8.

TABLE 1

ANTI-Ia ANTISERA WHICH DISTINGUISH PATIENTS WITH SJOGREN'S SYNDROME
FROM NORMAL CONTROLS

| Antiserum | Sjögren's syndrome | | Normal controls | | $\chi^2$ | p[a] |
|---|---|---|---|---|---|---|
| | + | − | + | − | | |
| Ia-172 | 24 | 0 | 68 | 116 | 31.7 | <.001 |
| Ia-AGS | 24 | 0 | 44 | 140 | 52.5 | <.001 |
| Ia-35 | 16 | 8 | 31 | 153 | 27.3 | <.001 |
| Ia-350 | 15 | 9 | 39 | 145 | 16.8 | .002 |
| Ia-590 | 14 | 10 | 37 | 147 | 14.8 | .007 |
| Ia-715 | 13 | 11 | 26 | 158 | 19.8 | <.001 |

[a] Corrected for the number of antisera tested


TABLE 2

SIGNIFICANT CORRELATIONS BETWEEN SPECIFICITIES IN NORMAL CONTROLS

| | Ia-35 | Ia-350 | Ia-715 |
|---|---|---|---|
| Ia-172 | .37[a] <.001[b] | .59 <.001 | |
| Ia-AGS | | | .36 <.002 |
| HLA-B8 | .45 <.001 | | |

[a] Coefficient of contingency
[b] P value


TABLE 3

SIGNIFICANT CORRELATIONS BETWEEN SPECIFICITIES IN PATIENTS WITH
SJOGREN'S SYNDROME

| | Ia-35 | Ia-350 |
|---|---|---|
| Ia-590 | .48[a] .007[b] | .39 .04 |
| Ia-350 | .43 .02 | |

[a] Coefficient of contingency
[b] P value

A similar analysis for the patients is presented in Table 3. Because of their 100 percent coincidence in SS, 172 and AGS are highly related to all the specificities examined and are not included. In the patient group, in a manner analagous to the controls, Ia specificities 35, 350, and 590 are associated with each other but not with 715. This could be due to either cross-reactivity or linkage disequilibrium.

Of the 24 SS patients, 19 have also been typed for HLA-DW3. The coefficient of contingency between DW3 and Ia-590 is .53, p=.007, and between DW3 and Ia-35 is .39, p=.07. DW3 did not correlate with Ia-350 or Ia-715.

Thus B cells of all our patients with sicca syndrome bear two Ia specificities, 172 and AGS, that are not associated in the normal population.

We have haplotyped several families and confirmed that the Ia specificities are linked to the major histocompatability complex. Patients have been documented who are heterozygous for either the Ia-172 or Ia-AGS so these alleles function codominantly in the pathogenesis of SS. We have not yet observed a patient who is simultaneously heterozygous for both so do not yet know whether they can function *trans* to each other.

Assuming that Ia-like B lymphocyte antigens in humans are coded by an Ir region, our results suggest that two immune response genes are involved in the pathogenesis of Sjögren's syndrome.

REFERENCES

1.  Moutsopoulos, H.M., Chused, T.M., Johnson, A.H., Knudsen, B. and Mann, D. (1978) Science, 1441-1442.

2.  Gershwin, M.E., Terasaki, P.I., Graw, R. and Chused, T.M. (1976) Tissue Antigens 6, 342-346.

3.  Ivanyi, D., Drizhal, I., Erbenova, E., Horejs, J., Salavec, M., Macurova, H., Dostal, C., Balik, J. and Juran, J. (1976) Tissue Antigens 7, 45-51.

4.  Fye, K.H., Terasaki, P.I., Moutsopoulos, H.M., Daniels, T.E., Michalski, J.P. and Talal, N. (1976) Arthritis Rheum., 19, 883-886.

5.  Chused, T.M., Kassan, S.S., Opelz, G., Moutsopoulos, H.M. and Terasaki, P.I. (1977) New Engl. J. Med. 296, 895-897.

6.  Hinzova, E., Ivanyi, D., Sula, K., Horejs, J., Dostal, C. and Drizhal, I. (1977) Tissue Antigens 9, 8-10.

# AUTOANTIBODIES TO THE INSULIN RECEPTOR: CLINICAL AND MOLECULAR ASPECTS

LEN C. HARRISON, JEFFREY S. FLIER, C. RONALD KAHN, DAVID B. JARRETT, MICHELE MUGGEO, AND JESSE ROTH
Diabetes Branch, National Institute of Arthritis, Metabolism, and Digestive Diseases, National Institutes of Health, Bethesda, Maryland, 20014 U.S.A.

## INTRODUCTION

Receptors are specific molecules whose primary function is to recognize and bind other molecules or ligands. This property of specific recognition, together with saturability and a high affinity for the ligand, establishes the existence of receptors for a large number of endogenous substances including hormones, neurotransmitters and antigens. (The efficacy of exogenous substances, e.g., drugs, depends on the presence of receptors, even though the endogenous ligands for these receptors may not be identified).

The concept of receptors originated in the theories of drug action early this century. However, it has only been within the last 10 years that the existence of receptors has been validated by measuring directly the binding of radioactive ligands. The pioneering studies were performed using radioiodinated, biologically-active peptide hormones and cells or cell fractions from appropriate target tissues. It was established that peptide hormones initiate their cellular effects by binding to receptors on the outer surface of the plasma membrane[1,2]. In most cases, however, it still remains to be shown whether the "message" resides in the ligand molecule, the receptor molecule, or their combination. For a number of hormones (e.g., beta catecholamines, glucagon, ACTH and gonadotropins), the ligand-receptor interaction results in activation of adenylcyclase and the generation of a "second messenger", cyclic AMP, which by activating protein kinases causes the phosphorylation of intracellular enzymes[3,4]. The post-receptor pathways in the case of other hormones (e.g., insulin, growth hormone, prolactin and alpha catecholamines) have not yet been characterized.

Studies originating on the insulin receptor have clearly shown that alterations in receptor concentration and/or affinity can determine, to a large extent, the responsiveness of the cell to the hormone[5,6]. This crucial role of the receptor was first demonstrated in the insulin resistance of obesity, the commonest form of hormone resistance. In this disorder, the circulating levels of insulin are elevated, but contrary to what would be classically predicted, the tissue response to insulin is impaired. Studies on obese syndromes in rodents[7] and later in humans[8,9,10] have clearly demonstrated that the resistance to insulin is directly correlated with a decrease in the concentration of insulin receptors. This decrease in receptor concentration was the first example of the general phenomenon of receptor regulation by its homologous ligand[11]. Other factors now known to modulate receptors and therefore the expression of the ligand include the stage of growth and differentiation of the tissue, the metabolic and ionic milieu, temperature, and exposure to heterologous ligands[2,3,5,6].

Receptors, especially those on the cell surface, e.g., receptors for peptide hormones, neurotransmitters and antigens, are logical candidates for immune-mediated disease. In endocrinology, the clustering of diseases such as insulin-requiring diabetes, adrenal insufficiency and thyroid disorders, and their association with "autoimmune" features, has been recognized for a number of years[12]. In these diseases, organ-specific autoantibodies against cell surface membrane components are demonstrable using immunofluorescence techniques. However, it was not until 1971 that Lennon and Carnegie suggested that antibody interference, specifically with hormone receptors, could be a general mechanism for disease[13]. Their concept was supported by the existence of LATS (Long-Acting Thyroid Stimulator), discovered by Adams and Purves in 1956[14] and subsequently shown to be IgG capable of stimulating the TSH receptor in Graves' disease[15,16,17]. The recent advent of methods for labeling ligands with radioisotopes and for purifying receptors facilitates the detection of receptor antibodies with a specificity and sensitivity far exceeding traditional immunofluorescence techniques. Receptor autoantibodies have now been identified in three disease states in man: to the TSH receptor in Graves' disease[16,17]; to the acetylcholine receptor in myasthenia gravis[18,19,20]; and to the insulin receptor in some patients with insulin-resistant diabetes[21]. In addition, antibodies against semi-purified receptors for insulin[22], acetycholine[23], prolactin[24], and growth hormone[25], have been raised in experimental animals. It is clear that receptor antibodies may act as either antagonists or agonists, or both, resulting in hypo- and hyper-functional syndromes.

DIABETES AND IMMUNITY

Diabetes is a disorder characterized by lack of insulin effect. This may be due to failure of secretion of insulin itself, an abnormality in the nature of secreted insulin or its transport, or a defect in the action of insulin at the tissue level. Less than one-quarter of diabetics are actually lacking in insulin (and require insulin therapy). The majority of diabetics actually secrete normal or greater than normal amounts of biologically-active insulin, but have tissue insensitivity to insulin, that is insulin resistance. In this latter category, the major group of diabetics, there is considerable evidence relating the insulin resistance to a decrease in insulin receptor concentration, affinity, or both[5,26].

Immune disturbances, especially of the humoral system, are implicated in the pathogenesis of the diabetic syndrome at three levels. At the level of the pancreas islet-cell antibodies may affect beta cell function and impair insulin secretion[27,28]; at the level of transport, insulin antibodies in the plasma may bind insulin and decrease its effective concentration[29]; and at the level of the target tissues insulin receptor antibodies may impair the binding and the action of insulin[21,30]. The role of islet cell antibodies will be discussed elsewhere. Circulating insulin antibodies are commonly seen in diabetics receiving insulin injections, but usually have a relatively minor, secondary role. Within the major group of diabetics characterized by insulin resistance and an associated receptor defect, there is a small subgroup with insulin receptor autoantibodies. We will review the clinical syndrome of insulin-resistant diabetes associated with insulin receptor autoantibodies, the detection and biological properties of these antibodies and their utility as probes of receptor function.

CLINICAL SYNDROME ASSOCIATED WITH INSULIN RECEPTOR AUTOANTIBODIES

In 1975, Jeffrey Flier described three patients with severe insulin-resistant diabetes and the skin lesion of acanthosis nigricans[21]. These patients had a marked, fixed defect of insulin binding to their circulating monocytes, which was mimicked on normal monocytes or other receptor preparations by incubating them in vitro with the patients' sera. The serum factors responsible for the specific binding defects in these patients are now known to be receptor autoantibodies. Fourteen cases of this syndrome have been reported or seen by us[30-35]. Their clinical features are summarized in Table I.

TABLE I

THE CLINICAL SYNDROME OF INSULIN RESISTANCE DUE TO INSULIN RECEPTOR
AUTOANTIBODIES

---

Incidence:  14 Documented cases (Oct., 1975 - June, 1978)
Sex:        10 Female / 4 Male
Age:        12 - 65
Race:       8 Black, 2 Japanese, 2 White, 1 Creole, 1 Mexican

Clinical Features:

1. Abnormal Glucose Tolerance and Insulin Resistance

   Usually severe, but may vary during disease.
   Some patients have received up to 25,000 units of
       insulin/day.
   3/13 patients had hypoglycemia at sometime in their
       disease (1 very severe).

2. Acanthosis Nigricans

   Present in 11/14, and very severe in some.
   Tends to follow clinical course of glucose tolerance
       abnormalities.
   Only 1 patient had an occult malignancy.

3. Immunologic Features

   In about one-third, some classic "autoimmune" syn-
       drome exists (SLE, Sjogren's, Ataxia telangectasia).
   In all cases, some signs    suggestive of autoimmune
       disease:
       ↑ ESR; ↑ γ-globulins
       + ANA; anti-DNA
       Leukopenia
       Alopecia
       Proteinuria → nephrotic syndrome
   May or may not follow clinical course of other features.

---

The immunoglobulin nature of the serum inhibitor was established
using standard immunochemical procedures. The antibodies are predominantly polyclonal IgGs, although at least one patient has also had IgM
antibodies[36]. Recent studies on a family of patients with ataxia
telangiectasia (a recessive disorder of progressive cerebellar ataxia,
telangiectasia, and diverse immune abnormalities) have uncovered a
more restricted low molecular weight IgM-λ chain specific, insulin
receptor antibody[37]. Interestingly, the mother of the propositus, who
was phenotypically normal but presumably a carrier, also had the same
insulin receptor autoantibody.

ANALYSIS OF THE BINDING DEFECT

Circulating monocytes are a convenient source of insulin receptors
and analysis of insulin binding to monocytes from affected patients
shows, typically, a marked defect in receptor affinity (Figure 1).

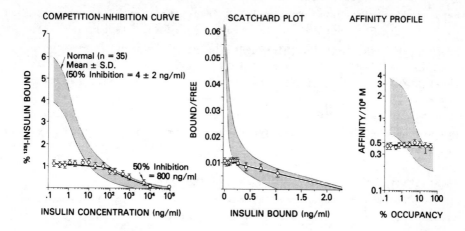

Fig. 1. $^{125}$I-insulin binding to circulating monocytes in a patient (B-6) with insulin receptor autoantibodies. The decrease in receptor affinity is indicated by the high concentration of unlabeled insulin required to inhibit tracer $^{125}$I-insulin binding by 50% (left panel), by the decrease in curvilinearity of the Scatchard plot (center panel), and by the flat affinity profile (right panel).

The binding of insulin is normally a negatively cooperative process; that is, as insulin occupancy of receptors increases the affinity of the interaction decreases[38]. In the presence of the antibodies the negative cooperativity, reflected by the concavity of the Scatchard plot, is no longer apparent. Although the antibodies appear to act mainly like classical competitive antagonists, some patients have had virtually no binding, indicating that the antibodies may also decrease the apparent receptor number. The binding defect on the patients' cells can be reversed by an acid-wash, a procedure which elutes surface immunoglobulin but does not alter insulin binding.

The binding defect is specific for insulin and is mimicked _in vitro_ by preincubating the patients' sera with normal receptors from a variety of tissues and species[21,39]. Some sera are equipotent on all receptors tested, whereas others have significantly higher inhibitory titers against human and rodent receptors than against avian and fish receptors. Kinetic studies have shown that the predominant effect of the sera to decrease affinity is due to effects on both the association rate and the dissociation rate of insulin.

The interaction of the antibodies with the insulin receptor has also been studied directly. Purified IgG fractions from anti-receptor sera have been labeled with [125]Iodine and adsorbed and eluted from cells containing insulin receptors to enrich for [125]I-receptor antibody[40]. This [125]I-antibody binds to a variety of cells in direct proportion to their insulin receptor concentration, and its binding is specifically competed for by insulin and insulin analogues[40]. Thus, there is both indirect and direct evidence that some of the antibodies bind to determinants on the receptor that are either in or very close to the binding site for insulin.

## BIOACTIVITY OF INSULIN RECEPTOR ANTIBODIES

The clinical picture associated with insulin receptor autoantibodies has usually been one of severe insulin resistance, consistent with the action of the antibodies to impair insulin binding. However, the

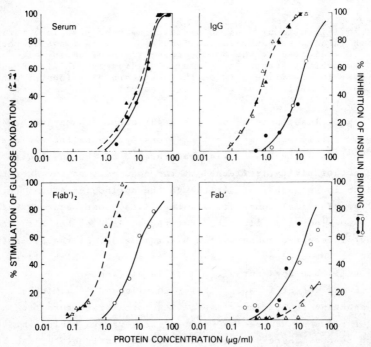

Fig. 2. Comparison of the effects of receptor antibody on insulin binding and glucose oxidation. Isolated rat adipocytes were incubated with buffer and serial dilutions of whole serum, IgG, F(ab')$_2$ or F(ab') from patient B-2. The inhibition of [125]I-insulin binding or the stimulation of [14]C-glucose oxidation caused by each fraction is expressed as a percentage of the maximum effect of insulin in each case. The small amount of bioactivity in the F(ab') is probably due to contamination by F(ab')$_2$.

acute effect of the antibodies _in vitro_ is to mimic the actions of
insulin[41] (Figure 2). In most cases, the bioactivity titer is higher
than the titer for inhibition of insulin binding, suggesting that some
antibodies bind to determinants which are not involved in insulin
binding, yet are still able to elicit insulin-like effects. The full
range of insulin-like effects of the antibodies is shown in Table II.


TABLE II

INSULIN-LIKE EFFECTS OF INSULIN RECEPTOR AUTOANTIBODIES.

FULLY MIMICS

  Adipocytes

      Stimulation of 2-deoxyglucose transport, glucose incorporation
         into lipid and glycogen, and metabolism to $CO_2$
      Stimulation of amino acid (AIB) transport
      Inhibition of lipolysis
      Activation of glycogen synthase (in the presence and absence
         of glucose)
      Inhibition of phosphorylase

  Muscle

      Stimulation of 2-deoxyglucose transport
      Stimulation of glycolysis

  Liver

      Stimulation of AIB transport

PARTIALLY MIMICS

  Muscle

      Stimulation of glucose incorporation into glycogen
      Activation of glycogen synthase (in the absence of glucose)

DOES NOT MIMIC

  Liver

      Stimulation of glycogen synthesis in fetal liver


The bioactivity of the antibodies is related to bivalency[42].
Purified IgG and F(ab')$_2$ inhibit insulin binding and mimic insulin
effects (Figure 2), but monovalent F(ab') fragments while able to
inhibit insulin binding, have very little bioactivity (Figure 2). The
bioactivity of the monovalent fragment can, however, be completely
restored by the addition of anti-IgG or anti-F(ab')$_2$ antibody[42]. This
shows that occupancy alone is insufficient and that cross-linking of
the insulin receptor is necessary for biological activation by antibody
(Figure 3).

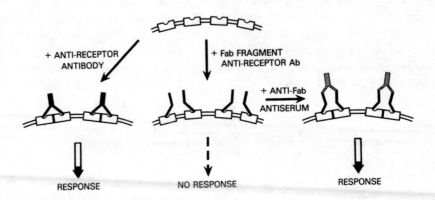

+ ANTI-RECEPTOR ANTIBODY

+ Fab FRAGMENT ANTI-RECEPTOR Ab

+ ANTI-Fab ANTISERUM

RESPONSE

NO RESPONSE

RESPONSE

Fig. 3. Schematic representation of the role of cross-linking in receptor activation.

Activation of other receptors by antibodies also appears to depend on cross-linking. Hence, IgE receptor-mediated mast cell degranulation requires cross-linking with either IgE and second antibody, antibodies to the receptor, or chemically cross-linked IgE dimers[43]. In addition, it has been shown that bivalent but not monovalent antibodies will accelerate degradation of the acetylcholine receptor in myasthenia gravis[44].

Our findings on the insulin receptor suggest that all the information for the acute biological effects of insulin is contained within the receptor and that any ligand, for example antibody, which perturbs the receptor can elicit the appropriate specific responses. Whether the antibodies can also mimic the chronic effects of insulin, such as mito-genesis and receptor "down-regulation", is presently unknown. That acid-washing of monocytes from affected patients restores binding nearly to normal argues against antibody-mediated down regulation of the receptor. The question remains: why do the patients have insulin resistance if the receptor antibodies are insulin-like? This apparent paradox may possibly be resolved by the recent work of F. Anders Karlsson in our laboratory, who showed that the insulin-like effect of the antibodies on 3T3-L1 fatty fibroblasts is transient and after several hours gives way to a state of imparied insulin responsiveness, consistent with the picture in vivo[45].

ASSAYS FOR INSULIN RECEPTOR ANTIBODIES

Receptor antibodies have been detected directly by their ability to inhibit insulin binding[21], immunoprecipitate the solubilized receptor[46], and cause insulin-like effects[41]. In addition, their presence may be inferred by reversal of the binding defect after acid-washing the patient's cells to remove surface immunoglobulins, and by uptake of [125]I-protein A into cells preincubated with anti-receptor serum. The two most useful assay techniques are illustrated in Figure 4. Immuno-

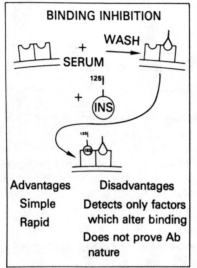
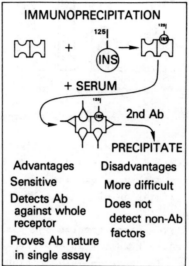

Fig. 4. Direct assays for insulin receptor antibodies.

precipitation gives the highest titer in every case, presumably because it detects antibodies against any determinant on the receptor, not only those which happen to impair insulin binding or evoke insulin-like effects[46] (Figure 5).

NATURAL HISTORY AND THERAPY IN PATIENTS WITH INSULIN RECEPTOR ANTIBODIES

The manifestations of autoimmunity in these patients characteristically show a variable pattern. Glucose intolerance and the titer of antibodies may remain constant over several years, but some patients have had spontaneous remission of glucose intolerance associated with a decrease in antibody titer[34,47]. In two cases treatment with glucocorticoids and cyclophosphamide was associated with remission, during which time antibody was not detected[31,47]. Two other patients have had a most unusual course. Despite having consistently high titers of

70

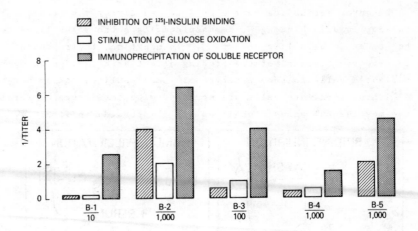

Fig. 5. Titers of receptor autoantibodies in five patients (B1-B5).

inhibitory antibodies, these patients have shifted from a state of
hyperglycemia and decreased insulin binding into one of hypoglycemia
and increased insulin binding[47]. The molecular basis of this appears
to be proliferation of the insulin receptor, but the precise mechanisms
are unknown.

Recently, we have treated one patient by serial plasma exchanges.
This procedure resulted in a marked decrease in antibody titer and a
concomitant increase in insulin binding to the patient's cells in
vitro; however the effect was short-lived and antibody levels quickly
rebounded. Logically, the procedure should be more successful if
combined with immunosuppressive therapy to prevent new antibody
synthesis.

APPLICATIONS OF RECEPTOR ANTIBODIES

Receptor antibodies are important probes of receptor function, as
illustrated above. In addition, receptor antibodies may be used as
reagents to both quantitatively assay and purify receptors. Immunoprecipitation of the solubilized receptor is highly specific[46] (Figure
4). Other receptors, including those for epidermal growth factor (EGF),
multiplication stimulating activity (MSA), growth hormone (GH), and
prolactin (PRL) are not precipitated. Furthermore, gel electrophoresis
shows that out of the multiple protein bands in solubilized placental
membranes, the antibodies precipitate only three (Figure 6). This

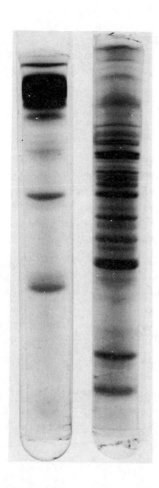

Fig. 6. Polyacrylamide gel electrophoresis of receptor-antibody complex
(left) immunoprecipitated from Triton-
solubilized human placental membranes
(right). (The dense band at the top
of the left-hand gel is IgG in the
immune precipitate.)

specificity of the antibodies has allowed us, firstly, to use them in
a radioimmunoassay for the receptor, capable of detecting $10^{-10}$M
insulin binding sites[48]. The radioimmunoassay adds a new dimension to
studies of receptors by allowing us for the first time to measure
receptors independently of their ligand binding. Preliminary findings
with the assay suggest that although the insulin binding site  on
the receptor has been tightly conserved throughout evolution receptors
from different organs or species may differ in other respects. Secondly, the receptor antibodies provide a unique affinity reagent for
purifying receptors. Antibody-affinity chromatography alone results
in an insulin-receptor preparation which is minimally 4-10% pure[46].

CONCLUDING REMARKS

The interference by antibody with ligand-receptor interactions is a mechanism potentially applicable to a large number of disease states. The studies which we have described illustrate some methodological approaches to the detection of receptor antibodies and demonstrate how receptor antibodies may be used as probes of receptor function. As far as diabetes is concerned, the prevalence of receptor autoantibodies appears to be very low, although a systematic survey using the recently developed and sensitive immunoprecipitation assay may change this view. A more fundamental question concerns the etiology and significance of receptor autoantibodies. Our studies suggest that the underlying insulin receptor may be normal. Thus, removal of the antibodies by acid-wash restores binding towards normal and binding during remission is normal. Most importantly, cultured lymphocytes and fibroblasts from affected patients have normal insulin binding and antibodies from the patients show no preferential blocking activity on their cells compared to control cells. Studies of HLA status and immune tolerance function in our patients are incomplete. In the light of Jerne's network hypothesis[49], it is important to note that insulin receptor antibodies have recently been raised as complementary idiotypes by immunization with antibodies to the natural ligand insulin[50]. Irrespective of whether receptor antibodies arise through defects in "recognition" or in "tolerance" they represent a specific subset of autoimmune phenomena whose functional sequelae can now be measured directly.

REFERENCES

1. Roth, J. (1973) Metabolism, 22, 1059-1073.

2. Kahn, C.R. (1976) J. Biol. Chem., 70, 261-286.

3. Catt, K.J., and Dufau, M.L. (1977) Ann. Rev. Physiol., 39, 529-557.

4. Exton, J.H., and Harper, S.C. (1975) Adv. Cyclic Nucleo. Res., 5, 519-532.

5. Roth, J., Kahn, C.R., Lesniak, M.A., Gorden, P., De Meyts, P., Megyesi, K., Neville, D.M., Jr., Gavin, J.R., III, Soll, A.H., Freychet, P., Goldfine, I.D., Bar, R.S., and Archer, J.A. (1975) Rec. Prog. Horm. Res., 31, 95-139.

6. Bar, R.S., Harrison, L.C., Muggeo, M., Gorden, P., Kahn, C.R., and Roth, J. (1978) Adv. Int. Med., 24, in press.

7. Kahn, C.R., Neville, D.M., Jr., and Roth, J. (1973) J. Biol. Chem., 248, 244-250.

8. Olefsky, J.M. (1976) J. Clin. Invest., 57, 1165-1172.

9.  Bar, R.S., Gorden, P., Roth, J., De Meyts, P., and Kahn, C.R. (1976) J. Clin. Invest., 58, 1123-1135.

10. Harrison, L.C., Martin, F.I.R., and Melick, R.A. (1976) J. Clin. Invest., 58, 1435-1441.

11. Gavin, J.R., III, Roth, J., Neville, D.M., Jr., De Meyts, P., and Buell, D.N. (1974) Proc. Natl. Acad. Sci., USA, 71, 84-88.

12. MacCuish, A.C., and Irvine, W.J. (1975) Clin. Endocrinol. and Metab. 4, 435-471.

13. Lennon, V.A., and Carnegie, P.R. (1971) Lancet, 1, 630-633.

14. Adams, D.D., and Purves, H.D. (1956) Proc. Univ. Otago Med. Sch., 34, 11-12.

15. Kriss, J.P., Pleshakov, V., and Chien, J.R. (1964) J. Clin. Endocrinol., 24, 1005-1028.

16. Smith, B.R., and Hall, R. (1974) Lancet, 2, 427-431.

17. Manley, S.W., Bourke, J.R., and Hawker, R.W. (1974) J. Endocrinol., 61, 437-445.

18. Almon, R.R., Andrew, C.G., and Appel, S.H. (1974) Science, 186, 55-57.

19. Abramsky, O., Aharonov, A., Webb, C., and Fuchs, S. (1975) Clin. Exp. Immunol., 19, 11-16.

20. Lindstrom, J.M., Seybold, M.E., Lennon, V.A., Whittingham, S., and Duane, D.D. (1976) Neurology, 26, 1054-1059.

21. Flier, J.S., Kahn, C.R., Roth, J., and Bar, R.S. (1975) Science, 190, 63-65.

22. Jacobs, S., Chang, K.-J., and Cuatrecasas, P. (1978) Science, 200, 1283-1284.

23. Lennon, V.A., Lindstrom, J.M., and Seybold, M.E. (1975) J. Exp. Med., 141, 1365-1375.

24. Shiu, R.P.C., and Friesen, H.G. (1976) Biochem. J., 157, 619-626.

25. Waters, M., and Friesen, H.G. (1978) Proc. Endo. Soc., 60th Ann. Meeting, June 14-16, Miami, Florida, p. 91.

26. Olefsky, J.M., and Reaven, G.M. (1977) Diabetes, 26, 680-688.

27. Bottazzo, G.F., Florin-Christensen, A., and Doniach, D. (1974) Lancet, 2, 1279-1283.

28. MacCuish, A.C., Barnes, E.W., Irvine, W.J., and Duncan, L.J.P. (1974) Lancet, 2, 1529-1531.

29. Berson, S.A., and Yalow, R.S. (1970) in Diabetes Mellitus: Theory and Practice, Ellenberg, M., and Rifkin, H. eds., McGraw Hill, New York, pp. 388-435.

30. Kahn, C.R., Flier, J.S., Bar, R.S., Archer, J.A., Gorden, P., Martin, M.M., and Roth, J. (1976) N. Engl. J. Med., 294, 739-845.

31. Kawanishi, K., Kawamura, K., Nishina, A.G., Okada, S., Ishida, T., Ofuji, T., Kahn, C.R., and Flier, J.S. (1977) J. Clin. Endocrinol. Metab., 44, 15-21.

32. Kibata, M., Hiramatsu, K., Shimizu, Y., Fuchimoto, T., Sasaki, N., Shimono, M., Miyake, K., Flier, J.S., and Kahn, C.R. (1975), Proceedings of the Symposium on Chemical Physiology and Pathology (Japan), 15, 58-63.

33. Pulini, M., Raff, S.B., Chase, R., and Gordon, E.E. (1976) Ann. Intern. Med., 85, 749-751.

34. Blackard, W.G., Anderson, J.H., and Mullinax, F. (1977) Ann. Intern. Med., 86,584-585.

35. Bar, R.S., Levis, W.R., Rechler, M.M., Harrison, L.C., Siebert, C. W., Podskalny, J.M., Roth, J., and Muggeo, M. (1978) N. Engl. J. Med., 298, 1164-1171.

36. Flier, J.S., Kahn, C.R., Jarrett, D.B., and Roth, J. (1976) J. Clin. Invest., 58, 1442-1449.

37. Harrison, L.C., Muggeo, M., Bar, R.S., Flier, J.S., Levis, W.R., Waldmann, T.A., and Roth, J., (manuscript in preparation).

38. De Meyts, P., Bianco, A.R., and Roth, J. (1976) J. Biol. Chem., 251, 1877-1888.

39. Flier, J.S., Kahn, C.R., Jarrett, D.B., and Roth, J. (1977) J. Clin. Invest., 60, 784-794.

40. Jarrett, D.B., Roth, J., Kahn, C.R., and Flier, J.S. (1976) Proc. Natl. Acad. Sci., USA, 73, 4115-4119.

41. Kahn, C.R., Baird, K.L., Flier, J.S., and Jarrett, D.B. (1977) J. Clin. Invest., 60, 1094-1106.

42. Kahn, C.R., Baird, K.L., Jarrett, D.B., and Flier, J.S. (1978) Proc. Natl. Acad. Sci., USA, in press.

43. Metzger, H. (1977) in Receptors and Recognition, Cuatrecasas, P., and Greaves, M. eds., Chapman and Hall, London, pp. 75-102, Vol. 4.

44. Drachman, D.B., Angus, C.W., Adams, R.N., Michelson, J.D., and Hoffman, G.J. (1978) N. Engl. J. Med., 298, 1116-1122.

45. Karlsson, F.A., Van Obberghen, E., Grunfeld, C., Kahn, C.R., and Roth, J., (manuscript in preparation).

46. Harrison, L.C., Flier, J.S., Kahn, C.R., and Roth, J. (1978) Proc. Endo. Soc., 60th Ann. Meeting, June 14-16, Miami, Florida, p. 331.

47. Flier, J.S., Bar, R.S., Muggeo, M., Kahn, C.R., Roth, J., and Gorden, P. (1978) J. Clin. Endocrinol. Metab., in press.

48. Flier, J.S., Harrison, L.C., Itin, A., Kahn, C.R., and Roth, J. (submitted).

49. Jerne, N.K. (1975) The Harvey Lectures, Series 70, 1974-75, 93-110.

50. Sege, K., and Peterson, P.A. (1978) Proc. Natl. Acad. Sci., USA, 75, 2443-2447.

DISCUSSION

ROITT:  Do I understand that you take the view, which I have heard
Dr. Pernis express, that the anti-idiotype provides the immunogenic
stimulus for autoantibody production rather than the tissue autoantigen itself?

HARRISON:  We considered that one possible cause of anti-receptor
antibody might be the development of complementary idiotypes, i.e.,
antibodies against antibodies which have specificities for the
binding region of the ligand.  This idea is based on Jerne's network
hypothesis.  Thus, one would speculate that some antibodies against
the ligand would be directed against its receptor binding region
and might therefore be images of the receptor.  Antibodies against
these antibodies might then be images of the ligand, i.e., anti-
receptor antibodies.  We wonder whether the natural history of
insulin resistance due to insulin antibodies in diabetics taking
insulin could be modulated by such a mechanism.
Dr. Anders Karlsson and I recently attempted, without much success,
to isolate antibodies directed against the receptor binding region
of insulin and to use them as immunogens to test this hypothesis.
In the meantime, Sege and Peterson have reported that anti-insulin
receptor antibodies can be generated in rabbits by immunizing with
rat anti-insulin antibodies.  This experimental evidence, if correct,
strongly favors the hypothesis.

VOLPE:  The model of immunoprecipitation of receptors by antibodies
is very attractive and important.  Have you any experience with the
TSH receptor or other receptors for which there are antibodies?

HARRISON:  We have no experience with other systems.  Our antibodies
do not appear to precipitate related receptors, e.g., for multiplication stimulating activity (MSA).  I think a problem with the TSH
receptor will be its affinity for normal IgG.

VOLPE:  It is of interest that patients with diabetes with acanthosis nigricans sometimes undergo spontaneous remissions.  Can you
tell us anything about the nature of such remissions?

HARRISON:  The course of any autoimmune disease seems to be characterized by fluctuations and the reasons, I believe, are unknown.
Remissions in these patients have been associated with a decrease
in titer of anti-receptor antibody and an improvement in glucose
tolerance.

TALAL:  The similar properties of anti-idiotypic antibodies and
anti-receptor autoantibodies suggest that the concept of an immunologic network might be expanded in the following sense:  The immune
system may have originated to survey  cell membrane receptors
and to participate in regulating the function of those receptors--
both immune and non-immune receptors.  Antibodies to the insulin
receptor may, therefore, be present in small amounts in normal
individuals.  Do you have any evidence for a physiologic role for
such antibodies?

HARRISON:  Such antibodies could modulate the insulin resistance
due to insulin antibodies in insulin-treated diabetics.  If receptor
antibodies are, in fact, present in small amounts or are normally
"suppressed" then perhaps we might expect to find them in certain
situations.  We have found low molecular weight IgM, lambda chain
specific, insulin-receptor antibodies in a family with ataxia-
telangiectasia, considered by some to be a model of aging.  Also

perhaps non-specific, polyclonal stimulation might uncover such antibodies.

COHEN: You suggest that the biological effects of antibody are due to cross-linking of insulin receptors. Is there any evidence that insulin itself acts by cross-linking the insulin receptors?

HARRISON: Although circulating insulin is probably monomeric, its local concentration on the cell may be high enough to induce self-association. However some insulins, e.g., guinea pig insulin, which do not dimerize still have some bioactivity. Ferritin-labeled insulin or fluorescein-labeled insulin has been shown to "patch" on the cell surface, but the relationship of this phenomenon to insulin action is still unclear. Recently Dr. Ron Kahn has exposed fat cells to submaximal doses of insulin followed by low concentrations of anti-insulin antibody, and has found an enhancement of insulin's effect, suggesting that cross-linking of insulin-receptor complexes may be important in insulin action.

LENNON: Your in vitro experiments suggested that binding of bi-valent antibodies to any part of the insulin receptor exposed on living cells leads initially to stimulation of receptor function (i.e., mimicry of insulin's action). A few hours later there followed inhibition of insulin binding (presumably due to modula-tion of receptors), which was analogous to the highly insulin resis-tant state that prevailed in the majority of the patients you de-scribed. My question is really a philosophic one to the audience. If this mechanism of action of anti-receptor antibodies is a general one, then when bivalent thyroid stimulating autoantibodies bind in vivo to TSH receptors of patients with Graves' disease, why is the TSH receptor not modulated off the membrane within a few days and why then does the clinical picture not transform functionally to one of hypothyroidism? Is it a question of an intrinsic difference in the TSH receptor, e.g., its structural integration in the thyroid epithelial cell membrane or its turnover rate, or is the continued stimulation of the TSH receptor perhaps a reflection of the class of antibody involved (e.g., its possible inability to activate complement)?

HARRISON: You raise an interesting point; I don't know of any detailed studies on the mechanism of action of TSH receptor antibodies in vitro. Insulin and TSH receptors may respond differently since they don't appear to activate cells by the same mechanisms. The TSH receptor is coupled to adenylcyclase and initiates the release of a "second messenger", cyclic AMP. A "second messenger" for insulin, if it exists, has not been identified. I also think it likely that the TSH receptor may mediate trophic effects in the thyroid.

ROITT: Since the patient's receptors are recovered by acid wash of their cells, it would appear that receptors are not lost by endo-cytosis or shedding.

THE IMMUNOLOGY AND GENETICS OF AUTOIMMUNE ENDOCRINE DISEASE

W. JAMES IRVINE

Endocrine Unit/Immunology Laboratories & University Department of
Medicine, Royal Infirmary, Edinburgh, United Kingdom

INTRODUCTION

   Included among the endocrine disease are many examples of the organ-
specific autoimmune group of disorders.  The example par excellence of
an organ-specific autoimmune disease is idiopathic (autoimmune)
Addison's disease, which is associated with a particularly high prevalence of the other diseases within this group (Table 1).

TABLE 1

THE MAIN ASSOCIATED CONDITIONS IN 383 PATIENTS WITH ADDISON'S DISEASE

|                              | Idiopathic | TB | Other |
|------------------------------|:----------:|:--:|:-----:|
| Ovarian failure              |     59     | 1  |       |
| Thyroid disease              |            |    |       |
|    Thyrotoxicosis |  22    | 1  |       |
|    Hypothyroidism |  25    | 53 |       |
|    Hashimoto  |     6      | 0  |       |
| Diabetes mellitus            |     35     | 1  |       |
| Hypoparathyroidism           |     18     | 0  |       |
| Pernicious anaemia           |     12     | 0  |       |
| Number of patients affected  |    199     | 4  |   0   |
| Total number of patients     |    313     | 66 |   4   |

Source:  Irvine, W.J. (1978) Medical Immunology.  Teviot, Edinburgh.

   In contrast, patients with adrenal destruction due to tuberculosis
have a low prevalence of other disorders affecting the endocrine glands,
and a low prevalence of pernicious anaemia.  The autoimmune aspects of
the endocrinopathies (and pernicious anaemia) have been reviewed elsewhere[1,2].  The following account refers to the main aspects only, pay-
ing particular attention to the recent HLA studies and the possible
role of immune complexes in pathogenesis.

78

ADDISON'S DISEASE

Patients with idiopathic (autoimmune) Addison's disease are characterised by the presence of adrenocortical antibodies in some 70% of
cases, while these antibodies are absent in patients with tuberculous
destruction of the adrenal[3,4]. The majority of these autoantibodies
react only with the adrenal cortex but others also react with steroid
producing cells within the gonads and placenta. The presence of antibodies cross-reacting with steroid producing cells in the gonads shows
a strong correlation with clinical and laboratory evidence of primary
ovarian failure[5]. Antibodies reacting solely with the adrenal cortex
and antibodies cross-reacting with steroid producing cells in the
gonads consist of a group or family of antibodies reactive with a
variety of antigens in these tissues. These antibodies, which are all
IgG, tend to persist in the serum for many years (even for several
decades) after the onset of adrenocortical and/or gonadal failure
(Figure 1). Being IgG they are transported across the placenta into
the foetal circulation, but there is no evidence that these antibodies
on their own cause damage to the foetal adrenal[6].

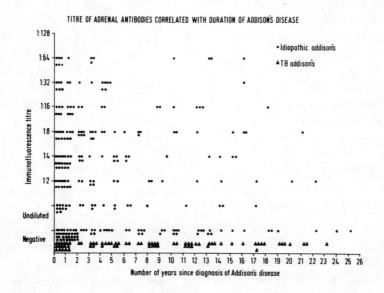

Fig. 1. Titre of IgG antibodies to human adrenal gland cortex in 261
patients with idiopathic Addison's disease and 64 patients with tuberculous destruction of the adrenals correlated with the number of years
since the diagnosis of Addison's disease. Source: Irvine and Barnes
(1975) Clinics in Endocrinology, Vol. 4, No. 2, Saunders, London.

Evidence for cell-mediated immunity to adrenocortical antigens in patients with idiopathic (autoimmune) Addison's disease has come from migration inhibition tests using the peripheral blood leukocytes of patients compared to those of controls[7]. At the time of writing studies on immune complexes in Addison's disease have not been reported.

Table 2 summarises the data currently available on HLA studies in Addison's disease. There would appear to be little doubt that there is a very high prevalence of B8 in caucasians with idiopathic (auto-immune) Addison's disease. The findings of Ludwig et al.[9] cannot be readily explained. Only small numbers of patients with tuberculous

TABLE 2

A SUMMARY OF THE LITERATURE ON HLA AND IDIOPATHIC ADDISON'S DISEASE

|  | Numbers | A1% | B8% |
|---|---|---|---|
| Thomsen et al., 1975[8] | 32 | 53 | 69 |
| Ludwig et al., 1975[9] | 20 | 20 | 20 |
| Irvine et al., 1978[10] | 50 | 56 | 80 |

Addison's disease have been HLA typed, but there is no tendency for B8 to be increased in these patients. Dunlop[11] reported that patients with Addison's disease who had relatives with Addison's disease always had the idiopathic form of the disorder, indicating that the autoimmune variety of adrenocortical failure has a genetic component. The prevalence of B8 in idiopathic (autoimmune) Addison's disease is essentially the same whether or not it is associated with other clinical autoimmune organ-specific disorders (Table 3).

TABLE 3

HLA-B8 IN IDIOPATHIC ADDISON'S DISEASE WITH OR WITHOUT ASSOCIATED DISEASES AS INDICATED IN TABLE 1

|  | No. | B8% |
|---|---|---|
| Idiopathic Addison's disease | | |
|   With associated AID | 29 | 83 |
|   Without associated AID | 21 | 76 |
| Total | 50 | 80 |
| Control | 300 | 28 |

Note: Data from Irvine, W.J., Feek, C.M., Morris, P.J., and Ting, A.

In contrast to findings of Thompson et al.[8], we[10] have found no correlation between adrenocortical antibodies in the serum and the presence of HLA-B8 (Table 4). There is the possibility that the criteria for diagnosis of idiopathic (autoimmune) Addison's disease is different in the two studies. In Addison's disease the situation does not pertain where the presence of the antibody or the clinical course is transient in some patients but persistent in others, so it would be particularly difficult to resolve this question. Extensive D or DR typing has not yet been reported in Addison's disease, but Svejgaard (personal communication) has indicated that the presence of adrenocortical antibodies in the serum of patients with idiopathic (autoimmune) Addison's disease is closely correlated with DRW 3.

TABLE 4

STUDIES CORRELATING THE OCCURRENCE OF ADRENAL ANTIBODIES AND HLA-B8 IN PATIENTS WITH IDIOPATHIC ADDISON'S DISEASE

| Adrenal antibody | Thomsen et al., 1975[8] | |
|---|---|---|
| | B8 positive | B8 negative |
| Positive | 21 (p < 0.002) | 1 |
| Negative | 1 | 5 |

| Adrenal antibody | Irvine et al., 1978[10] | |
|---|---|---|
| | B8 positive | B8 negative |
| Positive | 28 (NS) | 6 |
| Negative | 12 | 4 |

THYROID AUTOIMMUNE DISORDERS

Immunology. There are a variety of immune mechanisms that may be involved in the pathogenesis of thyroid disease, and the relative role of each may be different in relation to different disorders in different patients and with time in the same patient. Some of these mechanisms[12] are shown in Figure 2. Thus, in the thyrotoxicosis of Graves' disease (with diffuse involvement of the thyroid) TSH receptor stimulating antibodies (TSH R-Sab) are the probable cause (Figure 3). On the other hand, circulating thyroid IgG antibodies or those produced in the thyroid are unlikely by themselves to be the cause of Hashimoto goitre or primary atrophic hypothyroidism. Such IgG antibodies may play a role as part of immune complex mediated damage or as part of an antibody-dependent cell-mediated cytotoxic mechanism (Figure 4).

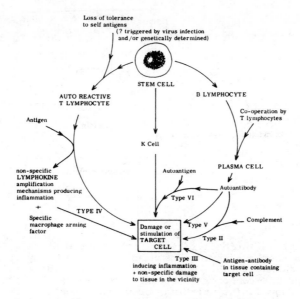

Fig. 2. Diagram representing the interactions of immune mechanisms Types II-VI in the production of autoimmune disease. The different immune mechanisms may have greater importance in some diseases than in others. Thus Types II, IV and VI may be predominant in Hashimoto's thyroiditis, while Type V may be more relevant in thyrotoxicosis. The lymphokine and complement systems may be regarded as nonspecific amplification mechanisms in order to boost the effect of a small number of specifically sensitized cells or antibodies interacting with the antigen at the target site. Source: Irvine, 1978. Medical Immunology, Teviot, Edinburgh.

Immune complexes can be detected in the serum of patients with autoimmune thyroid diseases and are present in the greatest amount in Hashimoto goitre patients, moderate amounts in patients with primary atrophic hypothyroidism and in lesser amounts (but still significant) in patients with thyrotoxicosis[13]. Moreover, antigen-antibody complexes in the presence of antibody excess may "arm" K-cells so as to render these specifically cytotoxic to target cells that have the corresponding antigen on or part of their surface membrane[14] (Figure 4). Since Hashimoto goitre and primary atrophic hypothyroid patients tend to have such complexes in their sera in the presence of thyroid antibody excess and since it is likely that some of these antibodies may be reactive with thyroid cell membrane antigens, one may surmise that antibody-dependent cell-mediated cytotoxicity could be a potent mechanism for immunological damage in these conditions. The leukocyte migration inhibition test gives some indication that T lymphocytes may be sensi-

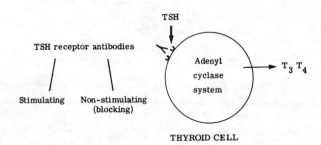

Fig. 3.  Thyroid antibodies react with TSH receptors (or with sites adjacent to them) on thyroid membranes in such a way as to interfere with the binding of TSH.  These antibodies should be collectively referred to as TSH-receptor antibodies and may be classified according to whether they do (stimulating) or do not (non-stimulating or blocking) trigger the adenyl cyclase system of the thyroid cell with production of increased amounts of thyroid hormones (T3 and T4).  They are also characterised by their species specificity according to whether they react with mouse, guinea pig or human thyroid.  Source:  Irvine, 1978.  Medical Immunology, Teviot, Edinburgh.

**TYPE VI IMMUNE REACTIONS**

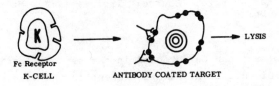

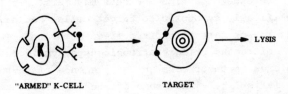

Fig. 4.  The mechanism of antibody-dependent cell-mediated cytotoxicity, involving K cells that may be "armed" in the presence of antigen-antibody complexes in antibody excess to give them antigen-specific cytotoxicity.

tised to the antigen in question. Using this method there is some
suggestion that T lymphocyte-mediated mechanisms may be involved in
Hashimoto goitre and to a lesser extent in primary atrophic hypothyroidism and to a small extent in thyrotoxicosis. However, the analysis
of the lymphoid cells infiltrating Hashimoto goitres suggest that the
ratio of B to T lymphocytes is higher in such goitres than in the peripheral blood[15,16].

The follow up of thyrotoxic patients over several decades reveals
that a substantial minority become subclinically or clinically hypothyroid, although the only thyroid treatment these patients may have
had was a non-destructive course of antithyroid drugs[17]. Presumably
in these patients the balance of the autoimmune mechanisms has altered
from stimulating antibodies to mechanisms that are damaging to the thyroid cells. It is also apparently possible that TSH receptor antibodies that are blocking may have become stimulating, explaining the
occasional patient who goes in the reverse direction from having primary atrophic hypothyroidism to developing thyrotoxicosis[18].

HLA studies. When all cases of Graves' disease are considered together, irrespective of their clinical course or the outcome of different forms of treatment, the prevalence of HLA-B8 is between 44% and
51% compared to 21% and 28% in the control population respectively[19,20].
However, when patients with Graves' disease are compared according to
whether they relapsed or remained in remission after a prolonged course
of antithyroid drugs, it was found that those who relapsed had a higher
prevalence of B8 than those who remained in remission[21] (Figure 5). In

## HLA IN THYROTOXICOSIS

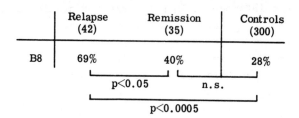

Fig. 5. There is a significantly higher prevalence of HLA-B8 in thyrotoxic patients (Graves' disease) who relapse after a prolonged course
of anti-thyroid drugs than in patients who remain in remission for at
least four years after a similar course of anti-thyroid drug therapy.
Source: Irvine, 1978, Medical Immunology, Teviot, Edinburgh.

those who had a lasting remission the prevalence of B8 was not significantly different from that in the control population. The difference
between the patients who showed a relapse and those who showed lasting
remission was statistically significant. Similar findings have been
reported by others, including D typing[22].

So far no studies directly correlating the titre and the persistence
of TSH receptor antibodies and HLA type have been done. Such a study
should take into consideration whether these receptor antibodies are
stimulating or blocking. However, it may be reasonably inferred from
what is already known that persistence of thyroid stimulating immunoglobulins is likely to be related to HLA-B8 and DW3 and therefore
presumably to the immune response (Ir) gene/s in linkage disequilibrium
with these HLA antigens (see Figure 10.)

Of all the thyroid conditions one would think from a prior argument
that Hashimoto thyroiditis with goitre would be among the conditions
showing the strongest association with the HLA system. It is not the
case; at least on present evidence. Hashimoto thyroiditis has the
most conspicuous lesion in terms of density of lymphocytes and plasma
cell infiltration and quantitatively the titres of autoantibodies in
serum are probably the highest in the whole range of organ-specific
autoimmune disease. Moreover, these titres tend to be persistent unless subtotal thyroidectomy is performed. Yet no association with any
HLA antigen has yet been described for goitrous patients with Hashimoto
thyroiditis, irrespective of whether the patient is euthyroid or hypothyroid (Figure 6). However, when primary atrophic hypothyroidism is
studied on its own, excluding patients with goitre, there is a significant association with HLA-B8. The difference between the goitrous
Hashimoto patients and the non-goitrous atrophic hypothyroid patients
is statistically significant[23]. The patients in this study were almost
all adults and information is not currently available in terms of the
HLA association with juvenile autoimmune thyroiditis which may be
transient.

The fact that Hashimoto thyroiditis with goitre stands out as the
exception in the spectrum of autoimmune endocrine disorders may be due
the fact that it is the only one of the various forms of organ-specific
autoimmunity with failure of the function of the target gland that is
associated with such gross enlargement of the gland. Clearly there
must be some other factor operating to produce a goitre in Hashimoto
patients which is absent in the patients with primary atrophic hypothyroidism. Hashimoto goitre may therefore be more polygenic than
simple atrophy of the adrenal or thyroid. There is no clear relation-

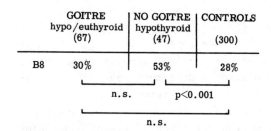

**HLA IN HASHIMOTO GOITRE AND PRIMARY
ATROPHIC HYPOTHYROIDISM**

|  | GOITRE<br>hypo/euthyroid<br>(67) | NO GOITRE<br>hypothyroid<br>(47) | CONTROLS<br><br>(300) |
|---|---|---|---|
| B8 | 30% | 53% | 28% |

n.s.          p<0.001

n.s.

Fig. 6. The increased prevalence of HLA-B8 in primary atrophic hypothyroidism and not in Hashimoto goitre compared to controls.

ship between titres of antibodies to thyroglobulin or to thyroid micro-
somes and HLA-B8, even although these antibodies tend to persist in the
serum for many years[23]. Extensive D or DR typing has not yet been reported in patients with Hashimoto thyroiditis or primary atrophic
hypothyroidism.

DIABETES MELLITUS

Immunological studies pioneered the way to the recognition that
there are two main types of primary diabetes which for the moment are
best referred to as type I and type II (Table 5)[24]. Thus, the clinical

TABLE 5

THE TWO MAIN TYPES OF PRIMARY DIABETES

| Type I |
|---|
| Insulin-dependent juvenile-onset diabetes |
| Insulin-dependent maturity-onset diabetes |
| ICAb positive diabetics initially controlled by OHA |

| Type II |
|---|
| Insulin-independent maturity-onset diabetes |
| Some insulin-independent juvenile-onset diabetes |
| (ICAb) negative at diagnosis) |

Source: Irvine, 1977, Lancet, i, 638.

association of diabetes mellitus with the members of the group of autoimmune diseases (Table 1) indicated that there must be a common factor in their pathogenesis. The fact that the association is almost exclusively with ketosis-prone insulin-dependent diabetes[25] indicates that the common pathogenic factor of autoimmunity is relevant only to insulin-dependent diabetes and not to the more common form of diabetes that in the past has been referred to as the "maturity onset" type. Furthermore, the age at onset of insulin-dependent diabetes associated with idiopathic Addison's disease is very wide, making the use of terms such as "juvenile-onset" and "maturity-onset" quite inappropriate[25]. Such terms infer that diabetes can be effectively subdivided into two main categories on the criterion of age of onset alone. Secondly, when diabetics without any present or past history or clinical features of thyroid disease or pernicious anaemia were studied for the prevalence of thyroid and of gastric autoantibodies compared to age and sex matched controls, only insulin-dependent diabetics, and not diabetics who could be adequately controlled by diet or oral hypoglycaemic agents (insulin-independent), had a significantly increased prevalence of thyroid cytoplasmic and gastric parietal cell antibodies[26]. Thirdly, evidence for cell mediated immunity to pancreatic antigens was shown to be confined to insulin-dependent diabetics using the leukocytes from the peripheral blood of diabetics compared to those from control subjects[27,28]. All this evidence had accumulated before an association between HLA and diabetes had been initially suggested by Singal and Blajchman[29] and substantiated by Nerup et al.[30] and confirmed by Cudworth and Woodrow[31].

Since 1974 the association of both B8 and B15 in caucasians with juvenile-onset diabetes as opposed to "maturity-onset" diabetes has been extensively analysed, but only recently have the HLA studies concentrated on the analysis of insulin-dependent versus insulin-independent diabetes irrespective of age at onset. Thus, though HLA studies have fully confirmed the description of diabetes shown in Table 5, they have done little that is innovative in forming that classification. When the age of onset was used as the criterion to sub-divide diabetes into two types, much confusion results in terms of genetic analysis[32]. When the criterion of insulin-dependency was used, it became a simple matter to obtain strongly suggestive evidence that type I diabetes (insulin-dependent) and type II diabetes (insulin-independent) breed true and, therefore, are presumably the outcome of separate disease processes within the syndrome of primary diabetes[33].

In 1974 islet cell antibodies (ICAb) were described in insulin-dependent diabetes associated with polyendocrine autoimmune disease[34,35]. Subsequently it was shown that there was a wide range in terms of the duration of ICAb following diagnosis and that diabetics with polyendocrine disease tended to have ICAb in their serum for many years, while the majority of insulin-dependent diabetics only had ICAb transiently[36]. The prevalence of ICAb at diagnosis in insulin-dependent diabetics is of the order of 60-70% in terms of statistically meaningful figures. The prevalence of ICAb in the diabetics only requiring diet is similar to that in the general population at 0.5% while the prevalence of ICAb in diabetics controlled by oral hypoglycaemic agents (OHA) is significantly higher, especially in those who have associated autoimmune disease or a family history of it (Figure 7).

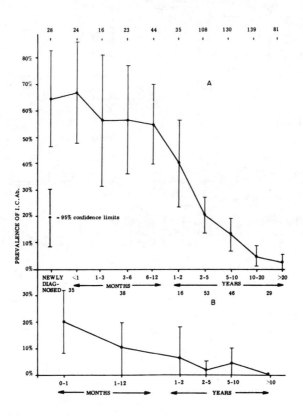

Fig. 7. The prevalence of ICAb according to the duration of diabetes. A) 628 insulin-treated diabetics; B) 217 diabetics requiring oral hypoglycaemic agents (OHA). Source: Irvine et al., 1977, Diabetes, 26, 138.

The follow up of ICAb positive diabetics controlled by OHA shows that such patients have a strong tendency to become insulin-dependent in subsequent years[37] (Figure 8). For that reason ICAb positive diabetics controlled by OHA are thought to have an earlier stage of type I diabetes that culminates in insulin-dependency (Table 5). This is important in relation to further family studies to ensure that such subjects are not wrongly classified as belonging to type II diabetes and the value of the study subsequently vitiated.

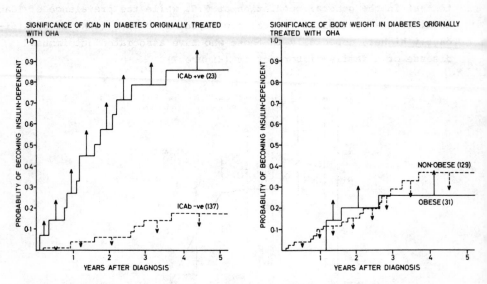

Fig. 8. The value of ICAb as a predictor of secondary OHA failure using actuarial statistics (Fig. 8a), compared to obesity (Fig. 8b). Source: Irvine et al., 1978, J. Clin. Lab. Immunol., 1, in press.

In terms of the pathogenesis type I diabetes, antibodies reactive with the cytoplasm of islet cells are unlikely to be an important factor on their own. For one thing they react with the cytoplasm of the glucagon, somatostatin and polypeptide producing cells as well as that of the insulin producing B cells. It is not established whether microsomal antigens are also present in or on the cell membranes of the islet cells. Antibodies reactive with antigens in or on the cell membrane of the islet cells have recently been described using isolated rat islet cells in culture[38], and it is probable that these are different antibodies from those reactive with islet cell cytoplasm[39]. As in the case of thyroid autoimmune disease, surface antigens are highly relevant to the mechanism of antibody-dependent cell-mediated cytotoxicity (Figure 4). Immune complexes have been detected in type I diabetics using the Clq and the Raji assays (for technical details see WHO report[40]), and these show a high prevalence at the time of diagnosis of type I diabetes and correlate well with the occurrence of islet cell cytoplasmic islet cell antibodies[41] (Figure 9). Although these complexes may simply be a consequence of islet cell damage, it is conceivable that such complexes in the presence of antibody excess may be important in causing islet cell damage by the mechanism of antibody-dependent cell-mediated cytotoxicity.

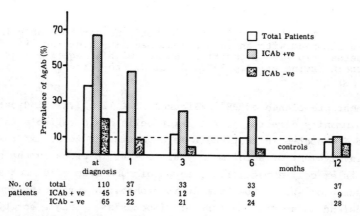

| No. of | total | 110 | 37 | 33 | 33 | 37 |
| patients | ICAb + ve | 45 | 15 | 12 | 9 | 9 |
| | ICAb - ve | 65 | 22 | 21 | 24 | 28 |

Fig. 9. The prevalence of soluble immune complexes in the serum (as detected by the Clq solid phase assay) in Type I diabetics according to the time since diagnosis and according to the presence or absence of islet cell cytoplasmic antibodies. Source: Irvine et al., 1978, J. Clin. Lab. Immunol., 1, in press.

HLA antigens. There are three groups of HLA antigens in linkage disequilibrium with three immune response (Ir genes) that influence the risk of a caucasian individual becoming an insulin-dependent diabetic[42,43] (Figure 10). The first of these groups (which includes A1, B8 and DW3) is the one which has an increased prevalence in idiopathic (autoimmune) Addison's disease, persistent thyrotoxicosis and atrophic hypothyroidism as described above. This is also the group that is associated with the autoimmune aspects of type I diabetes. Indeed an interesting gradation can be demonstrated from weak evidence of autoimmunity (e.g., negative or transient ICAb and low prevalence of thyrogastric antibodies) and a low prevalence of HLA-B8, through moderate evidence of autoimmunity (i.e., persistent titres of ICAb and higher prevalence of thyrogastric antibodies) with increased prevalence of B8, to strong evidence of autoimmunity (i.e., polyendocrine autoimmune disease with very high prevalence of thyrogastric antibodies) and the

90

HLA in TYPE I DIABETES

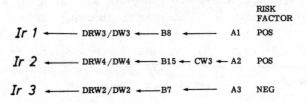

|  |  |  |  | RISK FACTOR |
|---|---|---|---|---|
| Ir 1 ←——— DRW3/DW3 ←—— B8 ←——— | | | A1 | POS |
| Ir 2 ←——— DRW4/DW4 ←—— B15 ← CW3 ← A2 | | | | POS |
| Ir 3 ←——— DRW2/DW2 ←——B7 ←——— | | | A3 | NEG |

Fig. 10. The three series of HLA antigens each probably in linkage disequilibrium with an immune response (Ir) gene associated with Type I diabetes; two with a positive and one with a negative risk factor. Source: Irvine et al., 1978, J. Clin. Lab. Immunol., 1, in press.

highest prevalence of B8[44] (Figures 11 & 12). This indicates that autoimmunity directed at the islet cells has a variable role within type I diabetes in that there is a continuing spectrum from minimal or no involvement of autoimmunity up to what is probably the sole cause for islet cell damage (i.e., an autoimmune form of diabetes). In the presence of such a spectrum it is probably a worthless argument to try to defend an autoimmune form of diabetes as a single entity, but rather to regard such a state as one extreme of a spectrum of interacting factors only one of which is autoimmunity (Figure 13).

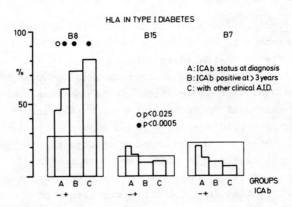

Fig. 11. The positive association of the autoimmune aspects of Type I diabetes with HLA-B8 and the lack of it with B15 and with B7. Source: Irvine et al., 1978, J. Clin. Lab. Immunol., 1, in press.

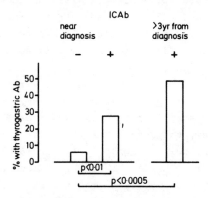

Fig. 12. The relationship between ICAb and thyroid/gastric autoantibodies in Type I diabetes. Source: Irvine et al., 1978, J. Clin.
Lab. Immunol., 1, in press.

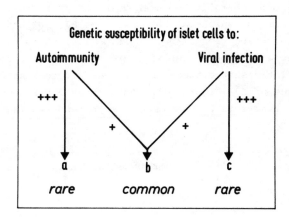

Fig. 13. The possible interaction between autoimmunity and pancreatotropic viral infection in the pathogenesis of Type I diabetes, based
on genetic susceptibility to either or to both. Source: Irvine, 1978,
Lancet, i, 638.

Type I diabetes is different from other organ-specific autoimmune
disorders in that it has a second group of antigens positively associated with an increased risk of getting the disease; namely, A2, B15,
Cw3, Dw4 and DRw (Figure 10). As shown in Figure 11 this group or
series of HLA antigens shows no correlation with the tendency towards
autoimmunity[44]. It must therefore operate through some other mechanism
(e.g., susceptibility to pancreatotropic viral infection or to chemical

damage). The fact that studies to date have been unconvincing in relation to any correlation between viral antibodies at diagnosis of
type I diabetes and B15 is perhaps not surprising since the nature of
the virus that might cause islet cell damage in man has not yet been
defined. All the evidence for a viral pathogenesis for type I diabetes
in man is circumstantial or by inference using the rat or mouse as an
experimental model. If we knew what the virus was in type I diabetes
in man, it would then be possible to determine whether a correlation
exists between susceptibility to it and a particular HLA phenotype. In
the author's view, the second HLA series is coding for genetic susceptibility to islet cell damage by a factor in the environment, be it
virus or chemical. The fact that this second HLA series is concerned
with immune response to certain foreign antigens (i.e., not autoantigens) is demonstrated by the increased humoral immune response that B15
and Cw3 individuals mount against heterologous insulin compared to that
of A1 and B8 individuals[45-47] (Figure 14). This also demonstrates that
the association between $IR_1$ and organ-specific autoimmunity is not simply due to a generally heightened immune response to all antigens[47].

The third group of HLA antigens (A3, B7, DW2 and DRW2) is associated
with a decreased risk of getting type I diabetes[48] (Figure 10). When
populations of type I diabetics are studied there tends to be an inverse relationship between the strength of the autoimmune component and
prevalence of B7 (Figure 11). An increased prevalence of cutaneous
allergic reactions to insulin has been reported in diabetics who are

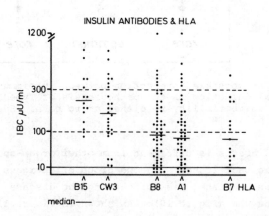

Fig. 14. The association of higher titres of insulin antibodies with
HLA-B15 and CW3 in insulin-treated diabetics. Source: Irvine et al.,
1978, J. Clin. Lab. Immunol., 1, in press.

B7[49]. To this extent the third HLA series is linked to a third immune response gene ($IR_3$). However, it must be borne in mind that the three HLA series may be in linkage disequilibrium with other genes relevant to an individual's susceptibility to type I diabetes that are not concerned with immune response but possibly with the regenerative capacity of islet cells following damage from whatever cause. Thus, the resistance of B7, DW2 and DRW2 individual to type I diabetes could be due to his/her resistance to mount an autoimmune reaction against islet cells, low susceptibility to pancreatotropic viral infection or to a high ability to regenerate the B cells of the islets if damaged. Nevertheless, the B7, DW2 and the DRW2 individual will probably still get diabetes if the assault by unknown factor x is strong enough.

In the Japanese HLA-B8 is much less common in the general population compared to caucasians[50]. Figure 15 shows that in insulin-dependent Japanese diabetics the prevalence of cytoplasmic ICAb is only 17% within the year following diagnosis compared to some 50% in caucasian insulin-dependent diabetics, and that in the Japanese insulin-dependent diabetics no cytoplasmic ICAb was observed more than 4 years from diagnosis. Thus there is little evidence so far of persistent ICAb in Japanese type I diabetics. A weak autoimmune component in type I Japanese diabetes may contribute to the lower prevalence of insulin-dependent diabetes in that ethnic group.

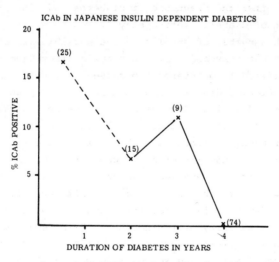

ICAb IN JAPANESE INSULIN DEPENDENT DIABETICS

Fig. 15. The prevalence of ICAb in insulin-dependent Japanese diabetics. Note the lower prevalence of ICAb at diagnosis than in the Caucasian population shown in Fig. 7 and the absence of any persistant ICAb at more than 4 years after diagnosis. Data from W.J. Irvine and K. Nagaoka.

Leslie and Pyke[51] have reported on a large series of diabetic twins in Britain. Although total ascertainment of such twins was far from being achieved, the difference in the concordant/discordant ratio between the maturity-onset non-insulin-dependent twins and the juvenile-onset insulin-dependent type provides further evidence that these types of diabetes are genetically distinct. The finding that many insulin-dependent twin pairs were discordant for diabetes is in keeping with the concept that this type of diabetes cannot be entirely genetic in origin.

Complications. The vascular complications in terms of microangiopathy constitute the most serious aspect of diabetes. Thus the commonest cause of blindness in young and middle-aged people is diabetes[52]. Although Ortved Anderson[53] has reported a positive association between severe proliferative retinopathy and insulin antibodies, my colleagues and I have argued that insulin-anti-insulin complexes are unlikely to have much biological activity[54]. Insulin is a bivalent antigen and is therefore unlikely to form complexes that will effectively fix complement. On the other hand, we have shown that there is a positive association between the prevalence of soluble complement-fixing immune complexes and the severity of retinopathy, irrespective of whether the diabetic patient had received insulin or not. This association holds in relation to the development of diabetic proliferative retinopathy many years after the diagnosis of diabetes and also in relation to the early onset of diabetic retinopathy[54].

Although a number of studies on the possible association between certain HLA antigens and diabetic complications have been reported[55-58] many are difficult to interpret because of lack of controls in relation to the duration of diabetes and because of the grouping together of patients with different degrees of retinopathy. Larkins et al.[59] reported that the unusual pattern of B8 positive A1 negative was significantly increased in diabetics with proliferative retinopathy. This was not confirmed by Cudworth et al.[42] whose preliminary data is shown in Figure 16. Clearly, detailed studies are required carefully matching diabetics with similar disease and with similar degree of metabolic control (as far as this is possible) and of similar duration, but differing in the severity of retinopathy. This requires to be done in conjunction with other laboratory measurements, such as the measurement of immune complexes in the serum as described above.

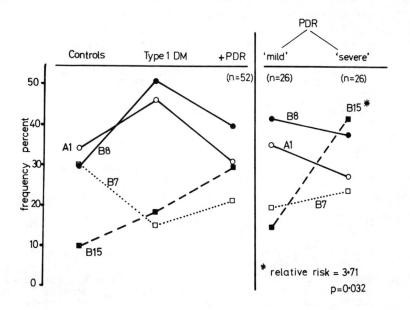

Fig. 16. HLA frequencies in relation to Type I diabetes mellitus and retinopathy. Source: Cudworth, Bottazzo and Doniach, 1978, in The Immunology of Diabetes Mellitus, Irvine, ed., Teviot, Edinburgh.

REFERENCES

1. Irvine, W.J. (1975) in Clinics in Endocrinology, Irvine, W.J. ed., Vol. 4, No. 2, Saunders, London.

2. Irvine, W.J. (1978) in Medical Immunology, Irvine, W.J. ed., Teviot, Edinburgh, Chapter 6.

3. Irvine, W.J., Stewart, A.G. and Scarth, L. (1967) Clin. exp. Immunol., 2, 31.

4. Nerup, J. (1974) Acta Endocrinologica 76, 142.

5. Irvine, W.J. and Barnes, E.W. (1974) J. Reprod. & Fert. 21, 1-31.

6. Gamlen, T.R., Aynsley-Green, A., Irvine, W.J. and McCallum, C.J. (1977) Clin. exp. Immunol., 28, 192-195.

7. Nerup, J. and Bendixen, G. (1969) Clin. exp. Immunol. 5, 341-353.

8. Thomsen, M., Platz, P., Anderson, O., Christy, M., Lyngsøa, J., Nerup, J., Rasmussen, K., Ryder, L.P., Staub-Nielsen, L. and Svejgaard, A. (1975) Transpl. Rev., 22, 125-147.

9. Ludwig, H., Mayr, W.R., Pacher, M. et al. (1975) Z. Immunitaets-forsch., 149, 423-427.

10. Irvine, W.J., Feek, C.M., Morris, P.J., and Ting, A. (1978) J. clin. lab. Immunol., 1, in press.

11. Dunlop, D.M. (1963) Brit. med. J., 2, 887-891.

12. Irvine, W.J. (1974) Proc. roy. soc. Med., 67, 548-555.

13. Al-Khateeb, S.F. and Irvine, W.J. (1978) J. clin. lab. Immunol., 1, 55-58.

14. Calder, E.A., McLeman, D. and Irvine, W.J. (1973) Clin. exp. Immunol., 15, 467-470.

15. Totterman, T.H., Mäenpää, J., Gordin, A., Mäkinen, T., Taskinen, E., Andersson, L.C. and Häyry, P. (1977) Clin. exp. Immunol., 30, 193.

16. Whitmore, D. and Irvine, W.J. (1979) J. clin. lab. Immunol., 1, in press.

17. Irvine, W.J., Gray, R.S., Toft, A.D., Seth, J., Lidgard, G.P. and Cameron, E.H.D. (1977) Lancet, ii, 179-181.

18. Irvine, W.J., Lamberg, B.-A., Cullen, D.R. and Gordin, R. (1979) J. clin. lab. Immunol., 1, in press.

19. Grumet, F.C., Payne, R.O., Konishi, J., Kriss, J.P. (1974) J. Clin. Endoc., 39, 1115-1119.

20. Thorsby, E., Segaard, E., Solem, J.H. and Kornstad, L. (1975) Tissue Antigens, 6, 54-55.

21. Irvine, W.J., Gray, R.S., Morris, P.J. and Ting, A. (1977) Lancet, ii, 898-900.

22. Beck, K., Lumholtz, B., Nerup, J., Thomsen, M., Platz, P., Ryder, L.P., Svejgaard, A., Siersbaek-Nielsen, K., Hansen, J.M. and Larsen, J.H. (1977) Acta endocr. (Copenh.), 86, 510-516.

23. Irvine, W.J., Gray, R.S., Morris, P.J. and Ting, A. (1978) J. clin. lab. Immunol., 1, in press.

24. Irvine, W.J. (1977) Lancet, 1, 638-642.

25. Irvine, W.J., Feek, C.M., Morris, P.J. (1978) J. clin. lab. Immunol., 1, in press.

26. Irvine, W.J., Clarke, B.F., Scarth, L., Cullen, D.R. and Duncan, L.J.P. (1970) Lancet, 2, 163-168.

27. Nerup, J., Anderson, O.O., Bendixen, G., Egeberg, J., Gunnarsson, R., Kromann, G. and Poulsen, J.E. (1974) Proc. roy. soc. Med., 67, 506-513.

28. MacCuish, A.C., Jordan, J., Campbell, C.J., Duncan, L.J.P. and Irvine, W.J. (1974) Diabetes, 23, 693-697.

29. Singal, D.P. and Blajchman, M.A. (1973) Diabetes, 22, 429-432.

30. Nerup, J., Platz, P., Ortved Anderson, O., Christy, M., Lyngsoe, J., Poulsen, J.E., Ryder, L.P., Staut Nielsen, L., Thomsen, M. and Svejgaard, A. (1974) Lancet, 2, 864-866.

31. Cudworth, A.G. and Woodrow, J.C. (1975) Diabetes, 24, 345-349.

32. Smith, C., Falconer, D.S. and Duncan, L.J.P. (1972) Ann. human Genet., 35, 281-299.

33. Irvine, W.J., Toft, A.D., Holton, D.E., Prescott, R.J., Clarke, B.F. and Duncan, L.J.P. (1977) Lancet, 2, 325-328.

34. Bottazzo, G.F., Florin-Christensen, A. and Doniach, D. (1974) Lancet, 2, 1279-1283.

35. MacCuish, A.C., Barnes, E.W., Irvine, W.J. and Duncan, L.J.P. (1974) Lancet, 2, 1529-1531.

36. Irvine, W.J., McCallum, C.J., Gray, R.S., Campbell, C.J., Duncan, L.J.P., Farquhar, J., Vaughan, H. and Morris, P.J. (1977) Diabetes, 26, 138-147.

37. Irvine, W.J., McCallum, C.J., Gray, R.S. and Duncan, L.J.P. (1977) Lancet, i, 1025-1027.

38. Lernmark, A., Freedman, Z.F., Kanatsuna, T., Patzelt, C., Rubenstein, A.H. and Steiner, D.F. (1978) in The Immunology of Diabetes Mellitus, Irvine, W.J. ed., Teviot, Edinburgh.

39. Irvine, W.J., Feek, C.M., Freedman, Z.F., Lernmark, A., Huen, A., Steiner, D.F. and Rubenstein, A.H. (1978) in The Immunology of Diabetes Mellitus, Irvine, W.J. ed., Teviot, Edinburgh.

40. Lambert, P.H. et al. (1978) J. clin. lab. Immunol., 1, 1-15.

41. Irvine, W.J., Di Mario, U., Guy, K., Gray, R.S. and Duncan, L.J.P. (1978) J. clin. lab. Immunol., 1, in press.

42. Cudworth, A.G., Bottazzo, G.F. and Doniach, D. (1978) in The Immunology of Diabetes Mellitus, Irvine, W.J. ed., Teviot, Edinburgh.

43. Irvine, W.J. (1978) in The Immunology of Diabetes Mellitus, Irvine, W.J. ed., Teviot, Edinburgh.

44. Irvine, W.J., Di Mario, U., Gray, R.S., Feek, C.M., Ting, A., Morris, P.J., Duncan, L.J.P. (1978) J. clin. lab. Immunol., 1, No. 2, in press.

45. Bertrams, J., Jansen, F.K., Gruneklee, D., Reis, H.E., Drost, H., Beyer, J., Gries, F.A. and Kuwert, E. (1976) Tissue Antigens, 8,13.

46. Ludvigsson, J., Säfwenberg, J., Heding, L.G. (1977) Diabetologia, 13, 13.

47. Irvine, W.J., Di Mario, U., Gray, R.S., Morris, P.J. (1978) J. clin. lab. Immunol., 1, No. 2, in press.

48. Ludwig, H., Schernthaner, G. and Mayr, W.R. (1976) New Engl. J. Med., 294, 1066.

49. Bertrams, J. and Gruneklee, D. (1977) Tissue Antigens, 10, 273.

50. Wakisaka, A., Aizawa, M., Matsuura, N., Nakagawa, S., Nakayana, E., Itakura, E., Okuno, A. and Wagatsuma, Y. (1976) Lancet, ii, 970.

51. Leslie, R.D.G. and Pyke, D.A. (1978) in The Immunology of Diabetes Mellitus, Irvine, W.J. ed., Teviot, Edinburgh.

52. Kohner, E.M. and Dollery, C.T. (1975) in Complications of Diabetes, Keen, H. and Jarrett, J. eds., Edward Arnold, London, p. 7.

53. Ortved Andersen, O., Vejtorp, L., Christy, M., Platz, P., Svejgaard, A., Thomsen, M., Reersted, P. and Nerup, J. (1975) Diabetologia, 11, 329.

54. Irvine, W.J., Di Mario, U., Guy, K., Iavicoli, M., Pozzilli, P., Lumbroso, B. and Andreani, D. (1978) in The Immunology of Diabetes Mellitus, Irvine, W.J. ed., Teviot, Edinburgh.

55. Barbosa, J., Noreen, H., Emme, L., Goetz, F., Simmons, R., deLeiva, A., Najarian, J. and Yunis, E.J. (1976) Tissue Antigens 7, 233-237.

56. Becker, B., Shin, D.H., Burgess, D., Kilo, C. and Miller, W.V. (1977) Diabetes, 26, 997-999.

57. Chuck, A.L. and Cudworth, A.G. (1977) Diabetologia, 13, 387.

58. Moller, E., Persson, B., and Sterky, G. (1978) Diabetologia,14, 155-158.

59. Larkins, R. G., Martin,F.I.R.and Tait,B.D. (1978) Brit. Med. J., 1,1111.

DISCUSSION

CANTOR: The major problem with the view that D. mellitus is an autoimmune disorder - accompanied by anti-islet cell antibodies, or even anti-insulin receptor antibodies - is that neither autoantibody accounts for the major clinical feature of the disease - angiopathy and renal failure. The autoimmune view would be strengthened considerably if there were good evidence that these central clinical features of diabetes were associated with antibody-antigen complexes in small vessels and glomeruli, particularly if these complexes included anti-islet cell Ig. Is there any data to this effect?

IRVINE: The causation of beta cell failure in the islets may have little in common with the causation of diabetic complications, although both may have immune components. Thus, I would not expect islet cell antibodies, nor insulin-receptor antibodies, to be involved in a major way in diabetic complications. This would not provide an argument against their involvement in the development of islet cell failure or insulin resistance in the first place.

My hypothesis is that immune mechanisms may play a contributory role in diabetic microangiopathy through the accumulation of immune complexes involving a very wide range of antigens (not necessarily autoantigens). This is quite different from the immune mechanisms and the antibody specificities that may be involved in the pathogenesis of beta cell failure in the pathogenesis of diabetes itself.

ROITT: In the majority of cases of autoimmune insulin dependent diabetes, the antibodies disappear within a year of onset yet you describe circulating immune complexes sometime later. Could you comment on this?

IRVINE: The methods we used for measuring soluble immune complexes in the sera of diabetics do not give any indication of the nature of the antigen(s) involved. There is no reason why the antigens involved close to the time of diagnosis of diabetes should be the same as those involved many years after treatment. Thus, we have demonstrated a close correlation between the presence of the immune complexes and cytoplasmic islet cell antibodies at and shortly after diagnosis, but no such overall correlation exists many years after diagnosis. Thus, in the majority of type I diabetics we believe that immune complexes are present at diagnosis and then go away, only to return again, but involving different antigens. We think that most of the immune complexes in the sera of diabetics of long duration are not related to insulin as detected by the Clq and Raji assays. For example insulin is a bivalent antigen so that insulin-anti-insulin complexes could only be expected to bind complement poorly and they may not be very good at binding to Fc receptors. There is not a good correlation between the presence of insulin antibodies as determined by an insulin binding assay and evidence of immune complexes in long standing insulin treated diabetics. Furthermore, diabetics of long duration but never treated with insulin have plenty of soluble immune complexes in their sera. The nature of the antigens involved in long duration diabetes could be anything; i.e., antigen antibody complexes that would normally be efficiently cleared from the serum but which accumulate in the sera of these patients.

ROITT: Contrasting the rapid difference of islet cell antibodies in the majority of cases of autoimmune insulin dependent diabetes with the persistence of antibodies in those cases associated with other "primary" autoimmune disorders, one might postulate that the first group arises by viral infection inducing an anti-islet cell immune response which destroys the target and eliminates the antigen whereas if the second group involved stimulation of autoantibodies by anti-idiotypes (comparable in shape to antigens), the stimulus for autoantibody production would remain even after destruction of the appropriate pancreatic cells.

IRVINE: That is an interesting suggestion, but as far as I know this has not been investigated as yet with regard to the persistence of islet cell antibodies. If so then the formation of anti-idiotypes must presumably be correlated with HLA-B8.

DUPONT: I wonder if the present controversy with regard to the HLA-linked genetic factors in diabetes to some extent is caused by a lack of uniformity with regard to the diagnostic criteria for juvenile onset diabetes, insulin-dependent diabetes, and islet cell antibody positive diabetes?

IRVINE: I don't think there is serious controversy in relation to HLA-linked genetic factors in diabetics within comparable ethnic groups. Where controversy has arisen in the past, it has been due as you suggest to poor diagnostic criteria and especially to the inclusion of diabetics of different ethnic origins within a single group.

HARRISON: I'd like to add to your comments on the role of immune complexes in the vascular complication of diabetes. Lesions identical to the Kimmelstiel-Wilson nodules in diabetic kidneys have been seen in immune complex diseases and reported in SLE.
Also, would you comment on the recent report by Larkins of an absence of association between B8 and A1 in those diabetics with retinopathy.

IRVINE: The recent report by Larkins was presented and discussed at the International Symposium on the Immunology of Diabetes that has just been held in Edinburgh (Immunology of Diabetes, W.J. Irvine (ed.), Teviot Scientific Publications, Edinburgh, 1978). His findings have not been the experience of others to date; confirmation by a large and carefully controlled study taking into account type of diabetes, duration, and quality of control and nature of the retinopathy is required. Indeed, it would be fair to say at the present time that most if not all of the published papers on the subject of HLA and diabetic complications suffer from very serious defects in terms of clinical design making their conclusions highly dubious. However, I am sure that future work following on these early beginnings may establish something of considerable interest in terms of genetic susceptibility to diabetic complications and the mechanisms involved.

MACKAY: Your use of the notation Ir1-Ir3 might create confusion in relation to their use in mouse genetics. Can you identify genetic components other than HLA in type I diabetes?
The B8 effect seems to be exerted through long persistence of autoantibody. Can you comment on this in terms of known function of immune response genes in the mouse?

IRVINE: Presumably some prefix could be used to specify whether one was referring to immune response (Ir) genes in mouse or man. I am not aware that other genetic components amenable to in vitro testing have been identified in relation to type I diabetes.
Immune response genes in mice are known to exert their influence at several levels of the immune system. Until now there has not been good evidence in man that the HLA system is directly linked to immune response genes in terms of autoantibody formation. What the data that I have presented show is that persistence of islet cell autoantibodies tends to occur in B8 type I diabetics. Type I diabetics with negative or transient islet cell antibody titers at diagnosis do not appear to be any different from those with persistent antibody titers, apart from the HLA phenomenon and the formation of other autoantibodies. Preliminary studies done in conjunction with Dr. Arthur Rubenstein of Chicago have indicated that it is not the persistence or otherwise of C peptide secretion

that determines the continued autoantibody response. Thus, it would appear that the immune response to islet cell autoantigen is determined genetically through a gene linked with the HLA system. By my way of thinking that must reflect the presence of an immune response gene. We have to be careful how we argue from animal models to man because the induction of experimental auto-immune disease in animals may, according to the technique used, produce a disease that does not in fact reflect the human situation.

VOLPE: It is of interest that the atrophic hypothyroid form of auto-immune thyroiditis had a significant correlation with HLA-B8, whereas goitrous Hashimoto's thyroiditis did not, similar to several previously reported studies. This continues to exercise me, since goitrous Hashimoto's thyroiditis has been the prototype of autoimmune organ-specific disease, with very high levels of auto-antibodies - in contrast to the low levels of antibodies (or negative antibodies) in the atrophic myxodematous form. Moreover, Hashimoto's disease is so closely related to Graves' disease (where there is a relationship to HLA-B8 and Dw3); one can have monozygote twins, one with Graves', the other with Hashimoto's disease. Yet in goitrous Hashimoto's disease, there is no proven HLA association. Would you care to speculate about this?

IRVINE: My own experience of twins and Hashimoto's disease is that of concordance, i.e., both had Hashimoto's. However, the occurrence of Graves' disease in one member and Hashimoto's in the other to which you refer emphasizes that there is a common genetic factor underlining the spectrum of thyroid autoimmune disease and suggests that an environmental factor may be operative in partially determining what form of thyroid autoimmune disease develops. Although I do not recall the details of these particular cases, it is also possible that the one with Hashimoto's disease had an earlier thyrotoxic stage that was not diagnosed. As I mentioned in my paper, it may be that Hashimoto's disease with goiter is more polygenic than either thyrotoxicosis or primary atrophic hypothyroidism. This may mask a B8 association in Hashimoto goiter. I may add, however, that Addisonian pernicious anemia which is another member of the organ specific autoimmune group also shows no association with B8, unless the pernicious anemia occurs in patients with autoimmune endocrine disease. It should be emphasized also that persistence of thyroid cytoplasmic antibodies and antibodies to thyroglobulin and antibodies to gastric parietal cells do not show a clear correlation with B8.

WARNER: I would tend to agree with your response to Dr. Mackay's question and would like to take this further by asking whether the difference in persistence of autoantibodies is related to the Ig class or subclass of the antibodies? Data from mice would suggest that one level of Ir control is through T cell regulation of antibody response, and that differences are observed in the degree to which the different Ig classes are so regulated.

IRVINE: I have no data on the subclasses of organ-specific autoantibodies according to their transcience or persistence. Thus I cannot say whether B8 comprise a different subclass of islet cell autoantibodies. This study should be done.

HLA AND AUTOIMMUNE DISEASE WITH SPECIAL REFERENCE TO THE GENETICS
OF INSULIN-DEPENDENT DIABETES

A. SVEJGAARD, M. CHRISTY*, J. NERUP*, P. PLATZ,
L.P. RYDER AND M. THOMSEN

Tissue-Typing Laboratory of the State University Hospital
(Rigshospitalet) of Copenhagen, and
*Steno Memorial Hospital of Copenhagen, Denmark

ABSTRACT

The relationships between HLA and some autoimmune diseases are
reviewed. Most of these associations primarily concern HLA-D/DR
antigens, and it is concluded that *Ir* genes are probably responsible for the relationships. Recent studies on HLA-D types in insulin-dependent diabetes (IDD) have shown an absence of Dw2 in IDD
and confirmed previously observed increases of Dw3 and Dw4. Results
of HLA-D genotyping indicate that IDD may be heterogeneous and that
more than one HLA factor are involved in the susceptibility to IDD.

INTRODUCTION

For many years, the usual way to study the genetic control of
diseases was to investigate the frequencies of affected relatives
of a group of propositi. Such studies were fully sufficient when
simple dominant or recessive Mendelian traits with complete penetrance were concerned. In most cases, however, family studies
merely revealed increased frequencies in various groups of relatives as compared to the frequency in the general population and
this lend support to the assumption that genetic elements play a
role in disease susceptibility. In this way, it was for example
shown that there is a genetic background of juvenile diabetes, but
the detailed mode of inheritance could not be established[1,2]. The
diseases were called multifactorial, i.e. both genetic and environmental factors play a role, and the genetic background was considered polygenic, i.e. due to the interaction of many different
genes[2,3]. This situation changed significantly when it was discovered that some of these diseases occur preferentially in individuals carrying certain HLA antigens (see ref. 4 for review). These
antigens are genetic markers associated with disease susceptibility
and this makes it possible to study the mode of inheritance into
more detail. The purpose of this paper is to review the relation-

ships between HLA and autoimmune diseases and to discuss how this information may be used to gain insight into the genetic control of disease as examplified by insulin-dependent diabetes. First, however, a few notes on the HLA system may be appropriate.

THE HLA SYSTEM

The HLA system[5,6,7] is the major histocompatibility complex (MHC) in man, and like other MHCs in other species, it controls three different sets of characters: (i) alloantigens, (ii) some components of the complement cascade, and (iii) some immune responses. The antigens fall in two categories: the HLA-ABC antigens which are present on all cells except red cells and controlled by three different loci (*HLA-A*, *B* and *C*), and the HLA-D and DR antigens which are only present on some cells (macrophages and B lymphocytes among others); the HLA-D antigens are detectable by cell culture techniques while the HLA-DR antigens (like the ABC antigens) are recognized by serological methods, but it is possible that the HLA-D and DR antigens are present on the same molecules and controlled by the same locus (*HLA-D/DR*). The complement components controlled by HLA include properdin factor Bf of the alternative pathway and the second (C2) and fourth (C4) components of the classical pathway. The immune response (*Ir*) genes are not yet very well defined in man, but in other species these genes are known to be closely related to the homologues of the *HLA-D/DR* genes in man, and it is possible that *Ir* and *HLA-D/DR* genes may be identical. In animals, *Ir* genes are involved in the specific cell-mediated immune response and in the formation of IgG antibody production to thymus-dependent antigens (see ref. 8 for review).

The HLA system is extremely polymorphic at each locus (except those controlling complement components): between six and 30 alleles are known at each of the loci, *HLA-A*, *B*, *C*, and *D*. There is pronounced linkage disequilibrium between these various loci, *i.e.* some *HLA* genes and antigens occur much more frequently in the same individual than would be expected from their individual frequencies. This phenomenon is of crucial importance for the understanding of the associations between HLA and disease. For example, the antigens, HLA-B8 and Dw3, are positively associated in the general Caucasian population and when one of these antigens occurs with increased frequency in a disease, there is a secondary increase of the other. Occasionally, it is difficult to decide which antigens show the strongest association. Moreover, still unknown HLA factors are likely to show linkage disequilibrium too, and thus we do not know whether it is the action of such hypothetical HLA factors which are truly responsible for the associations presently known.

HLA AND AUTOIMMUNE DISEASES

We shall not attempt to define 'autoimmunity' in this place, but have chosen to include a variety of disorders where autoimmune phenomena seem to be involved.

The vast majority of the studies seeking correlations between HLA and these diseases are population (association) studies, and table 1 summarizes the results of combined calculations on the data on Caucasians known at the HLA and Disease Registry in Copenhagen on some autoimmune diseases. Complement deficiencies are discussed by others elsewhere in this volume.

TABLE 1

ASSOCIATIONS BETWEEN HLA AND SOME 'AUTOIMMUNE' DISEASES (CAUCASIANS)

| DISEASE | HLA ANTIGEN | RELATIVE RISK |
|---|---|---|
| Juvenile diabetes | B8 | 2.4 |
| | B15 | 2.1 |
| | B18 | 1.7 |
| | Dw3 | 3.8 |
| | Dw4 | 3.5 |
| Graves' disease | Dw3 | 4.4 |
| Idiopathic Addison's disease | Dw3 | 8.8 |
| Pernicious anaemia | B7 | 1.5 |
| Hashimoto's thyreoiditis | | |
| Subacute thyreoiditis | Bw35 | 13.7 |
| Sicca syndrome | Dw3 | 19.0 |
| Chronic autoimmune hepatitis | Dw3 | 6,8 |
| Myasthenia gravis | B8 | 4.1 |
| | Dw3 | 2.3 |
| Dermatitis herpetiformis | Dw3 | 13.5 |
| Coeliac disease | Dw3 | 73.0 |
| Systemic lupus erythematosus | B8 | 2.1 |
| Rheumatoid arthritis | Dw4 | 3.9 |

The relative risks are combined estimates based on data available at the HLA and Disease Registry of Copenhagen.

Studies including HLA-D and/or DR antigens are much less numerous than those involving the classical HLA-A and B antigens. For example, it is not known whether the increase of B8 in systemic lupus erythe-

matosus (SLE) is secondary to a primary increase of Dw3/DRw3. However, when HLA-D antigens have been studied they have usually shown
stronger associations than the HLA-B antigens. Exceptions to this
rule are the stronger increases of Bw35 than Dwl in subacute thyreoiditis and possibly of B8 than Dw3 in myasthenia gravis. Studies of
DR antigens are even more sparse, but in general these antigens seem
to show the same or a higher degree of association than the D antigens[7].

It is a striking fact that Dw3 is involved in most diseases
showing 'primary' HLA-D associations, and this may lead to speculations that it is the same gene which is responsible for the HLA controlled susceptibilities to all these disorders.

It has sometimes been suggested that the associations concerned
specific haplotypes and not just single HLA factors. However, it may
be calculated that the increases of B8 is usually of the magnitude
which would be seen when all Dw3 positive individuals have the same
risk irrespective of the presence of B8.

Insulin-dependent diabetes, idiopathic Addison's disease, Graves'
disease, Hashimoto's thyreoiditis, and pernicious anaemia, have
often been grouped as 'organ-specific autoimmune disorders'[9]. However, the observation that only the former three are B8 and Dw3
associated indicate that they have at least partially different
genetic backgrounds[9,10].

Dermatitis herpetiformis (DH) and coeliac disease (CD) involve
both an element of hypersensitivity to gluten, but it has always
remained an enigma why some DH patients develop CD while others do
not. We observed that DH patients with simultaneous CD had an increased frequency of Dw3 and an insignificant increase of Dw7, whereas Dw3 alone was associated with uncomplicated DH[11]. Together with
L. U. Lamm, we recently looked into the possibility that the development of CD in DH is secondary to an association with one of the
$Bf$ alleles. A normal frequency of the $Bf^F$ allele was found in DH+CD
while the frequency of $Bf^F$ was insignificantly decreased in DH without CD. It seems likely, however, that this decrease is secondary
to a decrease of Dw7 and B12 in this group of patients because $Bf^F$
is associated with Dw7 and B12. However, more data are needed to
clarify the situation.

The concept that it is HLA-D associated $Ir$ genes which are
responsible for the disease associations[12,13] is supported by the
observations that the presence of anti-adrenal antibodies in Addison's disease[14] is more frequent in Dw3 positive than in Dw3 neg-

TABLE 2

ASSOCIATION BETWEEN ANTI-ADRENAL ANTIBODIES AND HLA-Dw3 IN IDIO-
PATHIC ADDISON'S DISEASE

| | Anti-Adrenal Antibodies | | |
|---|---|---|---|
| | present | | absent |
| Dw3-pos. | 25 | (96%) | 1 |
| Dw3-neg. | 4 | (40%) | 6 |
| Fisher's p = .0007 | | | |

ative patients (table 2). However, it should be noted that the
pathogenetic significance of such antibodies is unclear; they may be
secondary to the disease process, but even then their association to
HLA-Dw3 could reflect *Ir* gene involvement.

It is worth noting that most of the above associations differ
between different ethnic groups. For example, Graves' disease and
juvenile diabetes are associated with HLA-D-OH and HLA-D-YT in
Japanese[15].

The mere association between a disease and an HLA factor indicates in itself that there is a genetic element in the disease
susceptibility but it does not necessarily imply the exact mode of
inheritance. However, the HLA markers provide valuable tools in
the unravelling of the genetics. Insulin-dependent diabetes (IDD)
has a special situation because it is associated with not less
than three HLA-D antigens (cf. below), and in the rest of this paper
we shall discuss how this information may be used in combination
with family studies to gain more insight in the inheritance of IDD.

ON THE GENETICS OF INSULIN DEPENDENT DIABETES (IDD)

It appears from table 1 that the HLA-B8, B15, and B18 antigens
are highly significantly associated with juvenile diabetes in Caucasians. It is now fairly well established that the increases of B8
and B15 are secondary to stronger increases of Dw3 and Dw4, respectively. Moreover, we have recently found evidence that the increase
of B18 may also be secondary to the increase of Dw4: Nine of 11 B18
positive patients were Dw4 positive. Thus, juvenile diabetes is
strongly associated with two different HLA-D antigens. Moreover, it
was recently reported[16] that HLA-Dw2 has a very low frequency in

this disorder, - a finding we can confirm: none of 118 unrelated IDDM patients typed for Dw2 carried this antigen (table 3) and we have not found it in affected relatives either. The almost entire absence of Dw2 in IDD cannot be accounted for by the increases of Dw3 and Dw4, and it must be assumed that Dw2 is associated with a strong resistance to diabetes.

The association between three different HLA-D antigens and IDD does not necessarily imply that there are three different HLA factors involved in the genetic predisposition to this disorder. The findings could also be explained by the action of one as yet unknown IDD-predisposing gene positively associated with both Dw3 and Dw4, and negatively with Dw2. Indeed such a 'monogenic' hypothesis has been advanced by Thomson and Bodmer[17] and by Rubinstein et al.[18] on the basis of family studies which have shown that there is an excess of HLA identity among sibpairs both of whom suffer from IDD. This led to the suggestion that IDD is a recessive HLA linked disorder and Rubinstein et al.[18] further suggested that it had a penetrance of 50 per cent. However, this assumption is neither compatible with the observation[19] that the incidence of diabetes in siblings of IDD patients is only about 5 per cent nor with the results of HLA studies in unrelated patients.

We have earlier suggested[13] that the HLA controlled susceptibility to IDD is due to two different HLA factors each acting by its own mechanism. This hypothesis was based on the observation that *HLA-B8/B15* heterozygotes have a much higher risk of IDD than have individuals carrying only one of these antigens whether homozygous or heterozygous. However, with the finding of more strong associations with HLA-D than HLA-B factors[14], it became clear that more information could be derived by HLA-D typing the patients. In order to unravel the situation in more detail, we undertook a study of two groups of IDD[20]: (i) a prospective study of a sample of all newly diagnosed cases of IDD during a one year period in a geographical region in Denmark covering about a third of the total Danish population, and (ii) a study of sibpairs both of whom suffer from IDD. These studies will be reported in detail elsewhere[20], but the most crucial results are summarized here.

Table 3 shows the frequencies of HLA-Dw2, 3, and 4, in these two groups and in an earlier retrospectively studied group[14] of IDD patients ascertained from an out-patient clinic. It appears that Dw2 is absent in all three groups; all the other D antigens were present in at least some patients. In particular, Dw3 and Dw4 are increased in all three groups. However, the increase of Dw4 is significantly higher (p = .007) in the prospective group and the sibpairs (the eldest sib in each family has been used as propositus) than in the retrospective group. This may indicate heterogeneity between the groups, e.g. due to biased sampling or technical difficulties. Although we cannot exclude the latter possibility, we do not find it likely as we have used the same typing cells for all three samples. Accordingly, we consider true heterogeneity a more plausible explanation and it could have arisen by a sampling of relatively more cases with later onset in the retrospective group (cf. below).

TABLE 3

HLA-D ANTIGEN FREQUENCIES IN UNRELATED PATIENTS WITH INSULIN-DEPENDENT DIABETES

| HLA-D | FREQUENCY | | | Controls |
| | Group I (retrospective) | Group II (prospective) | Group III (familial) | |
|---|---|---|---|---|
| Dw2 | 0/20 ( 0.0) | 0/73 ( 0.0) | 0/25 ( 0.0) | 89/345 (25.8) |
| Dw3 | 26/52 (50.0) | 32/73 (43.8) | 14/25 (56.0) | 88/334 (26.3) |
| Dw4 | 32/79 (40.5) | 41/73 (56.2) | 18/25 (72.0) | 67/345 (19.4) |

Group I:     retrospectively ascertained patients
Group II:     prospectively ascertained patients
Group III:   familial cases (eldest affected sib only)
Figures are Nos. positive/Nos. investigated and percentages in brackets.
The frequency of Dw4 is significantly lower in group I than in groups II (p = 0.04) and III (p = 0.006).

To increase the amount of information, we tried to establish the *HLA-Dw3* and *Dw4* genotypes in the two most recently studied groups. This was done by typing for all the HLA-Dw1 through Dw8 antigens, by family studies, and when in doubt by using the patients' lymphocytes as stimulators in MLC (mixed leucocyte·culture) towards *HLA-Dw3/3* and *Dw4/4* homozygous typing cells as responders. Table 4 shows the preliminary results of this analysis. The genotype frequencies in the control population have not been determined in the same way, but were estimated on the basis of gene frequencies calculated by Bernṣtein's formula and assuming Hardy-Weinberg equilibrium.

The results (table 4) may be summarized as follows. Firstly, there is significant (p = .02) heterogeneity between prospectively

TABLE 4

HLA-D GENOTYPE FREQUENCIES IN INSULIN-DEPENDENT DIABETES

| HLA-D Genotype | Controls (%) | Group II (prospective) | | | | | | Group III (familial) | | |
|---|---|---|---|---|---|---|---|---|---|---|
| | | Onset ≤15 years | | Onset >15 years | | All | Risk | | | |
| | | N | (%) | N | (%) | N | (%) | N | (%) | Risk |
| *Dw3/3* | ( 2.0) | 3 } | (26) | 1 } | (27) | 4 | ( 5) 8.8 | 1 | ( 4) | 9.5 |
| *Dw3/X* | (21.4) | 8 | | 7 | | 15 | (21) 3.1 | 3 | (12) | 2.7 |
| *Dw3/4* | ( 2.9) | 8 | (19) | 5 | (17) | 13 | (18) 19.7 | 10 | (40) | 65.8 |
| *Dw4/X* | (15.5) | 19 } | (49) | 5 } | (23) | 24 | (33) 6.8 | 7 | (28) | 8.6 |
| *Dw4/4* | (1.o5) | 2 | | 2 | | 4 | ( 5) 16.8 | 1 | ( 4) | 18.2 |
| *Other* | (57.2) | 3 | ( 7) | 10 | (33) | 13 | (18) | 3 | (12) | |
| Total | | 43 | | 30 | | 73 | | 25 | | |

Genotype frequencies in controls were not observed but estimated on the
basis of gene frequencies assuming Hardy-Weinberg equilibrium.
Relative risks were calculated with 'other' genotypes as reference
(risk = one).
X indicates the presence of HLA-D antigens other than Dw3 and 4 (cf.
text).
There is significant heterogeneity between the two different age at
onset groups ($X_3^2$ = 9.9, p = .02), all the differences being due to
the last three genotypes.
The frequency of *Dw3/4* is significantly (p = .03) higher in the
familial than the prospective series.

ascertained patients with age-at-onset before and after the age of 16.
This difference is solely due to higher frequency of the genotypes
*Dw4/X* and *Dw4/4* and a lower frequency of 'other genotypes' in patients
with age at onset before the age of 16. Accordingly, the presence of
Dw4 without Dw3 seems mainly to increase the risk of IDD before the
age of 16, whereas Dw3 increases the risk at all the age groups
studied. This supports the assumption that Dw3 and Dw4 (or associated
factors) act by two different mechanisms[10,13] and agrees with the
observation that there are two peaks for the age at onset for IDD[21]:
one at about the age of ten and one in the twenties. Secondly, it
can be seen that there is a very high frequency of *Dw3/4* heterozygotes
in familial cases and there is significant (p = .03) difference
between this frequency and the corresponding frequency in the prospective study. This finding supports our earlier assumption[5] that
familial cases represent a special sample of all cases of IDD and
may be explained either by disease heterogeneity between single and
familial cases or more likely by a more frequent occurrence of
disease predisposing genes in familial versus non-familial cases. In
any case, this bias must be kept in mind when inference is made from

family studies to the genetics of IDD as such. It should be noted that
the prospective series cannot yet in a meaningful way be divided in
familial and non-familial cases because the siblings still are too
young. Finally, when inspecting the relative risks for the various
genotypes in both groups of patients in table 4, it can be seen that
*HLA-Dw3/4* heterozygotes have the highest risk, followed by *Dw4/4* and
*Dw3/3* homozygotes, whereas *Dw4/X* and *Dw3/X* heterozygotes have the lowest.
This indicates an interaction between *Dw3* and *Dw4* when present together
and some dose effect for each of these genes.

Like others, we found an excess of HLA identity in the affected
sibpairs, and it is of interest to note that this excess mainly concerned *HLA-Dw3/4* heterozygotes. This observation and the increase of
Dw4 in early-onset IDD is at variance with those reported by Rubinstein et al.[18] primarily because of a high frequency of Dw4 in one
series. However, these differences might not by themselves invalidate
the 'one-recessive gene' theory, but the gene in question would then
have to be associated with both *Dw3* and *Dw4*. In this case, the risks
for one of the homozygotes should then be the highest, while the
risk for *Dw3/4* heterozygotes should be intermediate between those of
the two homozygotes, which is not the case in any of the two groups
in table 4. However, it should be stressed that the data are still
of small size and more information is needed before a definite conclusion can be made. Nevertheless, it is our impression that our
findings as a whole are more compatible with the assumption that
there are two different HLA factors - one associated with Dw3 and
one with Dw4 which predisposes to IDD. The absence of Dw2 may be
due to a protective effect by a Dw2-associated HLA factor or (less
likely) it may reflect strong negative linkage disequilibrium between
both of the predisposing factors and Dw2.

ACKNOWLEDGEMENTS

This study was aided by grants from the Nordic Insulin Foundation,
the Danish Medical Research Council, and the Danish Blood Donor Foundation. Our thanks are due to Bodil K. Jakobsen and Elly Andersen for
expert technical and secretarial assistance.

110

1. Simpson, N.E. (1968) Ann. Hum. Genet., London, 26, 1.

2. Falconer, D.S. (1967) Ann. Hum. Genet., 31, 1.

3. Carter, C.O. (1969) Br. med. Bull. 25, 52.

4. Dausset, J. and Svejgaard, A. (1977) HLA and Disease, Munksgaard, Copenhagen, pp. 1-316.

5. Snell, G.D., Dausset, J. and Nathenson, S. (1976) Histocompatibility, Academic Press, New York, pp. 1-401.

6. Svejgaard, A., Hauge, M., Jersild, C., Platz, P., Ryder, L.P., Nielsen, L. Staub and Thomsen, M. (1975) Monogr. human Genet. 7, 1-103.

7. Histocompatibility Testing 1977 (1978) Munksgaard, Copenhagen (in press).

8. Ir Genes and T Lymphocytes (1978), Möller, G., ed. Immunological Reviews, 38.

9. Nerup, J., Cathelineau, Cr., Seignalet, J. and Thomsen, M. (1977) in HLA and Disease, Dausset, J. and Svejgaard, A. eds., Munksgaard, Copenhagen, pp. 149-167.

10. Platz, P., Ryder, L.P., Thomsen, M., Bech, K., Buschard, K., Nerup, J., Andersen, O.O. and Svejgaard, A. (1978) in Menarini Series on Immunopathology, Miescher, P.A. ed., Schwabe & Co., Basel (in press).

11. Thomsen, M., Platz, P., Marks, J., Ryder, L.P., Shuster, S., Svejgaard, A. and Young, S.H. (1976) Tissue Antigens 7, 60.

12. McDevitt, H.O. and Bodmer, W.F. (1974) Lancet i, 1269.

13. Svejgaard, A., Platz, P., Ryder, L.P., Nielsen, L. Staub and Thomsen, M. (1975) Transplantn Rev. 22, 3.

14. Thomsen, M., Platz, P., Andersen, O. Ortved, Christy, M., Lyngsøe, J., Nerup, J., Rasmussen, K., Ryder, L.P., Nielsen, L. Staub and Svejgaard, A. (1975) Transplantn Rev. 22, 125.

15. Sasazuki, T., personal communication.

16. Ilonen, J., Herva, E., Tiilikainen, A., Åkerblom, H.K., Koivuhangas, T. and Kouvalainen (1978) Tissue Antigens 11, 144.

17. Thomson, G. and Bodmer, W. (1977) in HLA and Disease, Dausset, J. and Svejgaard, A. eds., Munksgaard, Copenhagen, pp. 84-93.

18. Rubinstein, P., Suciu-Foca, N. and Nicholson, J.F. (1977) New Engl. J. Med. 297, 1036.

19. Degnbol, B. and Green, A. (1978) Ann. Hum. Genet. 42, 25.

20. Copenhagen Study Group of Insulin-Dependent Diabetes, in preparation.

21. Christau, B., Kroman, H., Andersen, O.O., Christy, M., Buschard, K., Arnung, K., Kristensen, J.H., Peitersen, B., Steinrud, J. and Nerup, J. (1977) Diabetologica 13, 281.

DISCUSSION

DUPONT: The simulation that you calculated for the frequency of Dw3/-; Dw4/-; Dw3/Dw3; Dw4/Dw4 and Dw3/Dw4 was made for a hypothetical penetrance of 50% for one recessive diabetes gene, and you showed that this hypothesis did not fit the experimental data obtained in the prospective study. How do the values change for the different Dw-- phenotype frquencies for different values of the penetrance-level?

SVEJGAARD: A decrease of the penetrance to, say 25% (which is more likely to reflect the true situation), would not change the estimates notably.

TERASAKI: Is the negative association of Dw2 found in diabetes the only association known for resistance to a disease rather than susceptibility?
Also is it known whether multiple sclerosis patients have diabetes? A negative association might be expected since Dw2 is high in MS and low in diabetes.

SVEJGAARD: I think it is true that the negative association between HLA-Dw2 and insulin-dependent diabetes is the most pronounced example of "protection" we know at the moment, and it is of course interesting that it appears to be dominant. Many of the diseases showing positive associations with HLA are bound to show negative associations too, namely in terms of the antigens controlled by genes allelic to those controlling the susceptibility.
Insulin-dependent diabetes and multiple sclerosis are both rare and I don't know whether they tend to "exclude" each other.

COHEN: Your observation of HLA restriction for killing of DNP-modified target cells involved individuals who were primed against DNP in vivo and then restimulated in vitro. In contrast, experiments carried out in Gene Shearer's laboratory demonstrated no HLA restriction for killing of target cells modified by TNP. This response to TNP-modified cells was induced as a primary response in vitro. Therefore, it is possible that the HLA restriction that you observed might have been acquired by priming against DNP in vivo. In any case, HLA restriction is a complicated phenomenon.

SVEJGAARD: I agree. These experiments were carried out by Dr. Dickweiss et al. (Nature, 1977), and I may add that not all DNCB-sensitized individuals developed detectable killer cells even after priming of their cells in vitro. However, when they were present, they showed HLA-ABC restriction, and HLA-A2 was always the "best" antigen.

LENNON: In answer to Dr. Terasaki's inquiry whether there has ever been noted a negative association between diabetes mellitus and multiple sclerosis (MS) may I cautiously relate an anecdote.
While I was engaged in neuroimmunology research in Melbourne (1968-1971), I was frequently called by an enthusiastic lay-researcher in MS who was attempting to solicit my cooperation in organizing glucose tolerance tests for a large group of MS patients. She insisted that from her "surveys" of MS patients in Melbourne she found a striking absence of history of diabetes. She believed that this observation should be investigated by doctors involved in MS research.

ROITT: Would you like to make any speculation on a possible role of HLA coded products on the development of susceptibility to auto-immune disease in the light of the studies of Zinkernagel and colleagues on the "education process" of T cells in the thymus.

SVEJGAARD: As we have discussed elsewhere (Clinics in Rheumatic Diseases, 3:239, 1977), there are many different ways to explain the HLA and disease associations. When the primary associations concern HLA-D antigens—as is the case with most of the autoimmune diseases—it seems most likely that immune response (Ir) or immune suppressor (Is) genes are involved. Primary HLA-ABC associations may lead the thoughts to the altered self or dual recognition theory of Doherty and Zinkernagel. However, there are certainly other possibilities, e.g., molecular mimicry, which should be kept in mind, and the observations reported by Dupont on the C4 polymorphism is also worth mentioning particularly because one of the "silent" alleles are associated with B8. It is of crucial importance to investigate whether it is more strongly associated with Dw3 and to find a method allowing detection of individuals heterozygous for the "silent" alleles because such heterozygotes might theoretically have an increased susceptibility to autoimmune disease.

DUPONT: Professor Jean Dausset from Paris has recently presented in Boston data on HLA-haplotype frequencies and genetic linkage disequilibrium in the French population. The data were solely obtained from family studies. They included also studies of C-2 polymorphism, Bf and GLO. One particular HLA-B-Bf-DR-haplotype was associated with insulin-dependent diabetes, and this haplotype was characterized by a rare Bf-allele, implying that the complement genes within HLA may play a role in this disease.

MACKAY: On expected population frequencies of diabetes mellitus (2/100) and multiple sclerosis (20/100,000), an expected prevalence of both diseases would be four per million population, but the prevalence of coexisting juvenile onset diabetes and MS would be quite less. Thus, examination for dissociation of DM and MS would require a large population survey.

# B CELL ALLOANTISERA AS PROBES OF THE CONTRASTING IMMUNOGENETICS OF THREE RHEUMATIC DISEASES

R.J. WINCHESTER, M.E. PATARROYO, A. GIBOFSKY,
J. ZABRISKIE AND H.G. KUNKEL
The Rockefeller University, New York, New York  10021

## SUMMARY

A panel of B cell alloantisera revealed that patients with rheumatic fever, systemic lupus erythematosus or rheumatoid arthritis were each characterized by a distinctive profile of alloantigens.  Patients with rheumatic fever had a high frequency of an unusual alloantigen that was unrelated to any defined DR specificity.  Patients with rheumatoid arthritis were characterized by a high frequency of a "public" allo-antigen associated with HLA-Dw4, w7 and w10.  Among patients with systemic lupus erythematosus, the alloantiserums giving the highest positive association was related to DRw2 and DRw3.  The reduction of frequency of certain alloantigens raised the possibility of resistance factors.  The value of using alloantisera with other than narrow DR specificities as probes for disease immunogenetics was apparent in this study.

## INTRODUCTION

The association of particular alleles of loci mapping within the major histocompatability complex to the occurrence of certain diseases is a topic of considerable current interest because it provides a method of analyzing the extent and significance that genetic factors have in determining susceptibility to these diseases.  The primary approach that has been widely used is exemplified by the disease associations found with the HLA-A and B determinants(1).  These resulted from the application of HLA tissue typing reagents that were extremely well defined in terms of their genetic specificity.

More recently the recognition of an intricate system of serologically defined alloantigens selectively expressed on B cells, and resembling the murine Ia molecules (2), afforded a new approach towards the recognition of meaningful clues regarding the genetic basis of certain diseases. Early studies revealed provocative associations such as the extremely high association of a particular B cell alloantigen with multiple sclerosis (3). These studies raised the question of the utility of a new principle that sought to establish the primary significance of a reagent in terms of its ability to identify a major percentage of individuals within a disease group compared to a normal population, and thus achieve a disease-associated genetic marker that was present in a very high percentage of the patient group.

The present study is concerned with an analysis of three rheumatic diseases each with features of autoimmunity, rheumatoid arthritis, systemic lupus erythematosus and rheumatic fever that illustrate the use of the B cell alloantigens as probles of histocompatibility linked disease associations.

MATERIALS AND METHODS

Reagent alloantisera were selected for their reactivity with B cell alloantigens and in certain instances were absorbed with platelets or T cells to remove HLA-A, B or C specificities. Cytotoxic assays using the Amos two stage method and 37° incubations each for 1/2 hour on isolated B cells, or fluorescence assays employing the indirect technique on poke week mitogen transformed peripheral blood lymphocytes were used to detect the B cell alloantigens. Lymphoid cell lines were initiated by transforming purified B lymphocytes with supernatants from B95-8 cell line.

RESULTS

## Reagent Delineation

The screening of a large number of pregnancy sera on peripheral
blood T and B cells, following poke weed mitogen stimulation, yielded
a number with specificity for alloantigens selectively found on B
cells. In some instances absorption with pooled platelets of T cells
were necessary to eliminate contaminating antibodies with specificities for HLA-A, B or C determinants.

In order to provide an orientation to the specificity of the B
cell alloantisera, they were tested using cytotoxic assays on a panel
of over 40 B cell lymphoblastoid cell lines derived from individuals
that were homozygous for HLA-Dw determinants by mixed lymphocyte
culture testing. Table I illustrates the reactivity of representative typings. A variety of different patterns were evident with
certain sera reacting in a pattern closely related to D locus determinants. For example serum 989 detected alloantigens only found on
cell lines from individuals that were HLA-Dw2. Other sera such as
ch or 1038 reacted with cell lines from individuals that included
those from more than one D locus specificity. Serum 883 is of
interest in that it did not react with cell lines derived from any
individuals homozygous for HLA-Dw1, 2, 3, 4, 5, 6, 7 or 10. The
specificity of the sera inferred from these reactions was listed at
the right of the Table. A plus implies that two separate antibody
specificities are involved, and "x" signifies a shared antigenic
specificity.

## Analysis of specificity

The characterization of the reagent alloantisera that reacted with
individuals of more than one D locus allele was carried out by means

Table I

Reactivity of B-Cell Alloantigen Typing Antisera Against Panel of Lymphoblastoid Lines
Derived from Individuals Homozygous for HLA-Dw Determinants

| Serum Designation | 1/1 | 1/1 | 1/1 | 2/2 | 2/2 | 2/2 | 3/3 | 3/3 | 3/3 | 5/5 | 4/4 | 4/4 | 4/4 | 4/4 | 7/7 | 7/7 | 7/7 | 10/10 | 10/10 | inferred Specificity DR |
|---|---|---|---|---|---|---|---|---|---|---|---|---|---|---|---|---|---|---|---|---|
| | | | | | | | HLA-Dw Type of Cell Line Donor | | | | | | | | | | | | | |
| 7 AM | + | + | + | 0 | 0 | 0 | 0 | 0 | 0 | 0 | 0 | 0 | 0 | 0 | 0 | 0 | 0 | 0 | 0 | 1 |
| 1239 | 0 | 0 | 0 | + | + | + | 0 | 0 | 0 | 0 | 0 | 0 | 0 | 0 | 0 | 0 | 0 | 0 | 0 | 2 |
| ch | 0 | 0 | 0 | + | + | + | + | + | + | 0 | 0 | 0 | 0 | 0 | 0 | 0 | 0 | 0 | 0 | 2+3 |
| 1033 | 0 | 0 | 0 | 0 | 0 | 0 | + | + | + | 0 | 0 | 0 | 0 | 0 | 0 | 0 | 0 | 0 | 0 | 3 |
| 1995 | 0 | 0 | 0 | 0 | 0 | 0 | 0 | 0 | 0 | + | 0 | 0 | 0 | 0 | 0 | 0 | 0 | 0 | 0 | 5 |
| 1038 | 0 | 0 | 0 | 0 | 0 | 0 | 0 | 0 | 0 | 0 | + | + | + | + | + | + | + | + | + | 4x7x10 |
| 1283 | 0 | 0 | 0 | 0 | 0 | 0 | 0 | 0 | 0 | 0 | + | + | + | + | + | + | + | + | + | 4x7x10 |
| 924 | 0 | 0 | 0 | 0 | 0 | 0 | 0 | 0 | 0 | 0 | + | + | + | + | 0 | 0 | 0 | + | + | 4x10 |
| 191 | 0 | 0 | 0 | 0 | 0 | 0 | 0 | 0 | 0 | 0 | + | + | + | + | 0 | 0 | 0 | + | + | 4x10 |
| 1903/2 | 0 | 0 | 0 | 0 | + | 0 | 0 | 0 | 0 | 0 | 0 | 0 | 0 | 0 | + | + | 0 | 0 | 0 | 7 |
| 1903 | 0 | 0 | 0 | + | 0 | 0 | 0 | 0 | 0 | 0 | 0 | 0 | 0 | 0 | + | + | + | 0 | 0 | 2+7 |
| 883 | 0 | 0 | 0 | 0 | 0 | 0 | 0 | 0 | 0 | 0 | 0 | 0 | 0 | 0 | 0 | 0 | 0 | 0 | 0 | undefined |

of absorptions with the B cell lines. Table II illustrates two different patterns of specificity. In the instance of serum 1283, the un-
absorbed serum reacted with cell lines from individuals that were
HLA-Dw4, 7 or 10. Absorption of serum 1283 by any of these cell lines
resulted in the complete and reciprocal removal of all reactivity with
any cells. Control absorption with, for example, Dw2 cells did not
remove any reactivity. As a result, serum 1283 was designated to
have a pattern of specificity 4x7x10 to indicate that a single antigen
was involved. The determinants 4, 7 and 10 appear to have a special
interrelationship because 9 additional sera recognize sharing of
common antigens among these specificities. The two dominant patterns
are 4x10 illustrated by serum 924, and 4x7x10 illustrated by 1283.
In contrast, serum 1903 illustrated a different pattern. The un-
absorbed serum reacted with all cell lines that were derived from
individuals that were neither HLA-Dw2 or 7. In this example, absorption by Dw2 cells removed the antibodies that reacted with the Dw2
cells but not those from Dw7 individuals, and vice versa. Accordingly,
the serum 1903 had an inferred specificity of DRw2+7, indicating that
separate antibody specificities for determinants HLA Dw2 or Dw7 exist.

Application to certain disease populations

Table III illustrates the B cell alloantigen typing of normal
individuals and three populations of patients with systemic lupus
erythematosus, or rheumatoid arthritis or rheumatic fever. Among the
patients, two types of statistically significant relationships were
evident: In one, illustrated by the reactivities of reagents with
specificities for the DRw 4-7-10 complex, there was a positive association between the occurrence of the alleles expressing antigens of
this DRw 4-7-10 complex, and the presence of the disease rheumatoid

arthritis. In the second, illustrated by the occurrence of systemic
lupus erythematosus and the presence of the alleles expressing antigens of the DRw 4-7-10 complex, the association is a negative one.
In the instance of systemic lupus erythematosus, the positive association was maximal with serum ch, with specificity DRw 2+3 and of a
slightly lower order with sera that had single specificities for
DRw2 or DRw3. These associations are calculated in terms of relative
risk in Table III.

Of interest was the finding that patients with rheumatic fever
lacked a significant increase in reactivity with any serum that had
a defined DR specificity; yet, these patients were characterized by
a highly significant increase in the frequency of reactivity with
serum 883.

Table II

Delineation of Reagent Specificity by Absorption

with Lymphoid Lines Derived from HLA-D Homozygous Individuals

| Serum/Absorbing Cell | HLA-Dw Homozygous Test Cell | | | |
|---|---|---|---|---|
| (HLA-Dw) | 2/2 | 4/4 | 7/7 | 10/10 |
| 1283 | − | + | + | + |
| 1283/2 | − | + | + | + |
| 1283/4 | − | − | − | − |
| 1283/7 | − | − | − | − |
| 1283/10 | − | − | − | − |
| 1903 | + | − | + | − |
| 1903/2 | − | − | + | − |
| 1903/4 | + | − | + | − |
| 1903/7 | + | − | − | − |
| 1903/10 | + | − | + | − |

Table III

Contrasting Activities of B Cell Alloantisera in

Patients with Systemic Lupus Erythematosus, Rheumatic Fever, or Rheumatoid Arthritis

| Reagent | HLA-DR Specificity | Normal (n=40) % | Systemic Lupus Erythematosus (n=24) % | $X^2$ | RR* | Rheumatic Fever (n=21) % | $X^2$ | RR | Rheumatoid Arthritis (n=45) % | $X^2$ | RR |
|---|---|---|---|---|---|---|---|---|---|---|---|
| 7A0 | 1 | 28 | 28 | 0.0 | 1.0 | 26 | 0 | 1.0 | 9 | 4.4 | 0.2 |
| 1239 | 2 | 28 | 46 | 5.5 | 3.0 | | | | 13 | 1.3 | 0.5 |
| 1146 | 2 | 26 | 50 | 4.8 | 3.1 | 25 | 0.1 | 1.9 | 12 | 1.2 | 0.5 |
| CH | 2+3 | 40 | 72 | 11.0 | 5.7 | | | | 34 | 0.0 | 1.0 |
| 1033 | 3 | 24 | 55 | 7.9 | 4.7 | 20 | 0.3 | 0.7 | 14 | 0.4 | 0.6 |
| 2134 | 3 | 26 | 56 | 7.9 | 4.7 | | | | 14 | 0.3 | 0.6 |
| 1283 | 4x7x10 | 28 | 12 | 2.9 | 0.3 | | | | 80 | 18.9 | 9.1 |
| 1038 | 4x7x10 | 24 | 6 | 3.1 | 0.3 | 24 | 1.4 | 1.5 | 76 | 16.8 | 8.4 |
| 191 | 4x10 | 22 | 4 | 3.5 | 0.2 | 26 | 0.0 | 1.0 | 60 | 7.9 | 5.1 |
| 1995 | 5 | 12 | 8 | 0.2 | 0.7 | 11 | 0.9 | 0 | 4 | 0.9 | 3.4 |
| 883 | undefined | 17 | 0 | 4.7 | 0 | 71 | 18.5 | 12.5 | 4 | 4.0 | 0.2 |

RR = Relative Risk = $\dfrac{\text{Patients with Alloantigen}}{\text{Patients without Alloantigen}} \times \dfrac{\text{Normals without Alloantigen}}{\text{Normals with Alloantigen}}$

DISCUSSION

The central finding of the present study was that three rheumatic diseases each had a distinctive pattern of B cell alloantigens significantly different from a normal control population, as well as from each other. This association implies that genetic factors are operating in each disease and affords an approach to their further analysis.

In each of the disease categories, the nature of the specificity of the B cell alloantisera that best recognizes the particular patient group appears to be different. In the instance of rheumatic fever, serum 883 recognizes a B cell alloantigen that does not relate to any of the available reference cells with defined D locus alleles. These included all specificities except DRw8 and DRw9. Support for the interpretation that this alloantigen is distinct from others with DR specificities comes from the absence of a decrease in the frequency of the recognized DR alleles in the rheumatic fever patient group. However, at present there is not sufficient information available to determine if serum 883 detects a new DR allele, a B cell alloantigen unrelated to the Ia system, or an alloantigen on the chain of the Ia molecule that does not bear the DR alloantigens.

The antisera 1283 and 1038 detect a similar or identical alloantigen that is present in 80% of patients with rheumatoid arthritis but occurs at a significantly reduced level in patients with systemic lupus erythematosus. The specificity of these sera appears related to a "public" specificity expressed in common on Ia molecules present on individuals who were either HLA-Dw4, w7 or w10. Two related sera detecting a public antigen shared by individuals who were HLA-Dw4 or w10 have the highest negative association with systemic lupus erythematosus patients and a significant positive association with rheumatoid arthritis. It appears possible that these public antigens imply an evolutionary relationship among these three alleles, and furthermore that the susceptability and possible resistance factors relate closely to this complex of alleles.

The serum ch that reacts in highest frequency with the patients with systemic lupus erythematosus appears to be detecting determinants that relate to both DRw2 and DRw3. Absorption studies provided evidence that two populations of antibodies were involved. Support for this was given by the moderate increase in the specificities DRw3 as well as DRw2 detected by other sera in the panel. This suggests that disease susceptability in systemic lupus erythematosus patients could be associated with two distinct alleles.

It is of interest to note that the highest disease associations in each category were provided by sera that are not narrowly specific for single DR specificities. This emphasizes the utility of screening patient populations with a wide variety of alloantisera, using them as probes for meaningful immunogenetic associations that may in turn also serve to assist in unravelling the intricate genetics of the B cell alloantigens.

## ACKNOWLEDGEMENT

This investigation was supported by U.S. Public Health Service grant CA20107 and Colombian National Science Foundation grant 10000-3-37-77. R.J. Winchester is the recipient of a career development award AI-00216.

## REFERENCES

1. Svejgaard,A. and Dausset, J. (eds) HLA and Disease. Munksgaard, Copenhagen, 1977.

2. Winchester, R.J., Fu, S.M., Wernet, P., Kunkel, H.G., Dupont, B., and Jersild,C. J. Exp. Med., 141: 924-926 (1975).

3. Winchester, R.J., Ebers, G., Fu, S.M., Espinosa, L., Zabriskie, J., and Kunkel, H.G. Lancet (ii): 814 (1975).

DISCUSSION

DUPONT:   The typical DRw4 antisera cross-react with DRw5.   We can confirm your findings that another group of B lymphocyte antibodies exist, which cross-react with the DRw4,7, and 10 specificities.   We have performed absorption/elution studies on HLA-D homozygous lymphoblastoid B cell lines and have shown that the cell lines with the specificities DRw4,7, and 10 will absorb the same antibody, which after elution will react with all three types of cell lines, but not with DRw5. This demonstrates, in our opinion, that the cell lines of the specificities DRw4,7, and 10 share a "public" B cell specificity that is different from the DRw4 specificity.

WARNER:   Do you have any information on the molecular distribution of these cross-reactive type specificities?   For example, if you first precipitate cell surface preparations with a "monospecific" Dw4 serum, would there be any reactivity remaining against a cross-reactive DRw4/7/10 type serum? Also, has this type of experiment been separately performed with normal versus patient derived lymphocytes?

WINCHESTER:   Noel, as you recall from the introductory slide contrasting the quantity of Ia antigens brought down by allo- versus heteroantisera, it is clear that there is room for several distinct categories of alloantigens bearing Ia molecules.   The experiments you propose are in fact a current topic of investigation, but the answers are not yet fully ready.

CANTOR:   You have clearly demonstrated that it currently makes good sense to define the "Ia-antigenic profile" of lymphocytes according to the serologic reagents used, rather than to the putative MHC gene products that the sera define.   Nevertheless, in view of the striking absence of reactivity of lymphocytes from SLE patients to the 924 reagent, can you tell us which Dw allelles are most likely defined by this reagent?

WINCHESTER:   As defined on the normal panel of D-locus homozygous cells, the 924 reagent reacts specifically with cells from individuals that are HLA-Dw4 or HLA-Dw10.   Thus, this reagent would be termed "DRw4 X 10".   It is interesting to note this serum does not react with cells from individuals that are Dw5 or Dw7.

WARNER:   One approach to the question of multiple molecular components, some of which may be specifically expressed on patient's lymphocytes, is to look for differential expression on cell types. Have you looked at the frequency of expression of these various antigens in patients versus controls, when cells of the myeloid series are used for typing rather than B lymphocytes?

WINCHESTER:   For obvious reasons the bulk of studies with Ia antigens performed from the perspective of differentiation have been performed with heteroantisera.   Moreover, the alloantisera used to examine paralle expression of, and allo- and heteroantigens were often selected for, their broad reactivity.   Within these constraints, there has always been a complete parallel between the allo- and the heteroreagents.   I am not fully satisfied that "narrow" allo-typing reagents have had adequate usage here.   Certainly, in view of the differences between the pre- sence of Ia antigens on murine and human granulocyte progenitors, it is possible that one set of alleles might not be joined on a particular lineage in man.

TALAL:   Do the autoantibodies to lymphocytes that occur in SLE patients interfere with your typing studies?  Can the addition of these anti-lymphocyte antibodies to normal lymphocytes reproduce any of your findings?

WINCHESTER:   That is a good question and we have placed special stress on the potential problem of interference.  In all instances, cells were cultivated overnight at 37º and rewashed in order to allow elution or shedding of any bound anti-lymphocyte antibodies.  Fluorescent studies of these cells did not reveal any evidence of persisting antibodies prior to testing with alloantisera.

Another line of evidence that the typing with alloantisera is not influenced by autoreactive anti-lymphocyte antibodies are the identical results provided by B type lymphoid lines obtained from the patients after several months of culture.  Therefore, we feel that the present methods avoid interference from the autologous antibodies.  In answer to the second part of your question, we have not performed those experiments, but I would anticipate that the results would be negative.

MACKAY:   Are there degrees or grades of reactivity with the antisera and individual B cells you have used?

WINCHESTER:   In cytotoxic assay on B cells, one occasionally observes reactions with low levels of killing, for example 30-40%.  These taken by themselves are ambiguous of course.  As was mentioned, parallel typing using immunofluorescence on pokeweed mitogen trans- · formed cells has proven to be an excellent aid to resolving these occasional problems.  Furthermore, the availability of cell lines from the patients allows additional confirmation of the typing reaction.  Using these two ancillary approaches, we are reasonably confident of the typing reaction as presented.

ROITT:   The differences you obtain with partial reactions in various individuals with given typing sera would more likely indicate the existence of different antigenic specificities (possibly even on different molecular species) than variable expressions of the quantity of a given antigen.

WINCHESTER:   I presume that you are referring to the typing reactions given by the cell lines derived from D locus homozygous individuals.  These were not partial reactions, but clear-cut negative reactions substantiated by absorption experiments, a point that again illustrates the utility of having the lymphoid lines available.  Thus, this pattern of reactions is termed "short" and a serum such as 191 that reacts with some Dw4 individuals and all Dw10 individuals is designated as having Is or DR short 4-10 specificity.

SCHWARTZ:   Can you tell us more about the family studies?  Were the patients homozygous or heterozygous for the factors that you detected?

WINCHESTER:   We have not yet done enough family studies to provide a comprehensive answer, but thus far we have examples of both heterozygosity as well as homozygosity for the 4-7-10 complex in patients with rheumatoid arthritis.

SCHWARTZ:   If the patients were heterozygotes, like their normal relatives, what other genes do you propose that would be involved in lupus?

WINCHESTER:   I can only speculate on an answer to a question that has dominated many discussions over the years, and I would rather hear your view of this. Certainly what we are speaking about is variable penetrance. Presumably environmental factors play at least a modifying role as would be judged from the published studies on monozygotic twins who are discordant for lupus or acquire the disease at different ages.

WARNER: A technical point concerning typing with sera of restricted specificity has recently been made by J. Howard, who used hybridoma-derived monoclonal antibodies, and showed in cytotoxicity studies that the quantitative expression of some MHC antigens can certainly influence the degree of lysis occurring. This was markedly influenced by the presence or not of other "natural" antibodies in the complement source.

WINCHESTER: Yes, perhaps the existence of a narrow specificity might even account for certain instances of the "cytotoxicity negative absorption positive" phenomenon. In the present studies, this kind of reaction, were it to occur, would be readily detected by the parallel immunofluorescent examination, measuring as it does primary antibody binding.

HLA-D RESTRICTED SUPPRESSOR T CELLS OF THE MIXED LYMPHOCYTE REACTION:
POTENTIAL CLINICAL APPLICATIONS

EDGAR G. ENGLEMAN
Department of Pathology, Stanford University School of Medicine,
Stanford, California 94305

INTRODUCTION

Immune response genes linked to the major histocompatibility complex
have been identified in a variety of animal species[1]. These genes
influence virtually every aspect of immunity[2], and despite the lack of
unequivocal evidence, they are presumed to exist in man. The possibility that HLA-linked immune response genes may affect susceptibility
to disease justifies our efforts to identify such genes in man. In this
regard, the HLA-D region is of particular interest not only because
certain HLA-D alleles are associated with susceptibility to rheumatoid
arthritis, juvenile onset diabetes mellitus and multiple sclerosis (and
many other disorders) but also because the HLA-D gene products are analogous to the murine Ia antigens in structure, tissue distribution, and
in the ability to stimulate in a mixed lymphocyte reaction (MLR)[3].

Recently, the analogy between HLA-D and the murine I region has been
strengthened by the discovery that T cell mediated suppression of the
MLR in man is controlled by genes in the HLA-D region. In the current
report, the evidence for this contention is summarized, and the data
which suggest a possible role for MLR suppressor cells in disease is
reviewed. Lastly, we speculate on the possibility that such cells or
their products might be used therapeutically in conditions in which
suppression of the immune response is desirable.

EVIDENCE THAT MLR SUPPRESSOR T CELLS RECOGNIZE HLA-D GENE PRODUCTS

Examples of antigen-specific and antigen-nonspecific suppression.
In 1976 Andrew McMichael and Takehiko Sasazuki, working in Hugh
McDevitt's laboratory at Stanford University, discovered that the T
lymphocytes from an individual who failed to respond in the MLR
suppressed the MLR responses of certain other individuals. Subsequent
studies by the author in collaboration with Drs. McMichael and McDevitt
indicated that MLR suppressor T cells may represent a distinct subset
of T lymphocytes, the effects of which are genetically restricted and
highly specific for the MLR. Within this subset of lymphocytes, however, there are important variations as described below.

MM is a 23 year old man whose lymphocytes respond normally to mitogens and soluble antigens but fail to respond to allogeneic cells in the MLR. His medical history is remarkable in that he received thymic irradiation as an infant. HLA typing revealed that he is homozygous for HLA A2, B12 and Dw4. When MM's lymphocytes were added to the responder lymphocytes of other persons homozygous for the same HLA antigens, their responses to allogeneic cells but not to mitogens were suppressed by 50 to 95%[4]. This suppressive effect of MM lymphocytes is illustrated in Figure 1. Such inhibition was observed despite repeated washing of MM cells and was unchanged if MM cells cultured alone for 24 hours were used. Ordinarily, under the conditions of the MLR as performed in this laboratory, an increase in the number of responder cells results in an increased response. A time-course experiment ruled out the possibility that the apparent suppression at day 7 was, in fact, the descending curve of an earlier peak. When T and B lymphocytes from MM were separated, either on the basis of their differential affinity for antibody to immunoglobulin or by a sheep erythrocyte rosetting technique, the suppressor activity resided almost exclusively in the T cell fractions (Figure 2). The possibility that MM T cells were actually killer cells could not be completely excluded, but no evidence of killing by MM could be detected in a cell-mediated antibody-dependent cytotoxicity assay.

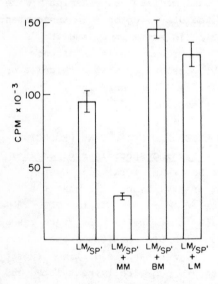

Figure 1. The effect of MM cells on the MLR between LM and SP'. LM is HLA-B,D identical to MM. SP, the irradiated stimulator cell, is an HLA-A2,29/B12,21.1/C-/Dw7,- individual selected at random from a panel of unrelated donor cells. Each column height represents the MLR response in counts per minute, with bars showing the standard error of the mean of six experiments. The first column depicts LM's response to SP'. The second column shows the same MLR in the presence of 50,000 MM cells. In the third column 50,000 cells from MM's HLA identical sibling BM have been added to the MLR. In the fourth column 50,000 additional LM responder cells have been added for a total of 100,000 LM responders. Reprinted with permission from Journal of Clinical Investigation (ref. 4).

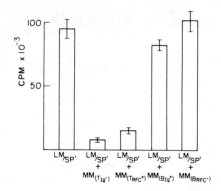

Figure 2. The effect of MM's T cells compared to MM's B cells on the MLR between LM and SP'. MM's T cells were separated from his B cells either by a technique in which B cells remained bound to immunoglobulin coated on a plastic surface ($B_{Ig+}$) and T cells eluted ($T_{Ig-}$) or by a technique in which rosette forming cells ($T_{RFC+}$) were separated from B cells ($B_{RFC-}$) on a Ficoll-Hypaque gradient. 50,000 T cells prepared by each technique, and 50,000 B cells prepared by each technique were co-cultured in the MLR LM/SP'. Each bar represents the mean of sextuplicate cultures. Reprinted with permission from Journal of Clinical Investigation (ref. 4).

Further studies of the requirements for MLR suppression revealed that only persons heterozygous or homozygous for the Dw4 antigen were inhibited by MM suppressor T cells (Table I)[4]. This effect was not altered by differences in the HLA-A, B or C antigens between the suppressor and responder. These data suggested the possibility, therefore, that genes in or near the HLA-D region code not only for the B cell antigens that elicit the MLR but also for structures on T cells or possibly macrophages, which are recognized by MLR suppressor T cells. Despite the high degree of specificity for the responder cell in the MLR, however, MM suppressor T cells were nonspecific with respect to the stimulating cell. That is, when MM cells were co-cultured in MLRs between Dw4 responders and allogeneic stimulator cells, the responses to all stimulator cells were suppressed. The MM suppressor cell is, therefore, similar to the MLR suppressor T cell described by Rich and Rich in the mouse[5]. In the murine system, a factor released by MLR suppressor cells only inhibits the responses of cells sharing the same I-C subregion within the H-2 complex, but the responses to a variety of stimulator cells are inhibited to the same extent.

TABLE 1

THE MM SUPPRESSOR CELL: SPECIFICITY FOR HLA-Dw4 IN RESPONDER CELLS*

| | Responder HLA type | | | | Response to SP | | |
|------|------|-------|------|-----|-------------------------|----------------------|----------|
| | A | B | C | D | Without MM | With MM | Change |
| | | | | | | | % |
| LM | 2,11 | 12,12 | 5,- | 4,4 | $95,460^{\pm}5,824$ | $29,202^{\pm}2,190$ | -69.4** |
| MC | 2,2 | 12,12 | | 4,4 | $63,612^{\pm}3,153$ | $26,260^{\pm}1,588$ | -58.6** |
| EG | 3,30 | 7,15 | | 2,4 | $89,254^{\pm}4,675$ | $34,905^{\pm}2,984$ | -60.5** |
| SF | 2,24 | 13,27 | 2,- | 4,- | $105,669^{\pm}5,247$ | $62,433^{\pm}8,520$ | -40.9** |
| LH | 3,3 | 7,7 | 1, | 2,2 | $44,622^{\pm}5,814$ | $67,911^{\pm}7,012$ | +52.2 |
| JB | 1,1 | 8,8 | | 3,3 | $91,388^{\pm}8,270$ | $111,749^{\pm}7,461$ | +22.3 |
| MH | 2,29 | 12,27 | 1, | 1,7 | $29,639^{\pm}3,120$ | $53,150^{\pm}6,336$ | +79.6 |
| BC | 1,29 | 17,35 | 4,T7 | 5,6 | $56,411^{\pm}4,703$ | $82,970^{\pm}9,645$ | +47.1 |

* Each row describes a 7-day MLR between 50,000 irradiated stimulator
cells and 50,000 responder cells with or without 50,000 MM cells. The
same stimulator cell from an unrelated donor, SP (A3,29/B12,21.1/C-/
Dw7,-) was used in each reaction. The response in counts per minute
$^{\pm}$ standard error represents the mean of six experiments.
** Suppression by MM significant at P <0.001 by the two-tailed t test.

Studies of a second individual who failed to respond in the MLR
revealed that not all MLR suppression in humans is nonspecific with
respect to the stimulating alloantigen. JH is an HLA A2,3/B7,7/Dw2,
2/DRw2,2 mother of ten. She responds normally in the MLR to most
allogeneic cells but surprisingly not to her husband, WH. This was
an unexpected finding because WH (HLA A11,28/Bw35,35/Dw1-/DRw1,4) has
no HLA antigens in common with JH, and his cells behave normally as
stimulator cells when tested with a random panel of responder cells.
Furthermore, although JH made an anti-HLA Bw35 antibody during her
child bearing years, she has had no detectable HLA A or B antibody
since 1972 nor has she ever had detectable B lymphocyte alloantibody.

In experiments similar to those described for MM, the failure of
JH to respond to WH was shown to be due to suppressor T cells[6]. It
initially appeared that only responders identical at both HLA D loci
to JH were inhibited by JH suppressor T cells. Cells which lacked
Dw2 or which had two D antigens, one of which was Dw2 and the other
a different D antigen, were not suppressed. The apparent lack of
suppression of Dw2 heterozygotes, however, was due to the proliferation of cells from JH responding to D antigens other than Dw2 in

heterozygous cells because JH lymphocytes, depleted of MLR responder cells, inhibited both Dw2 heterozygotes and Dw2 homozygotes[7]. Selective depletion of JH responder cells was accomplished with α-irradiation, as shown in Figure 3. Thus, all MLR proliferation activity in JH was lost at an irradiation dose of 1,000 rads, whereas JH suppressor activity remained intact. When JH lymphocytes exposed to 1,000 rads were tested for their ability to suppress the MLR, all HLA Dw2 individuals were inhibited, regardless of the associated HLA A or B antigens (see, for example, Figure 4). Cells lacking Dw2 were not suppressed.

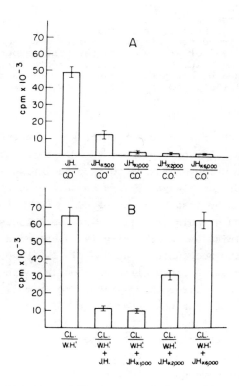

Figure 3. The sensitivity of JH MLR responder cells and suppressor cells to γ-irradiation. The upper panel (A) demonstrates the sensitivity of JH responder cells to increasing doses of γ-irradiation as measured in one-way MLRs between 50,000 JH cells and 50,000 irradiated (6,000 rads) cells from an allogeneic donor, CO. The responses represent the mean of six experiments at each radiation level. The lower panel (B) shows the sensitivity of JH suppressor cells to increasing doses of γ-irradiation as measured by the ability of 50,000 JH cells to inhibit the MLR between CL and WH'. Reprinted with permission from Journal of Experimental Medicine (ref. 7).

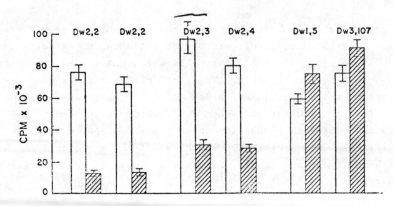

Figure 4: Effect of irradiated JH cells on the responses of Dw2 positive and Dw2 negative cells to WH. Each column represents the MLR response in counts per minute, with bars showing the standard error of the mean of six experiments. The responder's HLA-D type is shown above the columns. Open columns represent one-way MLR response to WH. Hatched bars represent the same MLRs in the presence of 50,000 JH cells exposed to 1,000 rads of γ-irradiation.

Only when WH or a few other cells were present as the irradiated stimulator was JH suppression of the MLR detectable. In general, these were cells to which JH failed to respond in a one-way MLR. Such cells sometimes, but not always, shared the HLA Bw35 specificity with WH. Thus, the JH suppressor cell appeared to recognize determinants in the irradiated stimulator cell as well as D locus products in the responder. This dual specificity is illustrated in Figure 5, in which JH inhibits the responses of an HLA B7, Dw2 homozygous responder to WH but not to three other allogeneic cells. The specificity of the JH suppressor T cell for both responder and stimulator cells is qualitatively different than the other human MLR suppressor T cell described above or the MLR suppressor T cell in mice, both of which lack specificity for the stimulator cell. Such results are consistent with the hypothesis that the JH suppressor cell arose <u>in vivo</u> as a result of sensitization to repeated fetal grafts with the same paternal antigens.

<u>Evidence for the existence of soluble T-cell derived suppressor factors</u>. With the knowledge that suppression by T cells of immune responses in mice can be mediated by soluble factors, we sought to determine whether such factors might also mediate T cell suppression of the MLR in humans. For preparation of suppressor factor, MLRs were carried out in loosely capped 15 ml plastic test tubes with $3 \times 10^6$ JH lymphocytes and $3 \times 10^6$ irradiated WH lymphocytes in a

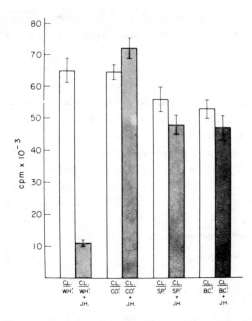

Figure 5. Specificity of the JH suppressor cell for the MLR stimula-
ting cell. Unshaded columns represent one-way MLR's between CL
(B7,Dw2 homozygous) and four different irradiated stimulator cells.
Shaded columns represent the same MLR's in the presence of 50,000
JH cells. Each result represents the mean of six experiments. Reprinted
with permission from Journal of Experimental Medicine (ref. 7).

volume of 6 ml. Control cultures consisting of JH cells alone or JH
plus irradiated cells other than WH were carried out simultaneously.
After 48 hours, the culture tubes were centrifuged and the supernatant
fluids withdrawn. These supernatants were then substituted for
medium in one-way MLRs carried out in the usual manner between 50,000
responder cells and 50,000 irradiated stimulator cells. The results
demonstrated that a soluble factor or factors suppressive of the MLR
were generated when JH lymphocytes were cultured with the irradiated
cells of her husband, WH[8]. Experiments with lymphocyte fractions
enriched for either T cells or B cells suggested that the source of
the JH suppressor factor is a T cell.

The remarkable features of the JH suppressor factor are its specificity both for the MLR stimulator cell and the MLR responder cell. The stimulator specificity paralleled that observed with JH cells in MLR suppression. The factor was generated in the combination JH/WH, and it was not generated when JH was cultured alone or with irradiated cells other than WH. As shown in Table II, the factor suppressed the response to WH and certain other stimulator cells although these cells did not have an HLA antigen in common. The factor also showed specificity for the HLA D products of the responding cell (Table III). The specificity was for HLA Dw2, which is the D type of JH. One cell which lacked Dw2 but was nonetheless suppressed was serologically positive for the Dw2 associated DRw2 antigen. These results confirmed the previous observation that sharing of D region products is required between the JH suppressor T cell and the target responder cell.

TABLE 2

SPECIFICITY OF THE JH SUPPRESSOR FACTOR FOR THE STIMULATING CELL IN THE MLR*

| | Stimulator | | | | Effect of the J.H. suppressor factor on mixed lymphocyte responses by: | | |
|------|------|------|------|------|------|------|------|
| HLA | A | B | Dw | DRw | TL (Dw2,2) | JL (Dw2,3) | EG (Dw2,4) |
| | | | | | | % | |
| W.H. | 11,28 | 35,35 | 1,- | 1,4 | −71 | −47 | −54 |
| S.D. | 1,31 | 17,12 | 7,- | ND‡ | −57 | −27 | −7 |
| S.N. | 2,24 | 21,35 | -,- | 5,4 × 7 | −46 | −11 | −32 |
| B.C. | 1,29 | 17,35 | 5,6 | 5,6 | −42 | +3 | −6 |
| S.F. | 2,24 | 13,27 | 4,- | 4,- | −58 | −3 | −10 |
| E.G. | 3,30 | 7,15 | 2,4 | 2,4 | −33 | +8 | ND‡ |
| L.W. | 1,11 | 8,35 | 3,- | ND‡ | −17 | +14 | +48 |
| H.K. | 1,2 | 8,7 | 3,- | 3,7 | +7 | +34 | +20 |
| W.B. | 3,29 | 7,12 | 2,7 | 2,7 | +13 | +18 | +10 |
| C.O. | 1,2 | 13,40 | 1,- | 1,5 | −17 | +35 | +40 |

Source: Journal of Experimental Medicine (ref. 8). Reprinted with permission.

* Supernate from a 48-hour culture between JH and irradiated WH cells was tested for the capacity to inhibit the responses of 3 Dw2-positive cells to a panel of 10 unrelated stimulator cells. Results are expressed as the % change (either inhibition {-} or enhancement {+} in baseline unidirectional MLRs.
**ND, not done.

TABLE 3

THE JH SUPPRESSOR FACTOR: REQUIREMENT FOR HLA-Dw2 IN THE MLR RESPONDER CELL

| | Responder | | | Response to WH in the presence of medium* | Response to WH in the presence of JH/WH super-nate* | |
|---|---|---|---|---|---|---|
| HLA | A | B | Dw | DRw | | |
| | | | | | | %Δ cpm |
| TI | 1,3 | 7,7 | 2,2 | 2,2 | $37,610 \pm 4,599$ | $11,112 \pm 908$ | −71 |
| TL | 3,3 | 7,7 | 2,2 | 2,2 | $81,883 \pm 6,729$ | $23,682 \pm 2,399$ | −71 |
| RL | 1,2 | 7,8 | 2,3 | 2,3 | $20,818 \pm 1,459$ | $10,363 \pm 1,368$ | −50 |
| JR | 1,10 | 8,18 | 2,3 | 2,3 | $66,953 \pm 7,386$ | $41,942 \pm 4,722$ | −37 |
| EG | 3,30 | 7,15 | 2,4 | 2,4 | $31,004 \pm 3,463$ | $14,260 \pm 2,847$ | −54 |
| CO | 1,2 | 13,40 | 1,- | 1,5 | $48,945 \pm 2,778$ | $44,413 \pm 2,060$ | −9 |
| HK | 1,2 | 8,7 | 3,- | 3,7 | $74,875 \pm 4,110$ | $67,550 \pm 4,552$ | −10 |
| CH | 1,2 | 8,40 | 4,6 | 4,6 | $37,025 \pm 5,905$ | $38,111 \pm 4,827$ | +3 |
| BC | 1,29 | 17,35 | 5,6 | 5,6 | $67,351 \pm 7,549$ | $73,590 \pm 5,162$ | +9 |
| RK | 1,28 | 5.2,14 | -,- | 2,5 | $72,214 \pm 5,350$ | $54,440 \pm 6,406$ | −25 |

* Responses in cpm represent the means of six experiments $\pm$ standard error.

<u>Evidence that the genes controlling suppressibility are not identical to HLA D or HLA DR</u>.  Although the data suggest that sharing of D region products by suppressor cells and responder cells is a requirement for suppression, the precise location of the gene coding for this product has not been determined.  In the mouse, T suppressor factor of the MLR only inhibits the responses of strains which share identity at a particular subregion within the immune response region of the major histocompatibility complex (H-2)[3].  By analogy, if the D region in man consists of several loci, it is possible that only one locus (not necessarily HLA D or DR) need be shared by suppressor and responder cells.  Therefore, we have recently attempted to identify a) cells which are of the suppressible D/DR type but are not suppressed or b) cells which lack the appropriate D/DR type and are suppressed. After screening more than 100 individuals, two such exceptional cells have been found[9].  One cell lacks Dw2 and DRw2 but is suppressed by JH.  A second cell has Dw2 and DRw2 but is not suppressed by JH. These data suggest that the gene product which must be present on

responder cells for suppression to occur is not the D/DR antigen(s) but rather the product of a linked gene. Family studies of these exceptional individuals are in progress in an effort to clarify this point.

The determinants on stimulator cells which are recognized by JH suppressor cells have not been characterized. It is clear that such determinants differ from the D region products on suppressible responder cells because WH, whose cells stimulate suppression, is not suppressed by JH. We have no evidence that the determinants on WH which elicit suppression are products of HLA genes because other cells which elicit suppression do not share a known HLA specificity with WH.


CLINICAL APPLICATIONS

Association of MLR suppressor cells and disease. To search for MLR suppressor cells in a large number of individuals, such as a group of patients with a particualr disease, would be extremely difficult with the techniques described. It is unlikely, for example, that healthy HLA D matched subjects would be routinely available for tests of suppression. Therefore, a screening procedure was sought which avoids the necessity of HLA D matching. Advantage was taken of the observation that MLR suppressor T cells resist a dose of irradiation which functionally eliminates responder cells from suppressor donor populations. Cultures were carried out between 50,000 responder lymphocytes and 50,000 irradiated stimulator cells with or without 50,000 irradiated (1,000 rads) autologous responder lymphocytes. By adding irradiated autologous cells, an attempt has been made to enrich the suppressor cell population. In the actual test, individuals were challenged by four different, normal stimulator cells. A test was considered positive if irradiated responder cells inhibited at least 2 of 4 responses by 40% or more.

When this test was applied to a group of 50 healthy volunteers, only one was found whose autologous irradiated cells suppressed her responses in the MLR[10]. This person's cells also suppressed the MLR responses of an HLA identical sibling but failed to suppress the responses of other siblings or unrelated persons. Further studies showed that suppression was mediated by a T cell, the behavior of which was similar to that described for MM.

Unlike healthy subjects, patients with Hodgkin's Disease typically mount weak responses in the MLR (Figure 6). When the MLR suppressor test was applied to a group of patients with this disorder, the findings were in striking contrast to those obtained with normal volunteers[11]. The result of one such test is shown in Figure 7. Note that in the presence of this patient's own irradiated cells, his responses to 3 of 4 stimulator cells are significantly reduced, suggesting the presence of suppressor cells.

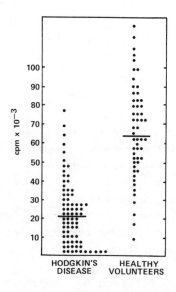

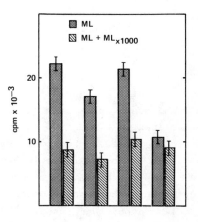

Figure 6. MLR responses of patients with Hodgkin's Disease and normal controls. Results represent mean one-way MLR responses against a panel of normal stimulator cells.

Figure 7. MLR suppressor activity in a patient with Hodgkin's Disease. Unshaded columns represent one-way MLRs between ML, a patient with Hodgkin's Disease, and four different irradiated stimulator cells. Shaded columns represent the same MLRs in the presence of 50,000 ML cells exposed to 1,000 rads. Each result represents the mean of three experiments.

The overall results of our screening efforts are summarized in Table 4. It is clear from these results that detectable MLR suppressor cells are rare in healthy volunteers but very common in Hodgkin's Disease. The Hodgkin's Disease patients were classified according to age, sex, histologic criteria, clinical stage, treatment and HLA type, but MLR suppression did not correlate with any of these parameters. This result is not necessarily surprising in view of the high frequency of MLR suppressor cells in these patients. Such cells appear to be absent in patients with mycoses fungoides, but the sample of non-Hodgkin's lymphoma patients is not yet large enough to determine the frequency of suppressor cells.

TABLE 4

THE FREQUENCY OF MLR SUPPRESSOR CELLS IN PATIENTS WITH HODGKIN'S
DISEASE AND OTHER LYMPHOMAS

|  | NORMALS | HODG DISEASE | NON-HODG LYMPHOMA | MYCOSIS FUNGOIDES |
|---|---|---|---|---|
| SUPPRESSION OF 2 OR MORE RESPONSES | 1 | 47 | 1 | 0 |
| SUPPRESSION OF 1 RESPONSE | 6 | 31 | 2 | 1 |
| NO SIGNIFICANT SUPPRESSION | 45 | 39 | 8 | 15 |

More recently, MLR suppressor tests have been performed with purified T and non-T cells from patients with Hodgkin's Disease.  T cells
were isolated by a modified sheep erythrocyte rosetting technique.  Of
12 patients tested, 6 had positive suppressor tests with unseparated
mononuclear leukocytes, but 10 had positive tests with purified T
cells; in only 4 of 12 cases, preparations enriched for B cells and
monocytes were suppressive of the MLR.  Thus, it appears that the use
of T cells rather than unseparated cells increases the sensitivity of
the suppressor test in patients with Hodgkin's Disease and that the
majority of such patients have MLR suppressor T cells.  Moreover, in
some patients suppressor T cells failed to inhibit the responses of
unrelated individuals, suggesting that these cells may be genetically
restricted in their effects.

Potential beneficial and deleterious clinical effects of MLR
suppressor cells.  Perhaps it is appropriate to conclude these
remarks with a brief discussion of the hypothetical "pros" and "cons"
of MLR suppressor cells.  We begin with the supposition (for which
there is as yet no direct evidence) that such cells may have significant effects in vivo.  In Hodgkin's Disease, for example, suppressor
cells might contribute to a state of tumor tolerance.  Of course,
the critical question of whether or not the occurrence of suppressor
cells precedes disease or follows it can  only be answered by prospective studies of patients and normals with and without suppressor
cells.  Nonetheless, in light of the recent demonstrations in mice
that suppressor T cells protect syngeneic tumors against host immune
defense mechanisms[12] and that antibodies to the same suppressor cells
enhance host immunity to these tumors[13], antisera to human suppressor
cells could conceivably be used to potentiate host immunity in man.

Conversely, suppressor cells or their products could be used to inhibit immunity in patients in whom such inhibition may be desirable, such as recipients of organ transplants and patients with autoimmune disorders. In this regard, our laboratory is currently studying recipients of HLA-mismatched renal and heart allografts to determine if long-term survivors have circulating suppressor cells. A more direct assessment of the possible clinical effects of suppressor cells will necessitate treatment of selected patients with suppressor cells or suppressor factor. We have already shown that suppressor factor can be generated in vitro by removing circulating suppressor cells and culturing them in the presence of appropriate stimulator cells. To be useful clinically, however, such genetically restricted factors must be prepared from donors who are HLA D compatible with the prospective recipient - a rather formidable task. Other investigators have reported that MLR suppressor cells can be induced, in vitro, from the lymphocytes of individuals who apparently lack detectable circulating suppressor cells[14,15]. Conceivably, such cells or their products could be autotransfused for their potential therapeutic effects.

A final alternative, induction of suppressor cells in vivo, is perhaps closer to being formally tested in humans. Strober and Slavin and their colleagues at Stanford University have found that pretreatment of mice and rats with total lymphoid irradiation (TLI) followed by histoincompatible bone marrow transplantation results in permanent acceptance of skin, heart or kidney allografts from the same histoincompatible donor[16]. These investigators have also used TLI to obtain permanent tolerance to soluble antigens, and their data suggest that radiation-induced suppressor cells are responsible for the observed tolerance[17]. Hopefully, such approaches as these will be applicable to man in the near future, and lead us to an age of more sophisticated immunologic intervention in the treatment of human disease.

SUMMARY

We have described suppression of the mixed lymphocyte reaction (MLR) in humans by thymus-derived (T) cells and soluble T-cell derived factors. Such suppression is highly specific for HLA D products on MLR responder lymphocytes, and in one example is also specific for determinants on MLR stimulator cells. Circulating MLR

suppressor cells are rare in healthy individuals, but common in
Hodgkin's Disease, a disorder characterized by markedly diminished
cell mediated immunity.  In this disorder and perhaps others, suppressor cells might have the undesirable effect of hindering host immune
defense mechanisms.  Conversely, suppressor cells or their products
could be used to inhibit immunity in patients with autoimmune disease
or recipients of organ transplants.

ACKNOWLEDGEMENTS

The author gratefully acknowledges the advice and encouragement
of Dr. Hugh McDevitt, in whose laboratory the majority of this work
was performed.  This work was supported by a grant from the Kroc
Foundation (Santa Ynez, Calif.) and a grant from the National Institutes of Health (AI 11313).

REFERENCES

1. McDevitt, H.O. (1976) Fed. Proc., 35, 2168.

2. Shreffler, D.C. and David, C.S. (1975) Adv. Imm. 20, 125.

3. McDevitt, H.O. and Engleman, E.G. (1977) Arth. Rheum., 20, 9-20.

4. Engleman, E.G. and McDevitt, H.O. (1978) J. Clin. Invest.,
   61, 828-838.

5. Rich, S.S. and Rich, R.R. (1976) J. Exp. Med., 143, 672-677.

6. McMichael, A.J. and Sasazuki, T. (1977) J. Exp. Med., 146, 368-380.

7. Engleman, E.G., McMichael, A.J., Batey, M.E. and McDevitt, H.O.
   (1978) J. Exp. Med., 147, 137-146.

8. Engleman, E.G., McMichael, A.J. and McDevitt, H.O. (1978)
   J. Exp. Med. 147, 1037-1043.

9. Engleman, E.G., unpublished observations.

10. Engleman, E.G. (1978) Transp. Proc., in press.

11. Engleman, E.G., Hoppe, R., Kaplan, H., and McDevitt, H.O. (1978)
    manuscript submitted for publication.

12. Fujimoto, S., Greene, M.I., and Sehon, A.H. (1976) J. Immunol.
    116, 791-799.

13. Greene, M.I., Dorf, M.E., Pierres, M., and Benacerraf, B. (1977)
    Proc. Natl. Acad. Sci. (U.S.A.) 74, 5118-5121.

14. Hirschberg, H. and Thorsby, E. (1977) Scand. J. Immunol. 6, 809.

15. Haynes, D., van Speybroeck, J. and Cochrum, K. (1978)
    Transp. Proc., in press.

16. Slavin, S., Reitz, B., Bieber, C.P., Kaplan, H.S., and Strober, S.
    (1978) J. Exp. Med., 147, 700-714.

17. Zan-bar, I., Slavin, S., and Strober, S. (1978) J. Immunol., in
    press.

DISCUSSION

ROITT: It could be deduced that the inhibition of MLR might be related to a failure to produce cells cytotoxic for virally infected cells; perhaps some way can be found to check this. If this proves to be so, the ability of this individual to handle viral infections normally would suggest that his delayed hypersensitivity responses that are not impaired, are acting through immune interferon and macrophage activation to combat viruses.

ENGLEMAN: Whether or not the MLR suppressor T cells also affect the generation of cytotoxic lymphocytes has not yet been examined. This is a possibility that we plan to explore.
What is clear is that these suppressor cells in MM and JH do not affect the majority of T cell functions measured in vitro in the absence of allogeneic cells. Also, MM and JH have normal Ig levels, and normal numbers and percentages of peripheral blood T cells, B cells, and monocytes. Although JH and MM are currently healthy, the fact that many patients with Hodgkin's disease have MLR suppressor cells may be cause for concern. I would stress, however, that we have yet to demonstrate any effects of suppressor cells on the clinical course of patients with Hodgkin's disease. In addition, it is not certain that MLR suppressor T cells in Hodgkin's disease and those in healthy persons are the same.

BIGAZZI: Did you check for the presence of lymphokines in the sera of your Hodgkin's patients? This would be interesting in view of previous reports of high levels of lymphokines in such patients and might mean that cells suppressing the MLR would not be suppressor cells for T lymphocytes producing lymphokines.

ENGLEMAN: We did not measure lymphokines in the sera of our Hodgkin's patients. Also, the target of the MLR suppressor cell and suppressor factor has not yet been identified. It is possible that a cell other than a proliferating T cell is the target; for example, another "induced" suppressor cell; a helper T cell; a macrophage; or even the stimulating B cell.

SCHWARTZ: Were the suppressor cells in the Hodgkin's patients "autosuppressors" and were they T cells?

ENGLEMAN: The screening assay for suppressor cells measures the ability of irradiated autologous mononuclear leukocytes to inhibit MLR, that is, to inhibit one's own response to allogeneic cells. In this sense, it is a test of "autosuppression". Our results would suggest that many patients with Hodgkin's disease have suppressor T cells of the MLR because when we add autologous irradiated T cells to the culture, the MLR is inhibited.
Nonetheless, it is likely that this is not the only explanation for the dysfunction of cell-mediated immunity in these patients. Twomey et al., for example, have reported that adherent cells (presumably monocyte macrophages) are suppressive of T proliferative responses in Hodgkin's patients. More recently, Goodwin et al. suggested that the suppressive effects of monocytes in Hodgkin's disease are mediated by prostaglandins and reversed by addition of indomethacin to the cultures. It is noteworthy, however, that the effects of prostaglandins were demonstrated in the response to PHA rather than in the MLR. In our experience, indomethacin has no effect on MLR suppression.

SCHWARTZ: It seems of interest that in mice, the graft versus host reaction is a powerful inducer of suppressor cells; and this reaction may terminate as a lymphoma that resembles Hodgkin's disease.

With regard to your presentation, have any family studies been performed on the Hodgkin's patients?

ENGLEMAN: Family studies of patients with Hodgkin's disease are just beginning in collaboration with Drs. Hoppe and Kaplan at Stanford. Hopefully, they will answer two key questions: Do MLR suppressor cells occur in the relatives of patients? If so, this might suggest that the occurrence of suppressor cells is under genetic influence. Secondly, do the patients' suppressor cells inhibit the responses of their relatives; that is, are these cells genetically restricted in their action?

I have had the opportunity to study identical male twins, one with and one without Hodgkin's disease. The affected twin had MLR suppressor cells that inhibited his twin's as well as his own responses in the MLR. The healthy twin lacked detectable suppressor cells.

GASSER: Have you looked at healthy pregnant women? Do you think these cells might appear during pregnancy and then in most cases go away after birth?

ENGLEMAN: We have not yet looked for MLR suppressor cells in pregnant women. We have looked for MLR suppressor cells in several multiparous women and have not found any except in the case of JH. However, all of these women were at least five years past their most recent pregnancy.

DUPONT: You have convincingly shown that some human T suppressor cells have an HLA-D restriction. Other investigators have also looked for this. Dr. Thomsen and coworkers from Copenhagen were the first to show human T suppressor cells which showed HLA-A restriction. Dr. M. Bean has recently published one case, where a male patient with bladder cancer had a T suppressor cell, which showed genetic restriction. This patient has never been transfused. We have studied this case with him and found that the T suppressor cell was active only on responder cells which carried HLA-B14. This was shown in family studies and in panel studies. It can be concluded that human T suppressor cells can show HLA-restriction. Some are restricted to the HLA-D region while others are restricted to the HLA-A and HLA-B loci.

GERSHON: I wasn't quite sure I understood Dr. Schwartz's question nor your answer to it. Dr. Schwartz, did your question about an etiologic agent ask whether Hodgkin's tumor is a tumor of suppressor cells?

SCHWARTZ: No.

GERSHON: Then let me ask that question; Dr. Engleman, have you considered the possibility that the suppressor cells you are measuring are being made by the Reed-Sternberg cell acting as a precursor?

ENGLEMAN: Not really. I hadn't considered it, but I wasn't aware that those tumor cells circulate.

GERSHON:  Well perhaps you won't find many of those Reed-Sternberg cells in the blood in Hodgkin's **disease**, but there is the possibility that Reed-Sternberg cells are precursor cells as are embryonal carcinoma cells which throw off a number of variants that differentiate normally and have some function.  I think that is something that you ought to consider.  I can't resist asking another question because looking directly at me with a wry pixy-like smile on his face is Dr. Wigzell.  Have you looked at the alpha-feto protein levels in the people with the spontaneous suppressor cells?

ENGLEMAN:  No, we haven't.

WARNER:  Studies at Stanford and in Boston appear to have established that the cultured Reed-Sternberg cell is in the macrophage lineage.  Furthermore, in the SJL mouse model that tumor also does not appear to be in the suppressor lineage.  In both instances the evidence is quite strong that the cell types studied are indeed the tumor cells themselves.

GERSHON:  **The nature of a science which relies on inductive logic is such that negative results usually have less significance than positive results.**  Thus, putting a precursor cell into a culture from which one cell type emerges does not preclude other cell types that can also emerge in other microenvironments.  Hodgkin's is really a very heterogeneous group of diseases and I don't think that the evidence we have is fully convincing that suppressor cells cannot be thrown off from precursor cells in some types of malignant tumors of stem cells.  Furthermore, I don't accept SJL tumors as models for Hodgkin's **disease**, certainly not the pleomorphic variety which is the type I was referring to.

SCHWARTZ:  Just to comment on that point, there is evidence, I believe, from Waldman's laboratory of at least one human tumor that functions in vitro as a suppressor cell.  I think this was the case of ALA and they had convincing evidence that this cell behaves like a suppressor.

MACKAY:  This interesting study clearly points the way that clinical immunology must follow in the future.
The frequent occurrence of hyporeactivity in DTH skin tests with ubiquitous antigens has previously led investigators to believe that the T cell system was damaged rather than suppressed.  Incidentally, were skin tests done in your patients?  In autoimmune diseases such as SLE, there is multiplicity of autoantibodies implying defective suppressor influence over B cell responses, yet hyporesponsiveness in tests for cell-mediated immunity implying augmented suppressor influence over one subpopulation of T cells.  It seems as if different functional subsets of suppressor T cell may exist in man.

ENGLEMAN:  Skin testing was not performed in most of these patients.  I certainly agree with your second comment, that is, that more than one population of suppressor cells may exist.  Tada and associates and Benacerraf and co-workers have described antigen-specific T cell suppression of antibody synthesis.  Suppression is mediated by factors which bear I-J determinants and according to Tada, the cells mediating suppression are extremely sensitive to x-irradiation.  In contrast, the murine MLR suppressor T cell of Rich and Rich releases a suppressive factor which bears products of the I-C subregion of H-2, and the suppressor cell is comparatively resistant to mitomycin-C and, I believe, x-irradiation. In humans, suppression by T cells of pokeweed mitogen induced B cell synthesis of Ig is sensitive to several hundred rads of x-irradiation, whereas the circulating MLR suppressor T cell is not. Thus there appears to be at least two distinct subsets of suppressor T cells: one for humoral immunity, and one for cell-mediated immunity. More recently, Drs. Cantor and Gershon have described still another T suppressor cell, the Ly 123 cell involved in feedback inhibition of antibody synthesis; and, of course, there are cells other than T cells which may have suppressive effects, particularly macrophages.

143

GENETICS OF AUTOIMMUNE DISEASES: PROSPECTS AND PERSPECTIVES

IVAN M. ROITT

Department of Immunology, Middlesex Hospital Medical School,
London, W1 United Kingdom

I wish to start by presenting some data on rheumatoid arthritis
which illustrate certain points that I want to make. One of the
fascinating things about this disease is that there is virtually no
correlation with any of the HLA-A,B,or C antigens but a strong association with D (Stastny, 1976) and DR (Stastny, 1978) specificities.
In conjunction with my colleagues M. Corbett, H. Festenstein, D. Jara-
quemada, C. Papasteriadis, F. Hay and L. Nineham, we have been carrying
out a prospective study on patients presenting within one year of the
onset of symptoms and followed up for five years or more. In confirmation of Stastny's report, we found 67% of our patients to be positive
for DRw4; this compares with an incidence of 25% in the controls and
gives a relative risk of 6.1. With respect to ARA classification,
16/31 (52%) possible or probable, 19/25 (76%) definite and 10/11 (91%)
classical cases were DRw4. There appears to be a relationship also to
the eventual severity of the disease in that bone erosions were seen
to a much greater extent in the DRw4 positive (26/34) than in the
negative (5/13) group.

It may be that the typing is showing up disease groups within the
rheumatoids who have essentially different basic characteristics or
groups likely to develop the disease in a milder or more severe form.
This suggested to us the possibility that these markers could provide
useful prognostic indices, the general idea being that if one can
recognize at an early stage whether a patient was liable to develop
severe erosive disease with irreversible changes in the joints, one
could institute aggressive therapy before such changes occurred. In
making our analysis, instead of looking at the final state of the
disease, we decided that we would try to estimate the severity of the
disease process by seeing how the doctor had titrated the patient.
If the patients were maintained in an adequate state by aspirin or no
treatment then they were assigned to Group 1; if they needed indomethacin or similar drugs then they were put in Group 2 and if they
required penicillamine, gold or steroids then they were in Group 3.
In other words, this represented the response of the doctor to the

degree of severity of the disease. Some may feel that there are limi-
tations to this approach but it does provide a more useful classification for those who have an active underlying disease process but
who are relatively well because they respond positively to treatment.
It has been known for many years that the classical rheumatoid factors
assessed by the SCAT and latex tests do correlate with poorer prognosis,
and in our studies, rheumatoid factor was positive within the first
year of presentation in 50% of patients in Group 1, 82% in Group 2 and
90% in Group 3. Combined tests for rheumatoid factor and DRw4 were
positive in only 27% of Group 1, whereas 72% of Group 2 and 70% of
Group 3 patients showed dual positivity.

There is now increasing evidence that IgG rheumatoid factor (or
antiglobulin) produced locally in the joint may be a dominant pathogenetic feature of the disease since IgG anti-IgG is both antigen and
complementary antibody in one; these molecules can self-associate to
form complexes which can be stabilized by IgM rheumatoid factor (itself
a polyvalent Fc binder) and then fix complement to initiate inflammatory processes. Our analysis of the high molecular weight material in
the synovial fluids of rheumatoid arthritis patients is entirely
consistent with this view since we find mainly IgG and IgM, lesser
amounts of IgA, Clq, C3 and C4 (the latter two accompanied by extensive
breakdown products) and trace amounts of Clr, Cls and factor B (Male,
Roitt & Hay, 1978). There are also split products of $\alpha_2$-macroglobulin
and fibrinogen present. These findings raise the following question:
does the HLA-linkage reflect in any way the ability to generate an
immune response to limited epitopes on the Fc portion of the IgG molecule which are critically placed, from the geometric standpoint, to
allow self-association to occur?

Turning now to the papers which were presented today, there seems
no doubt that typing for major histocompatibility specificities,
particularly those encoded by the D locus, will eventually provide
benefits in several aspects of autoimmune clinical disorders. The
first is concerned with the way in which tissue typing helps to define
homogeneous clinical entities within a heterogeneous group of patients
presenting with comparable symptoms e.g. the group of diabetics associated with autoimmunity and the subgroup of myasthenics with thymoma.
It is anticipated that this approach will have wide application and
of course there will be diagnostic benefits. Another aspect of autoimmune disease likely to be affected by typing is the application for
prognostic purposes and we have already mentioned our own studies in

rheumatoid arthritis and the possibility of designing prognostic
indices which would enable selected patients to receive the appropriate
therapy at an early stage in their disease.

Typing may also play a role in preventive medicine through genetic
counselling and by identification of individuals at risk.  Within
families of probands who already have a disease which is linked to HLA,
one could identify siblings who are at risk and follow them prospectively to see the evolution of the various parameters of the disease.

Looking at the technical aspects of typing there are still serious
constraints regarding the availability, complexity and specificity of
the reagents used.  The definition of D region specificities is
particularly imprecise and if the human resembles the mouse in terms
of the I subregions there must be a number of loci which have yet to
be identified.  To some extent we may derive better typing reagents
from the isolation of chemically discrete molecules either directly
from the surface of cells or indirectly through the cloning of relevant
genes.  However, I think it most likely that we will establish these
reagents through the cloning of antibody-forming cells either from the
Milstein technique of immunizing rats or mice and hybridizing them
with a cell line or from applying Klein's use of EB virus to produce
immortal cell lines from antigen or pokeweed mitogen activated lymphocytes obtained from a woman already sensitized to HLA alloantigens.
Another approach, which Bob Winchester has to some extent adopted, is
to ignore accepted typing reagents and instead screen sera for their
ability to discriminate between the lymphocytes from different disease
groups.  In a sense, Stastny was looking for disease related antigens
when he examined MLC reactivity between patients with rheumatoid
arthritis and found certain stimulators who tended to give very low
mixed lymphocyte responses;  he found a different specificity (T-Mo)
characteristic of juvenile RA.

What can we say about the mechanisms?  Everybody, I think, is
agreed that just one gene alone is unlikely to be responsible for
the susceptibility to the development of autoimmune disease particularly when one thinks of the tremendous complexity of the immune
response, the expression of self-reactivity and the production of
tissue lesions.  Certainly, the serious genetic studies on experimental
models of autoimmune disease have generally indicated multifactorial
control.  In the human, aside from clear-cut cases where there is a
deficiency in synthesis of factors which affect the immune response,
e.g. complement components, three main patterns of HLA-linked disorders

are seen. At one extreme, e.g. ankylosing spondylitis, there is almost total linkage to the HLA-B locus. Most of the organ-related autoimmune disorders show some linkage to both B and D loci, the majority of those studied being more strongly related to the DW3 specificity. Finally, a disease such as rheumatoid arthritis, which is not associated with B locus antigens, reveals a startling relationship to the D locus. The HLA region covers a very large number of genes and we have to consider that a number of different mechanisms might be operating. A central question is whether the mechanisms are antigen specific or antigen independent.

In organ-specific diseases such as Addison's disease, thyrotoxicosis and Hashimoto's thyroiditis, each individual makes autoantibodies to several different antigens which are probably unrelated and do not cross-react. This suggests that the mechanism concerned has more to do with the basic recognition of self, and in particular cell-surface autoantigens, than in selecting particular antigen specificities. A variety of epitope-independent mechanisms might operate. There is the question of antigen handling, its placing on the macrophage membrane and its connection or otherwise with surface Ia. Genes linked to steroid biosynthesis are present within the HLA complex and one could envisage an influence of steroid levels on the thresholds of different phases of immune responsiveness perhaps thereby raising the susceptibility of an individual to autoimmune disease.

Consideration should also be given to the role of complement components in mediating interactions between cells concerned in the immune response. Looking back phylogenetically at the primitive (non-antibody dependent) defence of the mucosal surfaces of the body by macrophages, co-operation with the alternative complement system would not be effectively provided by the fluid phase in which complement concentrations are low, and it would have seemed more helpful if evolutionary processes had started by putting the complement components on the macrophage plasma membrane itself. The adherence of macrophages to Sephadex G10 columns could be a result of activation of alternative pathway components on the macrophage with subsequent adherence to dextran-bound C3b. It is interesting to note that factor B and C4 have been demonstrated on the surface of lymphocytes and it is relevant to recall the formation of complement-induced clumps from spleen cells involving the interaction of T cells, B cells and macrophages (Pepys, 1974; Arnaiz-Villena & Roitt, 1975). Whatever the mechanisms and import of these complement-dependent interactions

between cells of the immune system finally prove to be, the location
of genes concerned with the synthesis of complement components within
the HLA region is intriguing.

There is a considerable body of evidence to implicate some derangement of non-specific T suppressor cells in certain autoimmune disorders
and it will be of interest to see whether some HLA association can be
identified.  So far we have not mentioned the question of thymic proc-
essing of immature T cells in which self H-2D and H-2K specificities
are imprinted so that the immunocompetent T lymphocyte can subsequently
recognize modified (e.g. virally infected) body cells (Zinkernagel et
al., 1978).  One supposes that the response of T cells to soluble antigens presented by macrophages in association with Ia will also involve
programming of the developing T cell  by Ia molecules on the thymic
epithelium (not yet demonstrated I believe at the time of writing).
A defect in this postulated process could predispose an individual to
the development of autoimmunity.  Certain abnormalities in the NZB
thymus have been reported and one recalls the histological studies of
de Vries & Hijmans (1967) describing, amongst other features, an early
deficiency of epithelial cells.

If it turns out that the major factors predisposing to autoimmunity
are not related directly to specific antigen epitopes, there must be
yet a further factor, or series of factors, which determine which organ
is affected.  For example, if one considers patients with Hashimoto's
thyroiditis, the close relatives are found to have thyroid autoimmunity
whereas relatives of probands with pernicious anaemia tend to have
gastric autoantibodies.  In other words, major histocompatibility genes
may predispose to the development of organ-specific autoimmunity in
general, but a further series of genes may be concerned with selection
of the target organ.  Separation of factors leading to autoimmunity
from those determining the autoantigen have been recognized in NZB
mice and their hybrids, some animals giving predominantly anti-red cell
responses and others anti-nuclear factors.

There could however be immune response genes within the HLA complex
controlling antigen specific responses.  In some instances molecular
mimicry between a pathogenetic agent and the HLA-specificity may be
involved.  One of the main protagonists of this viewpoint is Alan
Ebringer who has postulated a relationship between certain Klebsiella
species and HLA-B27 in ankylosing spondylitis;  really hard data on
cross-reactions of this kind are awaited with interest.  Immune response genes of the type described by Benacerraf, McDevitt, Sela and

Mozes for the mouse have not been unequivocally identified in Man. One major hurdle is the difficulty in studying _in vitro_ primary responses to any antigen, let alone ones with selected epitopes. However, antibody responses in sub-human primates should help to establish the general principle for higher animal species. Undoubtedly, the restricted epitopes which provoke autoimmunity should be good candidates for antibody responses controlled by Ir genes.

I would like to finish by extending my comments to genetic factors possibly outside the MHC region. I have already referred to genes which select the organ for attack. Another area in which we have little information relates to possible involvement of immunoglobulin structural genes coding for anti-idiotypic network control or auto-antigen simulation; Gm typing studies might be relevant. Lastly, we should mention the influence of the sex chromosomes. The female sex preponderance in most autoimmune diseases and the high incidence of thyroid autoantibodies in subjects with X-chromosome abnormalities are well-known. Whether the sex hormone pattern or a factor controlled by other genes on these chromosomes provides the dominant influence is, like so many other things, not entirely obvious.

REFERENCES

Arnaiz-Villena, A. and Roitt, I.M. (1975) Clin. exp. Immunol., 21, 115.

Pepys, M.B. (1974) Nature, 249, 51.

Stastny, P. (1976) J. clin. Invest., 57, 1148.

Stastny, P. (1978) N. Engl. J. Med., 298, 869.

Vries, M.J.de and Hijmans, W. (1967) Immunology, 12, 179.

Zinkernagel, R.M., Callahan, G.N., Klein, J. and Dennert, G. (1978) Nature, 271, 251.

# PART II

# SPONTANEOUS DISEASE IN ANIMALS

Noel L. Warner *and* Norman Talal, *Co-Chairmen*

LYMPHOCYTE COMMUNICATION AND AUTOIMMUNITY

HARVEY CANTOR

Harvard Medical School/Farber Cancer Institute, 44 Binney Street,
Boston, Massachusetts 02115 (U.S.A.)

Until recently, immunologists have generally held that the immune
system is comprised of clones of lymphocytes that, when activated by
antigen, would produce antibody or initiate a cell-mediated response
(such as an inflammatory response). According to this view, the duration and strength of a response depended only upon the number of lymphocyte clones in the host that carried receptors for a particular antigen
and that the absence of immunity to "self" reflected deletion at birth
of all lymphocytes bearing receptors for self antigens.

There is increasing evidence that this view of the immune system
is probably incorrect. A more accurate description of the immune system is that (a) it is composed of many sets of regulatory lymphocytes
which respond mainly to signals generated from within the system itself
and that for the most part these interactions inhibit both antibody
and cellular immune responses[1-5] and (b) these interactions may also
serve to prevent "B" lymphocytes from producing antibody against host
antigens. In other words, the absence of autoimmune reactions may be
due, in part, to continuous and active suppression rather than the
absence of self-reactive cells within the system.

One approach that has supported this view of the immune system involves the dissection and definition of T cells and the various immune
functions that they perform. Despite their uniform morphology, T cells
are by no means a homogeneous population; they comprise subclasses or
sets of lymphocytes with different and even seemingly opposing functions. Thus, one property of T cells called helper function is to
assist B cells to make antibody.[6] A second function of T cells had
been suspected from investigations of immunologic tolerance or unresponsiveness; these studies indicated that adoptive transfer of T
lymphocytes from an animal unresponsive to a given antigen to a normal
animal could render the recipient specifically unresponsive.[6] This
property of T cells is the generation of cells that are capable of
damaging or destroying cells recognized as antigenically foreign after,
for example, infection by a virus. This is associated with cytotoxic
or killer function of T cells. Yet another function of T cells involves
the ability of these cells to induce or activate other cells to
participate in inflammatory responses.

A crucial point arises: Are all these functions invested in a single set of T cells that have differentiated in the thymus and are these diverse responses governed entirely by extraneous conditions such as a mode or type of antigen stimulation? Alternatively, are these immunologic functions invested in distinct sets of T cells that have been programmed to respond in different ways during their differentiative history? According to the latter idea, thymus-dependent differentiation may give rise to a number of separate sublines of mature T cells, each genetically programmed to mediate one or another T-cell response.

In the mouse, this question has resolved itself into the practical problem of finding out whether it is possible to subdivide the T-cell population into different sets that, when confronted with antigens, are able to make only one or another of the possible T-cell responses. At the present time, the most effective technique for identifying and separating subpopulations of peripheral T cells has come from studies of the cell-surface components which become expressed on cells undergoing thymus-dependent differentiation. This classification is based upon the description of alloantisera to define a pattern of cell-surface differentiation components expressed on T cells.[7,1] Since these components have not been detected on cells of other tissues such as brain, kidney, liver or epidermal cells, they are evidently specified by genes expressed *exclusively* during T-cell differentiation. These have been called the Ly systems. The Ly1 component is coded for by a gene on chromosome 19 and the Ly23 components are both coded for by genes on chromosome 6. These last two are treated together tentatively because the two genes are tightly linked and these two systems have not so far exhibited any differences other than the fact that genetically they are coded for by distinguishable loci.

In general, the approach involves a complement-dependent assay. Lymphocytes exposed to, say, anti-Ly1 sera in the presence of complement are lysed; this lysis can be monitored by the use of trypan blue, which stains lysed but not living cells, or by the release of a radioactive label from the lysed cells.

The central problem to accurate and precise definition of cell-surface components is verification of the specificity of antisera used for this purpose. All antisera made against cells contain many antibodies to cell-surface components, in addition to those for which a particular antiserum is named. Usually, immunologists manage to overcome this problem by using the antiserum at a dilution that may mask contaminant antibodies. This approach is not satisfactory and usually produces data that are difficult to interpret. Ultimately, the use of monoclonal antibodies will avoid most of these problems.

Currently, the most stringent and reliable approach to verifying specificity comes from a comparison of the effects of the same anti-serum upon cells from congenic mice that differ *only* at the gene locus that codes for the relevant cell-surface component. For example, the ability of anti-Ly1.2 + complement to eliminate helper activity from a heterogeneous population of lymphocytes may indicate that T helper cells express Ly1. To establish this point, cells from B 6 mice (Ly phenotype 1.2,2.2,3.2) and congenic B6-Ly1.1 mice (Ly phenotype 1.1,2.2,3.2) are treated with anti-Ly1.2 + complement. If helper activity is eliminated from the former cell population but not the latter, it can be concluded that in this system T helper cells express the Ly1.2 cell-surface component (or a component coded for by a gene tightly linked to the Ly1 locus).

In addition, negative selection using antisera and complement requires that a large number of rabbit sera are screened for high complement activity and low background cytotoxicity. This means that different dilutions of each new batch of complement must be tested with a given antiserum to determine the precise dilution of antiserum and complement that will produce maximal specific cytotoxic effects. Finally, the specificity of antisera must be re-evaluated when *positive* selection techniques are employed. For example, one can select T cells expressing, say, the Ly2 cell surface component by coating purified T cells with anti-Ly2, washing, and passing the coated cells through a rabbit anti-mouse Fab2 column. Virtually all the cells eluted from the column express the Ly2.2 surface marker, as judged by immuno-fluorescence. However, extra care must be taken to evaluate the specificity of this selection since many antisera (including certain Ly reagents) contain antibodies that do not fix complement but which nevertheless will bind inappropriately to T cells. Thus, sera intended for positive selection must be screened for the presence or absence of reactivity against the appropriate Ly congenic partner both by fluor-escence and by their effect on functional activity of positively selected cells from appropriate congenic mice.

Analysis of this sort has revealed that the peripheral T cell pool contains *at least* three separate T-cell sets. We refer to them in shorthand as the Ly123 set, the Ly1 set, and the Ly23 set. They compose respectively 50%, 30% and approximately 5-10% of the peripheral T-cell pool. These findings indicate that, according to the criterion of selective expression of gene products on the cell surface, the T-cell

pool is divisible into three groups of cells, each following a different set of genetic instructions. The question then becomes whether
these individual differentiative programs include information that
decides what the function of each T-cell set should be. Evidence to
date indicates that cells of the Lyl set are genetically programmed to
help or amplify the activity of other cells after stimulation by antigen. Cells of this set are most aptly termed *inducer* cells since
they will induce or activate other cell sets to fulfill their respective
genetic programs: Lyl cells induce B cells to secrete antibody; they
induce macrophages and monocytes to participate in delayed type hypersensitivity responses; they can, under appropriate circumstances,
induce precursors of killer cells to differentiate to killer-effector
cells. Most recently, and perhaps most importantly, it has been found
that Lyl cells also induce a set of resting, non-immune T cells to
generate potent *feedback* suppressive activity.[2,3] Analysis of
isolated Lyl inducer cells from non-immune donors indicate that these
cells are already programmed for helper/inducer function *before* overt
immunization with antigen; this function is independent of the ability
of Lyl inducer cells to interact with antigen.[1]

By contrast, cells of the Ly23 set are specially equipped to develop
both alloreactive cytotoxic activity as well as to suppress both
humoral and cell-mediated immune responses following immunization.
Whether cytotoxicity and antigen-induced suppression are two manifestations of one genetic program or whether they represent the phenotype
of two separate genetic programs is not yet established.

Until recently cells of the Lyl23 set have been the least well-
defined of the various T-cell sets. The most likely possibility is
that at least some Lyl23 cells represent a store of receptor-positive
intermediary cells that regulate the supply and function of more
mature Lyl and Ly23 cells. This is based in part on experiments
showing that (a) after stimulation with chemically-altered syngeneic
cells, some Lyl23 cells give rise to Ly23 progeny[8,9] and (b) experiments indicating that purified populations of Lyl23 cells can give
rise to Lyl cells after polyclonal activation by Concanavalin A (ms. in
preparation). The notion that at least a portion of Lyl23 cells represents a precursor pool is also supported by earlier observations that
cells of the Lyl23 subclass are detectable in the spleens of mice
within the first week of life, while both Lyl and Ly23 cells do not
reach maximal numbers until adult life (8-12 weeks of age).

More recently, it has been demonstrated that antigen-stimulated Lyl
cells,[2] or supernatants of activated Lyl cells, in addition to inducing

B cells to secrete antibody, can induce or activate resting Ly123 T
cells to develop profound feedback suppressive activity. Feedback suppression is appropriate here since the degree of suppressive activity
exerted by a fixed number of non-immune Ly123 cells increases in direct
proportion to the numbers of antigen-activated Ly1 cells in the system.[2]
These findings also indicate that, like the formation of antibody, the
generation of immunologic suppression after stimulation by antigen is
not an autonomous function; both require induction by Ly1 cells. This
Ly1:Ly123 interaction has also been shown to govern the duration and
intensity of immune reactions in vivo.[3] These experiments have suggested that the following events ensue after stimulation of the immune
system by foreign materials: Activated antigen-specific Ly1 cells
induce B cells to form antibody and also induce resting Ly123 cells to
inhibit T helper cell activity. Reduction in T helper activity is
accompanied by decreased induction of B cells as well as progressively
decreasing induction of resting Ly123 cells; the net result is progressive decrease in both antibody formation and suppressor cell induction.

Thus, the ability of Ly1 inducer cells to activate various effector
cell systems on the one hand, as well as suppressive systems on the
other, probably plays a key role in the type and intensity of an antibody response. It is therefore important to know whether cells of the
Ly1 set that induce suppressive activity represent a specialized subgroup of Ly1 cells that differ from cells that, for example, induce
B cells to produce antibody. A direct approach to this question has
come from the finding that a portion of Ly1 cells also express a newly-
defined antigen called Qa1.[2,4,10-12] This antigen, or antigen system,
is coded for by genes that map between H2-D and the TL locus of the
mouse.[10] Studies of Ly1:Qa1$^+$ cells have shown that these cells are
responsible for induction of feedback inhibition and that Ly1:Qa1$^-$
cells are not.[4] In addition, these studies show that signals from both
Ly1:Qa1$^+$ and Ly1:Qa1$^-$ cells are required for optimal formation of
antibody by B cells. Thus, the ability of antigenic determinants to
induce a detectable antibody response may depend largely on the ratio
of Ly1:Qa1$^+$ and Ly1:Qa1$^-$ T-cell clones that bear receptors for that
determinant. It is also likely that the ability of Ly1:Qa1$^+$ inducer
cells to elicit strong suppressive responses is particularly important
in governing the duration and intensity of inflammatory reactions such
as delayed type hypersensitivity and IgE-mediated hypersensitivity.[4,13,14] Analysis of the cell-free products of homogeneous
populations of Ly1:Qa1$^+$ cells is therefore of considerable interest,
since these materials may well prove useful in strategies designed to

selectively suppress hypersensitivity or antibody responses to defined antigens.

In sum, these experiments have established that the genetic program
for a single differentiated set of cells, in this case immunologic
cells, combines information coding for a surface antigenic profile
that is associated with particularly physiologic functions. Second,
they have indicated that the majority of T cells are not effector
cells poised to respond to foreign antigen, but regulatory cells that
respond mainly to signals or messages generated from within the T-cell
system itself, and that detectable immune responses reflect perturbations of these signals after cells of the Ly1 system are stimulated
by antigen.

We have just begun to delineate the circuits involved in this regulatory system as summarized above. At the risk of oversimplification,
the picture that is beginning to emerge is that Ly1 cells act as
*sentinel* cells which screen the surfaces of other cells, particularly
macrophages for foreign material, associated with "MHC" molecules.
When activated, these sentinel cells can induce a variety of effector
cells (e.g., B cells which make antibody or macrophages and monocytes
which participate in inflammatory responses) to make a specific
immune response. In addition, they activate a *committee** of resting
T cells that are probably relatively immature. This committee of
cells emits inhibitory signals. The intensity of inhibition depends
mainly on the genetic background of the host, the nature of the
antigenic stimulus and the intensity of the inducing signal emitted
by the sentinel cells. The observed immune response depends upon the
relative potency and timing of feedback suppressive inhibitory signals
relative to the potency and timing of inductive signals that are
passed to effector cells (e.g., B cells).

What happens when this system goes wrong? There are, so far,
several examples of disorders of this immunoregulatory circuit; NZB
mice spontaneously develop an autoimmune disorder characterized by the
production of a variety of autoantibodies and a clinical syndrome
resembling human systemic lupus erythematosis. The major T-cell
deficit of NZB mice is the absence or malfunction of an Ly123 T-cell
set responsible for feedback inhibition.[3]

---

*The term committee is used here because it is virtually certain that
Ly123 cells are themselves a heterogeneous set, and perhaps not the
sole members of this system. Moreover, the response of a committee
is generally suppressive.

Another inbred mouse strain that we have just begun to study is the motheaten (ME) mouse.[15] This strain suffers from an autosomal recessive autoimmune disorder characterized by elevated immunoglobulin levels and deposition of IgM and IgG immunoglobulin complexes in kidney glomeruli. Comparison of the proportions of Ly$^+$ T-cell sets in 3-week-old me/me (fully recessive) mice and me/+ mice indicates that spleen cells from me/me mice have substantially increased proportions of Ly1 cells, and decreased proportions of Ly123 cells. Functional studies are underway to determine whether the abnormally high levels of Ly1 cells and low levels of Ly123 cells reflect a dysfunction within the feedback inhibitory circuit.

A second example comes from experiments by L. McVay-Boudreau in which Ly2$^+$ regulatory cells are deliberately eliminated from the host: mice depleted of all T cells are repopulated with either Ly1 inducer cells or all T-cell sets (including Ly2$^+$ regulatory cells). Within the first two weeks after repopulation, sera from the former but not the latter mice contain autoantibodies against erythrocytes and thymocytes. Thus, elimination of regulatory T cells which participate in feedback inhibition can be directly shown to result in the formation of autoantibodies.

How do Ly1:Qal$^+$ inducer cells communicate with resting T cells? There is good evidence that Ly1 cells that have been stimulated by antigen preferentially induce suppression of the in vitro response to that antigen and do not induce suppression to a second antigen even when both antigens are present in cell culture. These observations suggest that communication between Ly1:Qal$^+$ cells and resting Ly123: Qal$^+$ T cells is highly specific. One explanation for this specificity is that both inducer and effector cells bear receptors specific for the antigen and are brought into association by an "antigen bridge" (similar to the mechanism postulated for T-B interactions). Another possibility is that resting T cells bear receptors specific for idiotypic determinants carried on the Ly1 inducer population. There are several lines of evidence to support this latter view. First, experiments done by Elizabeth Bikoff in our laboratory have demonstrated that efficient induction of feedback inhibition in secondary antibody responses requires that both Ly1-inducer cells and non-immune T cells are obtained from donors that are identical at the Ig heavy chain locus; similar results have been obtained in primary anti-SRBC responses. The requirement for "matching" of identical Ig heavy chain gene clusters suggests that at least part of Ly1:Ly123 communication may involve immunoglobulin gene products; the contribution of the constant and variable portions of Ig heavy chain components is currently being investigated.

A second line of evidence that this inducer-acceptor circuit is mediated by Ig heavy chain associated gene products comes from an analysis of the induction of feedback inhibition in a system using a more well-defined antigen that elicits antibodies which carry identifiable idiotypic markers.[16-18] The particular idiotype we have studied has been identified by anti-idiotypic antibodies made in rabbits and has been extensively characterized over the past several years by Nisonoff and his colleagues. It is produced in A/J mice after immunization with para-azophenylarsonate ("Id") conjugated to a variety of proteins and accounts for 20-70% of the total anti-arsonate response. Lyl cells that bind to Id (and are potent inducers of $Id^+$ B cells[18]) can selectively induce id-specific suppression. By contrast, Lyl cells that lack idiotype-specific inducer activity do not induce feedback inhibition. These observations indicate that an important element in communication between T-cell sets may be based on shared or complementary cell surface $V_H$ structures.

These findings, taken together, suggest that the immune response is controlled in a highly precise way by signals continuously generated within the system itself and that the net effect of these interactions after perturbation of the system is to restore the homeostatic balance of the system. It is now important to examine the role of these regulatory circuits in the initial establishment of self tolerance at birth. There is certainly good evidence that B cells from normal animals are capable of producing autoantibodies.[19-21] A major question that is yet unresolved is whether the absence of autoantibody production reflects a continuous active process of feedback inhibition rather than neonatal deletion of T-inducer clones bearing receptors for self antigens.

Additional questions that remain unresolved concern the types of B cells that may be participating in autoantibody responses, in view of the demonstration that certain B-cell sets are more easily triggered than others; whether some types of autoimmune syndromes are induced to self antigens or whether they are initiated by viral modification of self antigen; and finally, whether the cascade of events that results in clinical disease is initiated by autoantibody-mediated elimination of a T-cell set that participates in this inhibitory cell circuit.

REFERENCES

1. Cantor, H. and Boyse, E. A. (1977) Lymphocytes as models for the study of mammalian cellular differentiation. Immunological Rev. 33:105.

2. Eardley, D. D., Hugenberger, J., McVay-Boudreau, L., Shen, F. W., Gershon, R. K. and Cantor, H. (1978) Immunoregulatory circuits among T-cell sets. I. T-helper cells induce other T-cell sets to exert feedback inhibition. J. Exp. Med. 147:1106.

3. Cantor, H., McVay-Boudreau, L., Hugenberger, J., Naidorf, K., Shen, F. W., and Gershon, R. K. (1978) Immunoregulatory circuits among T-cell sets. II. Physiologic role of feedback inhibition in vivo: Absence in NZB mice. J. Exp. Med. 147:1116.

4. Cantor, H., Hugenberger, J., McVay-Boudreau, L., Eardley, D. D., Kemp, J., Shen, F. W. and Gershon, R. K. (1978) Immunoregulatory circuits among T-cell sets. Identification of a subpopulation of T-helper cells that induces feedback inhibition. J. Exp. Med. (in press).

5. Gershon, R. K. (1974) T cell suppression. Contemp. Top. Immunobiol. 3:1.

6. Transplantation Reviews, Vol. 1. (1969) G. Möller, ed. Williams and Wilkins Co., Baltimore, MD.

7. Boyse, E. A., Miyazawa, M., Aoki, T. and Old, L. J. (1968) Ly-A and Ly-B. Two systems of lymphocyte isoantigens in the mouse. Proc. Roy. Soc. B. 170:175.

8. Cantor, H. and Boyse, E. A. (1977) Regulation of cellular and humoral immunity by T-cell subclasses. Proceedings of the 41st Cold Spring Harbor Symposium, Cold Spring Harbor, NY, p. 23.

9. Burakoff, S. J., Finberg, R., Glimcher, L., Lemonnier, F., Benacerraf, B. and Cantor, H. (1978) The biologic significance of alloreactivity. The ontogeny of T-cell sets specific for allo-antigens or modified self antigens. J. Exp. Med. (in press).

10. Flaherty, L. (1976) The Tla region of the mouse: identification of a new serologically defined locus, Qa2. Immunogenetics 3:533.

11. Stanton, T. H. and Boyse, E. A. (1976) A new serologically defined locus, Qa-1, in the Tla region of the mouse. Immunogenetics 3:525.

12. Stanton, T. H., Calkins, C. E., Jandinski, J., Schendel, D. J., Stutman, O., Cantor, H. and Boyse, E. A. (1978) The Qa-1 antigenic system. Relation of Qa-1 phenotypes to lymphocyte sets, mitogen responses, and immune functions. J. Exp. Med. (in press).

13. Askenase, P. W., Hayden, B. J. and Gershon, R. K. (1975) Augmentation of delayed type hypersensitivity by doses of cyclophosphamide which do not affect antibody responses. J. Exp. Med. 141:697.

14. Watanabe, N., Kojima, S., Shen, F. W. and Ovary, Z. (1977) Suppression of IgE antibody production in SJL mice. II. Expression of Ly-1 antigen on helper and non-specific suppressor T cells. J. Immunol. 118:485.

15. Shultz, L. D. and Green, M. C. (1976) Motheaten, an immunodeficient mutant of the mouse. II. Depressed immune competence and elevated serum immunoglobulins. J. Immunol. 116:936.

16. Nisonoff, A., Ju, S-T and Owen, F. L. (1977) A mouse cross-reactive idiotype. Immunol. Rev. 34:89.

17. Ward, K., Cantor, H. and Nisonoff, A. (1978) Analysis of the cellular basis of idiotype-specific suppression. J. Immunol. 120:2016.

18. Woodland, R. and Cantor, H. (1978) Idiotype-specific T helper cells are required to induce idiotype-positive B memory cells to secrete antibody. Eur. J. Immunol. (in press).

19. Möller, G. (1976) Mechanism of B-cell activation and self/non-self discrimination. Proceedings of the 41st Cold Spring Harbor Symposium, Cold Spring Harbor, NY, p. 217.

20. Dresser, D. W. (1978) Most IgM-producing cells in the mouse secrete autoantibodies (rheumatoid factor). Nature 274:480.

21. Steele, E. J. and Cunningham, A. J. (1978) High proportion of Ig-producing cells making autoantibody in normal mice. Nature 274:483.

DISCUSSION

DATTA:  How sensitive is the feedback $T_s$ to adult $T_x$ and to the natural thymotoxic ab of NZB mice?

CANTOR:  Approximately 50-90% of feedback inhibition activity is lost 6-10 weeks after removal of the thymus.  We have no information on the sensitivity of these cells to NTA.

ROSE:  Do you have information about the normal ontogeny of Ly 2,3 cells in regard to their time of appearance in the thymus and their migration to the periphery?

CANTOR:  No.  Recently we have developed more  sensitive systems for detecting Ly T cell sets in the thymus.  So far, this has allowed us to identify two populations early in ontogeny:  a major population of Ly 1,2,3 cells and a minor population of Ly 1 cells.  The low frequency of Ly 2,3 cells has made it so far extremely difficult to quantitate this T cell set.

COHEN:  I would like to comment about the relationship between the Ly 1 inducer cell you describe and the initiator T lymphocyte (ITL) that we have been studying for the past few years (Cohen and Livnat, Transplant. Rev. 29:24, 1976).  Unlike your system that involves activation of B cells to produce antibody, our studies relate to the capacity of ITL to recruit precursor T cells to become active cytotoxic T cells.  These recruited T lymphocytes (RTL) are specifically cytotoxic to target cells that bear the same H-2K or H-2D antigens that originally sensitized the ITL.  Despite the apparent differences in antibody production compared to generation of T cell cytotoxicity, it is noteworthy that the inducer or initiator T cells in both systems bears the Ly $1^+2^-$ phenotype (Ben-Nun and Cohen, submitted for publication).  This supports the notion that a unique class of T lymphocytes may act to initiate a network of different cell interactions.
Furthermore, we have found that the ITL might activate the RTL by a double signal:  products of the ITL's own H-2I region together with part of the allogeneic sensitizing antigen.

WEKERLE:  In your experiments using NZB mice you found decreases in feedback T cells as early as one month of age, a time point, where normally autoimmune symptoms are not found.  Could this mean that other immune defects, in addition to feedback T cell decrease, may be required for loss of self tolerance to occur in the NZB mice?

CANTOR:  No.  The delay in apparent autoantibody formation may simply reflect the accumulated effects of loss of normal immuno-regulation over a period of months.

WARNER:  Since the number of Ly 2,3 cells are normal to elevated in NZB mice, could the defect be in the induction of suppression? Have you been able to assess numbers of Ly $1^+$ Qa-$1^+$ cells in NZB mice?

CANTOR:  The NZB mouse does not express the appropriate Qa-1 phenotype.  However, the ability of three month old NZB Ly 1 cells to induce BALB/c Ly 1,2,3 cells to exert inhibitory effects is not grossly impaired.

SEX HORMONES AND AUTOIMMUNITY

N. TALAL and J. R. ROUBINIAN
Department of Medicine, University of California, San Francisco
and Section of Clinical Immunology, VA Hospital, San Francisco,
California 94121

ABSTRACT

The natural history of spontaneous autoimmune disease in NZB and
NZB/NZW $F_1$ mice is discussed with emphasis on an abnormal equilibrium
between T regulatory cells.  Recent attempts to develop newer forms of
immunotherapy are reviewed, and the possible therapeutic benefits of
male hormone are suggested.  Sex hormones modulate the development of
autoimmunity in NZB/NZW mice as determined by survival, concentration
of antibodies to nucleic acids, and severity of immune complex nephritis.  Androgens suppress and estrogens accelerate disease.  Mechanisms
responsible for these effects of sex hormones are currently under investigation.  It appears that both thymus-dependent and thymus-independent mechanisms are probably involved.

INTRODUCTION

Autoimmune disease represents a breakdown in the proper functioning
of the immunologic network.  When this breakdown occurs, the immune system behaves in a deranged manner.  Various sorts of autoantibodies are
produced, immune complexes are formed, and organs may undergo immunologic attack.

By using animal models for human illness, it is possible to study
disease in its preclinical manifestations.  Such opportunities are rare
in human medicine.  These studies can lead to valuable insights into
pathogenetic mechanisms underlying early events in disease, and offer
the hope of finding more specific and effective modes of prevention or
therapy.

Several animal models for systemic lupus erythematosus (SLE) now exist, including several newly-developed mouse strains and a canine model.
The best known and most extensively studied animal model for lupus is
the NZB/NZW $F_1$ (B/W) mouse, a hybrid of the NZB and NZW strains.  This
mouse spontaneously develops an autoimmune disease similar to human SLE
in three important respects: 1) the formation of antibodies to nucleic

acids, particularly to double-stranded DNA, 2) the deposition of DNA-containing immune complexes in the kidney leading to renal insufficiency and death, and 3) a sex factor which is manifested in the earlier onset of disease in females, who generally die before 1 year of age[1].

The parent NZB mouse also develops a spontaneous autoimmune disease characterized predominantly by Coombs'-positive hemolytic anemia. The parent NZB mouse is clinically normal for most of its life. Autoantibodies and mild nephritis appear in aged NZW mice, as they do in old mice of several normal strains. Genetic, viral and immunologic factors are all involved in the pathogenesis of autoimmunity in the NZB and B/W strains[2-4].

The various strains of New Zealand mice derive from randomly bred animals brought to New Zealand from Mill Hill over 40 years ago and subsequently bred for coat color. The genetic predisposition of NZB mice to develop autoimmune hemolytic anemia was appreciated 20 years ago. Despite fairly extensive genetic analysis in several laboratories, we still lack precise information as to the number of genes involved and how they might function. NZB mice have the allele $H2^d$. With regard to experimental antigens under the control of specific immune response genes, NZB mice make levels of antibody generally comparable to other $H2^d$ strains. There is evidence for modifying genes. For example, the mating of NZB with non-autoimmune strains often results in suppression of autoimmunity in the $F_1$ offspring. There is a variable expression of autoimmunity in the $F_2$ and backcross mice which is compatible with the action of modifying genes influencing a dominant pattern of inheritance.

With regard to viral factors, NZB and B/W mice contain abundant Type-C viral particles. High concentrations of gp70, the major envelope glycoprotein of this virus, are present in serum and tissues[5]. Immune complexes containing gp70 are found in the glomerular deposits of B/W mice along with the DNA-anti-DNA immune complexes. The possible role of Type-C viruses in normal processes of growth and differentiation, as well as in autoimmunity and neoplasia, is a subject of current biological interest[3]. Recent genetic studies suggest that virus expression and autoimmunity segregate independently. These results do not support a primary viral etiology for this murine model for systemic lupus erythematosus.

There is much evidence to suggest that normal mechanisms of immunologic regulation are disordered in NZB and B/W mice. Genetic factors probably underlie this regulatory disturbance. Abnormalities of B cells, T cells and macrophages have been described, as well as abnormalities of thymic epithelial function. For several years, interest focused on a loss of suppressor T cells with consequent escape of autoantibody producing B cell clones. Recent evidence for B cell abnormalities present at birth, and for defective interactions between T cell subpopulations, suggests that the defective suppressor cell concept is probably an oversimplification. A defect of splenic macrophages has also been observed using a system of in vitro immunization to foreign erythrocytes.

Our own recent studies on the sequential development of IgM and IgG antibodies to DNA and RNA suggest that the thymus, spleen and gonads exert major regulatory influences[6]. In general, these seem to reflect physiologic control mechanisms expressed on aberrant autoantibody responses.

NATURAL HISTORY OF NZB MICE

Investigative work of these New Zealand strains evolved in several phases. The first 10 years (from 1958 to 1968) were largely concerned with clinical and experimental pathology and detailed histologic descriptions of various tissue lesions[1]. Studies performed in different areas of the world indicated a general uniformity of disease expression. The development of splenomegaly and hemolytic anemia in the NZB, or LE cells and immune complex nephritis in the B/W, and of generalized lymphoid hyperplasia progressing at times to lymphoid neoplasia in both strains, were well described. Also documented were the ability of spleen cells from older Coombs'-positive mice to transfer autoantibody production into young Coombs'-negative syngeneic recipients, and the acceleration of disease that ensued following neonatal thymectomy.

The next phase of investigation (from 1968 to the present) was strongly influenced by a rapid burst of new knowledge in cellular immunology and lymphocyte biology. Experiments were performed measuring various immunologic responses in these autoimmune strains in hope that comparisons with non-autoimmune strains would bring insight into pathogenetic mechanisms. The major observations are listed in Table I. The New Zealand strains develop immunologic competence prematurely compared to normal strains[7]. Within the first week of life, they make antibody responses to sheep erythrocytes equivalent to that seen in adult NZB mice. Other strains require several weeks to achieve such immunologic maturity. This premature maturation of the immune system may also extend to cellular responses, since very young New Zealand mice can regress tumors more rapidly than age-matched control strain animals[8]. Newborn NZB spleen cells also synthesize excessive amounts of IgM and demonstrate a polyclonal hyperresponsiveness compared to age-matched control strain mice.

NZB and B/W mice make excessive antibody responses to many but not all experimental antigens, including foreign proteins, sheep erythrocytes and synthetic nucleic acids[9-11]. This hyperactivity is seen in young adult mice and is selective, since some antigens elicit responses which fall within the normal range[12]. Antibody responses tend to decline in older animals.

Adult New Zealand mice are relatively resistant to the induction and maintenance of immunologic tolerance to soluble deaggregated foreign proteins such as bovine gamma globulin[9,13,14]. Their ability to become tolerant declines at 1-2 months of age, in contrast to many control strains that manifest long-lasting tolerance. This resistance to tolerance is associated with T rather than B cells. A relative resistance to tolerance is also seen in other strains such as SJL and Balb/c mice, where it is under genetic control.

Older New Zealand mice have marked impairment of cellular immunity, demonstrated by decreased response to mitogens[15] and decreased ability to induce graft-vs-host disease[16] or to reject tumors and skin grafts. At this age, they show a decline in recirculating lymphocytes and in theta-positive lymphocytes. These alterations may be related in part to the aging process itself, and to the spontaneous appearance of an autoantibody cytotoxic to thymocytes and T lymphocytes[17]. This anti-T cell antibody occurs in virtually all NZB mice and in the majority of B/W mice.

The importance of suppressor T cells in immunologic regulation and tolerance has been emphasized recently[18]. NZB and B/W mice lose suppressor T cells[19-21] between 1 and 2 months of age, corresponding to the time when they become resistant to the development of immunologic tolerance. Moreover, neonatal thymectomy can result in an accelerated development of autoimmunity, suggesting that the thymus exerts a suppressing influence on disease expression[1].

The cause of the suppressor T cell deficiency is unknown. NZB and B/W mice demonstrate a premature decline in a thymic humoral factor which may function as a thymic differentiation hormone[22]. The administration of thymosin to these mice can restore some aspects of thymocyte and T cell function, but has no significant therapeutic effect[23].

Cantor, et al., have recently found a decreased number of Ly 123[+]

cells and a related impairment of feedback regulation in very young NZB mice[24] prior to the onset of autoimmune disease. This finding suggests a major derangement in the T cell network which may relate to a primary genetic abnormality predisposing to spontaneous autoimmunity.

The sequence of immunologic changes and disease manifestations in these mice evolves in a manner that suggests a causal relationship (Figure 1). The early decline in thymic epithelial cell function, suppressor T cells and T cell tolerance precedes the development of autoantibodies, tissue lymphocytic infiltrates, immune complex nephritis and hemolytic anemia. The more severe defects in cell mediated immunity occur at an age when malignant lymphomas and monoclonal macroglobulins may appear[25,26].

Several modes of therapy have been tried in NZB and B/W mice, some of which are new and highly promising (Table II). Conventional forms of treatment, such as corticosteroids and immunosuppressive drugs, have been used for many years. Cyclophosphamide is particularly effective but, as in humans, the development of lymphoid malignancy is enhanced by immunosuppression[27-29]. Various attempts to restore or maintain suppressor function have been tried. These have included repeated injection of spleen or thymus lymphocytes from young mice presumed to contain suppressor cells[30,31], thymus grafts from young mice[32], or various thymic humoral factors representing putative thymic hormones[23,33,34]. In general, treated mice have shown a delayed onset of autoantibodies and nephritis, consistent with the transfer of a suppressor mechanism. However, treated mice succumb to their disease often with little or no significant prolongation of survival. Results are best if treatment is started very early in life. Recent studies by Bach indicate that a thymic humoral factor actually increases levels of antibodies to DNA and proteinuria in B/W mice, although other parameters of disease may improve. These discordant results were seen in the same individual mice, suggesting that differential effects on T cell subpopulations could have potentially deleterious effects in humans.

Protein or calorie restriction delayed the onset of hemolytic anemia in NZB mice[35], and prolonged survival in B/W mice. An immunologic mechanism was suggested. The technique of hapten-specific carrier determined tolerance, using nucleosides of DNA covalently bound to isologous NZB IgG, has been shown to delay nephritis in B/W mice[36].

The ability of concanavalin A to induce a soluble immune response suppressor (SIRS) has been employed successfully to decrease autoimmunity and nephritis and prolong survival in B/W mice[37]. The New Zealand

mice lose the ability to produce SIRS themselves, but can respond when
this soluble suppressor factor is administered to them. Prostaglandin
$E_1$ administered once or twice daily to B/W mice[38] will also prolong
survival.

Autoimmunity develops earlier and with greater severity in female
B/W mice than in males. Our laboratory has recently been interested in
this latter point, and specifically in the possible therapeutic effects
of androgenic hormones.

SEX HORMONES AND AUTOIMMUNITY

The marked female predominance of various autoimmune diseases, including systemic lupus erythematosus, has long been apparent. The development of autoimmunity in Klinefelter's syndrome is further evidence
for a possible association with the hyperestrogenic state. Early experiments attempting to show that sex hormones influenced New Zealand
mouse disease were inconclusive. Recently, however, this point has
been clearly established with regard to several parameters of autoimmunity. Raveche et al. studied thymocytotoxic antibody in NZB and
$DBA/2$ $F_1$ mice. The incidence was greater in females than in males;
castration of the males brought the incidence up to female range,
whereas castration of the females had no effect.

In our laboratory, the modulation of autoimmune disease in B/W mice
by sex hormones has been studied by measuring antibodies to DNA, antibodies to polyadenylic acid, immune complex glomerulonephritis, and
mortality. Our first indication of an immunoregulatory difference between males and females was obtained in experiments involving neonatal
thymectomy. Male B/W mice were made worse by this procedure and died
of accelerated disease; female B/W mice lived significantly longer as a
result of neonatal thymectomy[39].

To investigate the sex difference further, we castrated males prepubertally at two weeks of age[40]. The castrated males died more rapidly
than the sham males and showed a survival rate identical to that of the
sham females (Figure 2). The castrated females showed no difference
from sham females: almost all were dead by the age of nine or ten
months. This experiment suggested a possible protective effect of androgenic hormones.

In a second experiment, prepubertal castration was combined with the
sustained administration of either male or female hormones[41]. There
was a prolongation of survival in mice given androgen. Thus, female
mice castrated and given androgen lived significantly longer than sham-

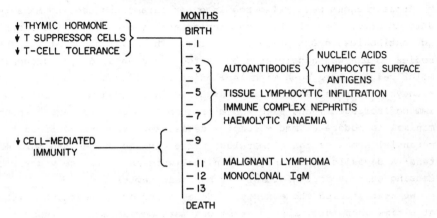

Fig. 1

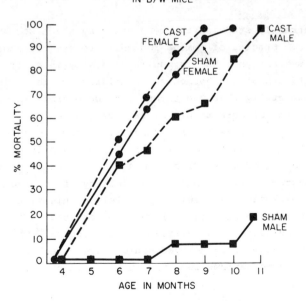

Fig. 2

operated females.  By contrast, there was a significantly decreased
survival of animals given estrogen.  In general, these results suggested that the genetic sex of the experimental animals made little difference.  The significant factor determining survival was the nature of
the sex hormones given to the mice.

These changes were reflected in other parameters used to measure
autoimmunity.  One of the more significant parameters is the amount of
IgG antibodies to DNA present in serum.  The development of IgG antibodies to DNA was accelerated by estrogen and delayed by androgen.  Similar results were seen with antibodies to polyadenylic acid.  The
kidneys from these animals were examined in a coded manner by light,
immunofluorescent, and electron microscopy.  Mice given androgen showed
minimal to moderate immune complex deposits which were confined to the
mesangial areas of the glomerulus.  Mice given estrogen had more extensive deposits of immunoglobulin that were more diffusely distributed,
causing changes in the glomerular basement membrane.

We next studied the effects of delayed androgen treatment initiated
at a time when autoimmune disease was more developed.  We found that
delayed androgen treatment started at three and six months of age also
prolonged survival in female B/W mice[42].  There was an additional interesting feature in these experiments performed in the older mice.
The androgen-treated animals had less immune complex nephritis even
though there was no significant reduction in levels of anti-DNA antibodies.

We are currently studying the mechanisms by which sex hormones exert
these effects.  Studies of immune responses to experimental antigens in
normal mice suggest that the thymus is necessary in order to demonstrate the effects of sex hormones.  Similarly, we found in B/W mice
treated prepubertally that the thymus must be present in order to observe some of the effects resulting from castration.

The mechanisms responsible for survival of mice given delayed androgen treatment may be quite complicated.  These animals live despite the
high concentration of antibodies to DNA.  Androgens might act on the
complement system or might improve the ability of treated mice to eliminate immune complexes.  A clinical parallel to this situation may be
patients with SLE who have large amounts of antibodies to DNA and yet
minimal or no evidence of significant renal disease.  These mice in
whom androgen treatment is delayed could serve as a laboratory model
for study of the mechanisms involved in the pathogenesis of lupus nephritis.

In summary, we have shown that sex hormones modulate the expression of murine lupus as determined by survival, concentration of antibodies to nucleic acids, and severity of immune complex nephritis. Androgens suppress and estrogens accelerate disease. These results probably explain the female predominance of human lupus and the increased incidence of autoimmunity in patients with Klinefelter's syndrome[43]. The immunoregulatory function of the thymus as well as the complement system may be involved in the action of male hormones. Our results support the thesis that lupus is a disorder of immunologic regulation and that the administration of androgens might create a more physiologic immunologic equilibrium. Androgen therapy for human autoimmune disease may be a possible outcome of this work.

TABLE 1

IMMUNOLOGIC ABNORMALITIES IN NZB AND B/W MICE

Premature Development of Immune Competence
Excessive IgM Production by Newborn B Cells
Excessive Antibody Responses to Many Antigens
Relative Resistance to Immune Tolerance
Impaired Cellular Immunity
Production of Thymocytotoxic Antibody
Decreased Suppressor Function
Decreased Thymic Humoral Factor
Decreased Ly 123+ Cells and Impaired Feedback Regulation

TABLE 2

MODES OF THERAPY IN NZB AND NZB/NZW $F_1$ MICE:
TREATMENT

Corticosteroids
Immunosuppression
"Suppressor" Lymphoid Cells
Thymus Grafts
Thymic Humoral Factors
Protein or Calorie Restriction
Nucleoside - Isologous IgG Tolerance
SIRS (Con A Induced Suppression)
Prostaglandin $E_1$
Androgens

ACKNOWLEDGEMENTS

Our research was supported by the Medical Research Service of the Veterans Administration, by U.S. Public Health Service grant AM 16140, by a grant from the Kroc Foundation, and a contract from the State of California.

REFERENCES

1. Howie, J.B., and Helyer, B.J. (1968)  The immunology and pathology of NZB mice.  Adv. Immunol., 9, 215-266.

2. Talal, N. (1970)  Immunologic and viral factors in the pathogenesis of systemic lupus erythematosus.  Arth. Rheum., 13, 887-893.

3. Levy, J.A. (1974)  Autoimmunity and neoplasia: the possible role of C-type viruses.  Am. J. Clin. Path., 62, 258-280.

4. Warner, N.L. (1973)  Genetic control of spontaneous and induced anti-erythrocyte autoantibody production in mice.  Clin. Immunol. Immunopath., 1, 353-363.

5. Yoshiki, T., Mellors, R.C., Strand, M., and August, J.T. (1974) The viral envelope glycoprotein of murine leukemia virus and the pathogenesis of immune complex glomerulonephritis of New Zealand mice.  J. Exp. Med., 140, 1011-1027.

6. Talal, N. (1976)  Disordered immunologic regulation and autoimmunity.  Transplant. Rev., 31, 240, 1976.

7. Evans, M.M., Williamson, W.G., and Irvine, W.J. (1968)  The appearance of immunological competence at an early age in New Zealand Black mice.  Clin. Exp. Immunol., 3, 375-383.

8. Gazdar, A.F., Beitzel, W., and Talal, N. (1971)  The age related response of New Zealand mice to a murine sarcoma virus.  Clin. Exp. Immunol., 8, 501-509.

9. Staples, P.J., and Talal, N. (1969)  Relative inability to induce tolerance in adult NZB and NZB/NZW $F_1$ mice.  J. Exp. Med., 129, 123-139.

10. Playfair, J.H.L. (1968)  Strain differences in the immune response of mice.  I. The neonatal response to sheep red cells.  Immunology, 15, 35-50.

11. Steinberg, A.D., Baron, S.H., and Talal, N. (1969)  The pathogenesis of autoimmunity in New Zealand mice.  I. Induction of anti-nucleic acid antibody in polyinosinic·polycytidylic acid.  Proc. Nat . Acad. Sci. U.S.A., 63, 1102-1107.

12. Cerottini, J.C., Lambert, P.H., and Dixon, F.J. (1969)  Comparison of the immune responsiveness of NZB and NZBxNZW $F_1$ hybrid mice with other strains of mice.  J. Exp. Med. 130, 1093-1095.

13. Weir, D.M., McBride, W., and Naysmith, J.D. (1968)  Immune response to a soluble protein antigen in NZB mice.  Nature (London), 219, 1276-1277.

14. Staples, P.J., Steinberg, A.D., and Talal, N. (1970)  Induction of immunological tolerance in older New Zealand mice repopulated with young spleen, bone marrow, or thymus.  J. Exp. Med., 131, 123-128.

15. Leventhal, B.G., and Talal, N. (1970)  Response of NZB and NZB/NZW spleen cells to mitogenic agents.  J. Immunol. 104, 918-923.

16. Cantor, H., Asofsky, R., and Talal, N. (1970) Synergy among lymphoid cells mediating the graft-vs-host response. I. Synergy in graft-vs-host reactions produced by cells from NZB/Bl mice. J. Exp. Med., 131, 223-234.

17. Shirai, T., and Mellors, R.C. (1971) Natural cytotoxic autoantibody and reactive antigen in New Zealand black and other mice. Proc. Nat. Acad. Sci. U.S.A. , 68, 1412-1415.

18. Gershon, R.K. (1974) T cell control of antibody production, in Contemporary Topics in Immunobiology, Vol. 3, Cooper, M.D. and Warner, N.L., eds., Plenum Press, New York and London, pp. 1-40.

19. Allison, A.C., Denman, A.M., and Barnes, R.D. (1971) Cooperating and controlling functions of thymus-derived lymphocytes in relation to autoimmunity. Lancet, 2,135-140.

20. Barthold, D.R., Kysela, S., Steinberg, A.D. (1974) Decline in suppressor T cell function with age in female NZB mice. J. Immunol., 112, 9-16.

21. Dauphinee, M.J., and Talal, N. (1973) Alteration in DNA synthetic response of thymocytes from different aged NZB mice. Proc. Nat. Acad. Sci. U.S.A. Part II, 70, 3769-3773.

22. Bach, J.F., Dardenne, M., and Salomon, J.C. (1973) Studies of thymus products. IV. Absence of serum thymic activity in adult NZB and (NZBxNZW) $F_1$ mice. Clin. Exp. Immunol. 14, 247-256.

23. Talal, N., Dauphinee, M.J., Pillarisetty, R., et al. (1975) Effect of thymosin on thymocyte proliferation and autoimmunity in NZB mice. Ann. N.Y. Acad. Sci. 249, 438-450.

24. Cantor, H., McVay-Boudreau, L., Hugenberger, J., Naidorf, K., Shen, F.W., amd Gershon, R.K. (1978) Immunoregulatory circuits among T cell sets. II. Physiologic role of feedback inhibition in vivo: absence in NZB mice. J. Exp. Med., 147, 1116-1125.

25. Mellors, R.C. (1966) Autoimmune and immunoproliferative diseases of NZB/Bl and hybrids. Int. Rev. Exp. Pathol. 5, 217.

26. Sugai, S., Pillarisetty, R.J., and Talal, N. (1973) Monoclonal macroglobulinemia in NZB/NZW $F_1$ mice. J. Exp. Med., 138, 989-1002.

27. Steinberg, A.D., Gelfand, M.C., Hardin, J.A., et al. (1975) Therapeutic studies in NZB/W mice. III. Relationship between renal status and efficacy of immunosuppressive drug therapy. Arth. Rheum., 18, 9-14.

28. Hahn, B.H., Knotts, L., Ng, M., et al. (1975) Influence of cyclophosphamide and other immunosuppressive drugs on immune disorders and neoplasia in NZB/NZW mice. Arth. Rheum., 18, 145-152.

29. Walker, S.E., and Bole, G.G., Jr. (1975) Selective suppression of autoantibody responses in NZB/NZW mice treated with long-term cyclophosphamide. Arth. Rheum., 18, 265-272.

30. Wolf, R.E., and Ziff, M. (1976) Transfer of spleen cells from young to aging NZBxNZW $F_1$ hybrid mice: effect on mortality, antinuclear antibody, and renal disease. Arth. Rheum., 19, 1353-1357.

31. Morton, R.O., Goodman, D.G., Gershwin, M.E., et al. (1976) Suppression of autoimmunity in NZB mice with steroid-sensitive X-radiation-sensitive syngeneic young thymocytes. Arth. Rheum., 19, 1347-1350.

32. Kysela, S., and Steinberg, A.D. (1973)  Increased survival of NZB/W mice given multiple syngeneic young thymus grafts.  Clin. Immunol. Immunopathology 2, 133-136.

33. Gershwin, M.E., Steinberg, A.D., Ahmed, A., et al. (1976)  Study of thymic factors.  II. Failure of thymosin to alter the natural history of NZB and NZB/NZW mice.  Arth. Rheum., 19, 862-866.

34. Bach, M.A., and Niaudet, P. (1976)  Thymic function in NZB mice. II. Regulatory influence of a circulating thymic factor on antibody production against polyvinylpyrrolidone in NZB mice.  J. Immunol., 117, 760-764.

35. Fernandes, G., Yunis, E.J., and Good, R.A. (1976)  Influence of protein restriction on immune functions in NZB mice.  J. Immunol., 116, 782-790.

36. Borel, Y. (1976)  Isologous IgG-induced immunologic tolerance to haptens: a model of self versus non-self recognition.  Transplant. Rev.,31, 3-22.

37. Krakauer, R.S., Waldmann, T.A., and Strober, W. (1976)  Loss of suppressor T cells in adult NZB/NZW mice.  J. Exp. Med., 144, 662-673.

38. Zurier, R.B., Sayadoff, D.M., Torrey, S.B., et al. (1978)  Prostaglandin $E_1$ treatment of NZB/NZW mice: prolonged survival of female mice.  Arth. Rheum., in press.

39. Roubinian, J.R., Papoian, R., and Talal, N. (1977)  Effects of neonatal thymectomy and splenectomy on survival and regulation of autoantibody formation in NZB/NZW $F_1$ mice.  J. Immunol., 118, 1524-1529.

40. Roubinian, J.R., Papoian, R., and Talal, N. (1977)  Androgenic hormones modulate autoantibody responses and improve survival in murine lupus.  J. Clin. Invest. 59, 1066-1070.

41. Roubinian, J.R., Talal, N., Greenspan, J.S., Goodman, J.R., and Siiteri, P.K. (1978)  Effect of castration and sex hormone treatment on survival, anti-nucleic acid antibodies, and glomerulonephritis in NZB/NZW $F_1$ mice.  J. Exp. Med., in press.

42. Roubinian, J.R., Greenspan, J., and Talal, N.  Delayed androgen treatment prolongs survival in murine lupus (in preparation).

43. Stern, R., Fishman, J., Brusman, H., and Kunkel, H.G. (1977)  Systemic lupus erythematosus associated with Klinefelter's syndrome. Arth. Rheum., 20, 18-22.

DISCUSSION

WARNER:  Given that sex hormones do not influence the production
of autoantibodies, but do affect complex  deposition, could you
speculate further whether MHC genes that control complement or
testosterone levels may be involved.

TALAL:  In the mouse, there are several complement components
which are induced by androgen (including $C_4$ and the binding protein
for $C_4$).  $C_4$ (also called Ss) is the product of a gene located within
the MHC.  It is possible that mechanisms involving the loci and male
hormone could be implicated with the pathogenesis of murine lupus.
 In man, the disorder known as hereditary angioneurotic edema is
due to a deficiency of the $C_1$ esterase inhibitor.  This inhibitor
can be induced by androgen.  The weakly androgenic material Danazol
brings inhibitor concentration into the normal range and markedly
reduces the frequency of attacks in this disease.

GERSHON:  Bob Zurier has treated NZB or B/W mice with prosta-
glandins.  This treatment has significantly prolonged survival and
lessened autoimmune disease, even though it did not seem to affect
titers of autoantibodies.  John Kemp in my lab has tested Zurier's
treated mice for feedback suppression (FBS) and finds that the
ability to deliver FBS does not return after treatment.  **Thus** we
would conclude, like yourself, that disease can be ameliorated by
blocking the inflammatory consequences of autoantibodies without
necessarily affecting their development or appearance.

VOLPE:  I am a little disturbed by your use of Klinefelter's
syndrome (47 XXY) as an example of the effects of increased estrogen
in inducing autoimmune disease, since despite the gynecomastia and
gynecoid pelvis found in that condition, to my knowledge, there is
not an actual increase in estrogen.  Moreover, how would you account
for the increased incidence of organ-specific autoimmune disease in
other chromosomal defects such as Turner's syndrome (Gonadal dysgenesis, 45 X0) where estrogen is very low, or Down's syndrome
(mongolism) where the defect is at the 21st chromosome (with no
change in estrogen)?

TALAL:  Patients with Klinefelter's syndrome manifest an abnormality
of the hormone metabolism which subjects them to a hyperestrogenic
influence.  We have recently reported a pair of monozygotic twins
with Klinefelter's syndrome, one of whom had clinical myasthenia
gravis and the other lupus.  They both had antibodies to the acetylcholine receptor, but only the twin with lupus had antibodies to
DNA, to polyadenylic acid, and to lymphocytes.  This observation
suggests a role for environmental as well as genetic and hormone
factors in autoimmunity.
 In the other conditions, generalized autoimmune disease doesn't
develop and hormonal mechanisms may not be important.

GERSHWIN:  Do you have any data on the influences of sex hormones
in NZB/W $F_1$ mice on their age dependent changes of immune responsiveness, in particular mitogens, response to antigens and ability to
be tolerized?
 Have you performed similar experiments and have any observations
in NZB mice been made?

TALAL:  These experiments are currently in progress.

SVEJGAARD:  Dr. Talal, you found fewer immune complexes in the
kidneys but the same amount of antibodies in the blood in the
androgen-treated animals.  Do you have evidence that there is less

antigen in these animals or that they have a more sufficient phagocytosis of immune complexes which could prevent these from reaching
the kidneys?

TALAL:  It is not possible to accurately measure antigen, or even
to know what antigens to look for.  There may be an improved clearance
of immune complexes in the mice given delayed androgen treatment.
We are currently examining this question.

SCHWARTZ:  How do these two facts relate to hormone effect by
peripheral action?

TALAL:  In mice given androgen from the age of two weeks, the
thymus plays a role and autoantibodies are reduced.
In mice given **delayed** androgen, the hormone effects may be peripheral
and independent of the thymus.

MACKAY:  Can Dr. Talal comment on therapeutic implications for
human disease of hormonal therapy in mice.  It might be noted that
cyclophosphamide, sometimes used in treatment of human lupus, may
have damaging effects on gonads.

TALAL:  The effect of cyclophosphamide is immunosuppressive and
cytostatic.  One hopes to avoid damage to the gonads when using this
agent.
Androgen therapy for human lupus offers a promising new approach
to management.  We are currently evaluating several hormone analogues
and antimetabolites in our mice.  We hope to arrive at a therapeutic
strategy that can be tried in patients.

MECHANISM OF AUTOIMMUNE DISEASE IN NEW ZEALAND BLACK MICE

THOMAS M. CHUSED, HARALAMPOS M. MOUTSOPOULOS, SUSAN O. SHARROW, CARL T.
HANSEN AND HERBERT C. MORSE, III
National Institutes of Health, Bethesda, Maryland 20014

ABSTRACT

The B-cells of New Zealand Black (NZB) mice undergo spontaneous
polyclonal activation detectable at one week of age.  The B-cell activation occurs in the absence of thymus-processed cells which suggests
an immunologic lesion intrinsic to B-cells.  IgM, presumably anti-T-
cell autoantibody, can be detected within, but not on, peripheral T-
cells by four weeks of age and may affect their function.

INTRODUCTION

NZB mice, and the other autoimmune strains which have recently
become available, offer us the opportunity to work out the cellular and
genetic mechanism of autoimmune disease in a model system which we can
freely manipulate.  Most of the current hypotheses concerning the
mechanism of the spontaneous autoimmune disease of NZB mice invoke a
loss or malfunction of suppressor T-cells leading to the failure to
control forbidden clones of autoreactive B-cells[1-3].  Experiments
conducted in our laboratory during the past year suggest, as an alternative explanation of the disorder of NZB mice, that they have a primary abnormality of their B-cells which causes their spontaneous activation[4].  This paper will draw on our own experiments and those of
other investigators in an attempt to contruct a model which provides an
explanation for many of the observations which have been made on NZB
mice.

Since autoantibodies are the immediate cause of the clinical manifestations of NZB mice, the strategy adopted was to compare their B-
cells, the source of the autoantibodies, with B-cells from non-
autoimmune mice.  The questions posed were: 1) Are B-cell abnormalities present? 2) If so, at what age could they first be detected? 3)
Could they be dissociated from T-cell effects?

MATERIALS AND METHODS

Mice.  NZB/BlN, NZW/BlN and NFS/N, a normal inbred Swiss line, were
from the colonies of the National Institutes of Health.  NZB *nu/nu*

mice, being bred by Dr. Carl T. Hansen, were at the fourth backcross generation. C57Bl/6J were obtained from Jackson Laboratory.

Antisera. New Zealand White rabbits were immunized with Fab fragments prepared from pooled mouse IgG, isolated kappa chains, or the IgM myeloma, MOPC 104E. The anti-kappa antisera was kindly provided by Dr. Rose Mage. The anti-Fab and anti-kappa were affinity purified on columns containing covalently bound mouse Fab fragments and the anti-IgM on columns containing MPOC 104E and pooled serum IgM. F(ab')$_2$ fragments were prepared by pepsin digestion and fluoresceinated to give molar F/P ratios of approximately three.

Flow microfluorometry. This was performed on a FACS-II with a 1000 channel pulse height analyzer (Becton, Dickinson FACS Division, Mountain View, CA). The resulting plots give the number of cells per channel along the ordinate and the fluorescence or light scatter intensity along the abscissa.

RESULTS

B-cell surface immunoglobulin. The fluorescence profiles of NZB and NFS spleen cells stained with anti-kappa are shown in Figure 1. The anti-kappa reagent stains all surface immunoglobulin, which in mouse spleen consists almost exclusively of IgM and IgD[5]. The NZB profile contains a much larger proportion of B-cells which have decreased amounts of total surface immunoglobulin than the control. Staining with anti-Fab gives fluorescence profiles essentially identical to anti-kappa (data not shown). This finding is quite consistent and unique to NZB and its F$_1$ hybrids. Unmanipulated NZW, C57Bl/6 and a large number of other normal strains give profiles very similar to that of NFS. In contrast with the anti-kappa, which stains both IgM and IgD, Figure 2 shows that there is little difference in the IgM profiles of the two strains. We conclude that NZB spleen cells have markedly decreased levels of surface IgD. Using surface iodination and poly-acrylamide gel electrophoresis Cohen has reached the same conclusion[6]. Askonas has recently shown that mitogens can produce the same change in normal mice[7]. For this reason, the effect of injecting C57Bl/6 mice with a potent antigen, sheep erythrocytes, was examined (Figure 3). Two injections of 10$^8$ erythrocytes seven and three days before sacrifice .aused the same change in the total surface immunoglobulin profile as occurs spontaneously in NZB mice. Thus the decreased IgD on NZB B-cells is consistent with their being activated.

B-cell activation in NZB spleen. Our laboratory recently developed a very sensitive immunoradiometric assay for murine IgM and found that

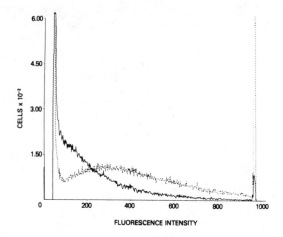

Fig. 1. Fluorescence distribution of total surface immunoglobulin of NZB (solid) and NFS (dashed) spleen cells stained with fluoresceinated, F(ab')$_2$ anti-kappa antibody. NZB spleen contains many more dull cells.

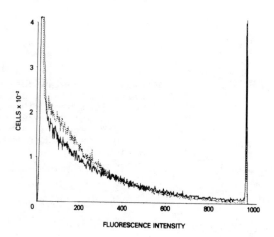

Fig. 2. Fluorescence distribution of surface IgM of NZB (solid) and NFS (dashed) spleen cells stained with fluoresceinated, affinity purified F(ab')$_2$ anti-IgM antibody. The two profiles are quite similar.

unstimulated spleen cells from NZB mice spontaneously release 40 to 100 times as much pentameric IgM as spleen cells from non-autoimmune controls during a short-term, four hour, *in vitro* incubation (Figure 4)[8]. Spontaneous IgM release by NZB spleen cells is greater than normal at birth and increases exponentially until six to eight weeks of age. NZB x NZW $F_1$ hybrids are less abnormal than NZB.

The fact that the secreted IgM is almost entirely pentameric suggested that it is the product of plasma cells rather than shed immunoglobulin receptors from the surface of B-cells. The high levels of NZB IgM secretion suggested that their B-cells may be nonspecifically, or polyclonally, activated. To test this the number of spontaneous direct

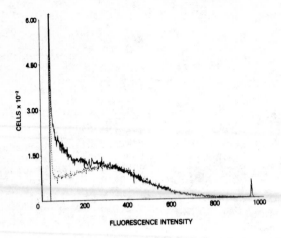

Fig. 3. Fluorescence distribution of total surface immunoglobulin of C57Bl/6J spleen cells stained with anti-Fab shows many more dull B-cells after sheep erythrocyte immunization (solid) than the saline injected control (dashed).

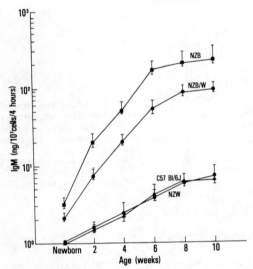

Fig. 4. Spontaneous secretion of IgM by spleen cells from NZB, NZB x NZW $F_1$, NZW and C57Bl/6J mice of different ages during a four hour *in vitro* incubation.

TABLE 1

SPONTANEOUS TNP-SPECIFIC DIRECT PLAQUE-FORMING CELLS
IN NZB AND NZW SPLEEN

| Mouse strain | Age (weeks) | | | |
|---|---|---|---|---|
| | 2 | 4 | 6 | 8 |
| | TNP pfc/$10^6$ spleen cells | | | |
| NZB | 4 | $17^a$ | $44^a$ | $40^a$ |
| NZW | 1 | 3 | 4 | 6 |

[a] Significantly different from NZW at p<.001

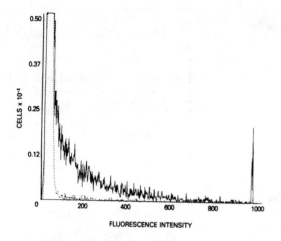

Fig. 5. Comparison of NZB (solid) and NFS (dashed) spleen cells stained with anti-Fab after fixation and examined by the FACS at low gain. The NZB spleen contained 2.64% positive cells and the NFS contained 0.24%. The negative cells are in the peak on the left.

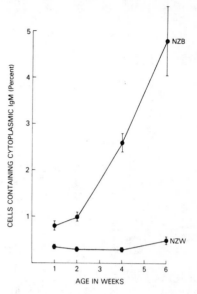

Fig. 6. Ontogeny of cells containing cytoplasmic IgM in the spleens of NZB and NZW mice. NZB spleen contains significantly more positive cells by one week of age.

plaque-forming cells to the hapten, trinitrophenol (TNP) was determined (Table 1). From the age of four weeks there is a significant increase in spontaneous TNP plaque-forming cells in NZB spleen. This has been confirmed by Izui and Dixon, who showed that anti-fluorescein plaque-forming cells are also increased[9].

Since immunoglobulin secreting cells have large amounts of cytoplasmic immunoglobulin, the increase in TNP plaque-forming cells led us to ask what proportion of NZB spleen cells contain cytoplasmic immunoglobulin. By greatly decreasing the gain of the FACS, only the very brightly stained cells containing large amounts of cytoplasmic immunoglobulin are detected. Such cells are not observed without fixation before staining. Fixed NZB and NFS spleen cells stained with anti-Fab are compared in Figure 5. The ontogeny of cells containing cytoplasmic IgM is shown in Figure 6. By one week of age NZB spleen contains significantly more plasma cells than the NZW control. By six weeks the adult levels of 5% in NZB and 0.5% in normals are attained. Similar results are obtained after staining with anti-Fab (data not shown).

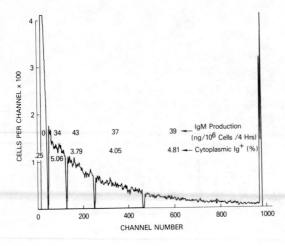

Fig. 7. NZB spleen cells stained with anti-IgM were sorted into an IgM negative pool and four pools with increasing amounts of total surface IgM. The IgM secretion and plasma cells content of each pool was determined.

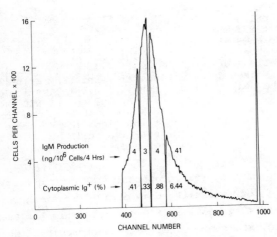

Fig. 8. NZB spleen cells were sorted by size, as determined by light scatter intensity, into four pools. Only the pool of large cells secreted IgM at a high rate and contained a high proportion of plasma cells.

This ten-fold increase in plasma cells in NZB spleen is the same order of magnitude as the increase in spontaneous TNP plaque-forming cells. We conclude that NZB B-cells are polyclonally activated by one week of age.

Properties of IgM secreting cells. The FACS was used to define the IgM secreting cells. Viable spleen cells from six week old NZB mice were sorted into a surface IgM negative pool and four pools with increasing amounts of surface IgM (Figure 7). Part of each pool was cultured for four hours and IgM secretion measured. Another portion was fixed and stained for plasma cells. No IgM was secreted by the surface IgM negative pool, while each IgM positive pool produced the same amount. Similarly, only 0.25% plasma cells were present in the

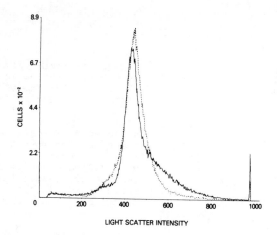

Fig. 9. Light scatter intensity distribution of NZB (solid) and NFS (dashed) immunoglobulin positive spleen cells. There are more large B-cells in the NZB spleen (channels 500 to 800).

surface IgM negative pool but each of the surface IgM positive pools contained four to five percent plasma cells. A similar experiment in which the cells were stained with anti-Fab gave the same result (data not shown). In contrast, Figure 8 demonstrates that when NZB spleen cells were sorted by size, as measured by light scatter, only the pool containing the largest cells secreted IgM and contained plasma cells. Thus the IgM secreting cells in NZB spleen are large but cannot be distinguised from non-secreting cells by the amount of surface immunoglobulin they bear.

Because the IgM secreting cells are large the size of surface immunoglobulin positive B-cells of NZB spleen was compared with NFS (Figure 9). There is a very noticable increase in the number of large B-cells in NZB spleen. In contrast, the surface immunoglobulin negative spleen cells, which are largely T-cells, have the same size profile in NZB and control (Figure 10). The lower peak is dead cells which have decreased light scatter.

Relationship of T-cells to B-cell activation. The fact that there is no evidence of T-cell blasts raises the very important question of the relationship of regulatory T-cells to the polyclonal B-cell activation. Are suppressor T-cells deficient? Is there excessive non-specific T-cell help? Or, are B-cells endogenously activated?

Two recent experiments performed in our laboratory address this issue. In the first, adult NZB and NZW mice were thymectomized, lethally irradiated and reconstituted with stem cells from fetal liver (Table 2). When examined 10 days later the spleens of the NZB mice

184

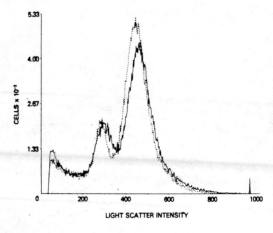

Fig. 10. Light scatter intensity distribution of NZB (solid) and NFS (dashed) immunoglobulin negative spleen cells (T-cells). There are no significant differences between the two strains. The lower peaks (channels 0 to 350) are debris and dead cells, respectively, which exhibit decreased light scatter.

TABLE 2

IgM PRODUCTION AND CYTOPLASMIC STAINING OF SPLEEN CELLS FROM
ADULT THYMECTOMIZED, LETHALLY IRRADIATED, NZB AND NZW MICE
RECONSITITUTED WITH SYNGENEIC FETAL LIVER CELLS

| Mouse strain | Treatment | Cytoplasmic Ig+ | IgM production |
|---|---|---|---|
| | | % | ng/$10^7$ cells/4 hr |
| NZB | Tx, 850R | 4.6 $\pm$ .6 | 183 $\pm$ 17 |
| | Sham Tx, 850R | 5.8 $\pm$ 1.2 | 199 $\pm$ 19 |
| | Sham Tx | 3.1 $\pm$ .4 | 206 $\pm$ 12 |
| NZW | Tx, 850R | .9 $\pm$ .1 | 8 $\pm$ 1 |
| | Sham Tx, 850R | 1.5 $\pm$ .3 | 9 $\pm$ 1 |
| | Sham Tx | .5 $\pm$ .0 | 11 $\pm$ 2 |

contained four to five times more plasma cells and secreted twenty
times as much IgM as the NZW controls. Thus, the B-cell activation of
NZB mice does not require the presence of radiosensitive T-cells. To
determine whether NZB B-cell activation requires thymus-processed
cells, in the second experiment homozygous nu/nu NZB mice were compared
with outbred NIH Swiss nu/nu mice (Table 3). NZB nu/nu mice have four
times as many plasma cells and secrete five times as much IgM. Thus
the spontaneous B-cell activation of NZB mice does not require the
presence of thymus-processed cells. Because the B-cell activation is

TABLE 3

IgM PRODUCTION AND CYTOPLASMIC STAINING OF SPLEEN CELLS FROM
*nu/nu* MICE WITH DIFFERENT GENETIC BACKGROUNDS

| Genetic background | Cytoplasmic Ig+ | IgM production |
|---|---|---|
| | % | ng/$10^7$ cells/4 hr |
| NZB | 8.1[a] | 247[b] |
| NIH | 1.7 | 44 |

[a] Significantly different from NIH at p=.01
[b] Significantly different from NIH at p=.03

independent of T-cells and is present as early as one week of age it is
likely to be intrinsic to the B-cells. A circulating B-cell stimulating factor has not been formally excluded, but the observation that
autoimmune disease is transferred by NZB marrow[10] implies that such a
putative factor must be produced by a marrow derivative. At this point
the most economical hypothesis is that NZB B-cells are intrinsically
activated.

Anti-T-cell autoantibody. What can be said about the T-cell abnormalities that have been described in NZB mice: the lack of graft-
versus-host suppressors[11] and of Ly 123+ cells[12], and the presence of
an *in vitro* primary cytotoxic response to minor histocompatability
antigens[13], among others? There are two possibilities. The first is
that NZB T-cells are deregulated or spontaneously activated in the same
way as B-cells. The second is that autoantibodies directed against T-
cell receptors may affect the function of intrinisically normal T-
cells. For example, Eichmann and Rajewsky have been able to prime T-
cells with anti-idiotype antibody[14]. It has been shown that NZB mice
spontaneously produce autoantibody to the Thy 1 surface antigen[15-17]
but it should be noted that the number of Thy 1 molecules on T-cells
appears to be be much greater than many other surface antigens and
receptors. In the presence of anti-Thy 1, autoantibodies directed
against other determinants which may critically affect T-cell function
may not be detected. Alternatively, the anti-Thy 1 may selectively
deplete suppressor cells, as Shirai, et al., have suggested[18].

Anti-T-cell autoantibodies do not attain appreciable titers in NZB
serum until relatively late in life. Could such autoantibody be
detected on the surface of T-cells with the FACS by staining with anti-
Fab and turning the gain of the instrument up to its maximum? Figure
11 shows that the answer is no. At this high gain, all the B-cells are

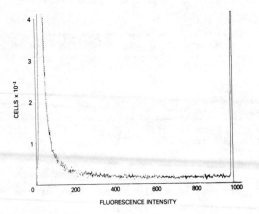

Fig. 11. Fluorescence distribution of spleen cells of NZB (solid) and NFS (dashed) stained with anti-Fab and examined with the FACS at maximum gain. No difference is observed. The B-cells are in the clipping channels on the right and the T-cells are in the peak on the left.

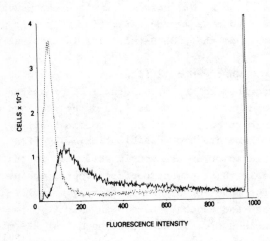

Fig. 12. Fluorescence distribution of fixed spleen cells of NZB (solid) and NFS (dashed) stained with anti-Fab and examined with the FACS at maximum gain. The T-cells are in the peaks on the left. NZB T-cells are shifted to right in comparison to the control.

so bright that they are off-scale and are collected in a peak in the clipping channels on the right while the T-cells remain on scale in the peak on the left. There is no difference between the NZB and control T-cells. Suprisingly, the result is different if the spleen cells are fixed before staining (Figure 12). Again, the T-cells are all in the peak on the left. The T-cells from six week old NZB spleen are much brighter than those from the control spleen. The same result is obtained after staining for mu-chains (data not shown). No immunoglobulin is found in or on NZB thymocytes. Thus peripheral T-cells in NZB mice contain detectable intracellular but not surface IgM. This is

POSTULATED MECHANISM OF AUTOIMMUNE DISEASE IN NZB MICE

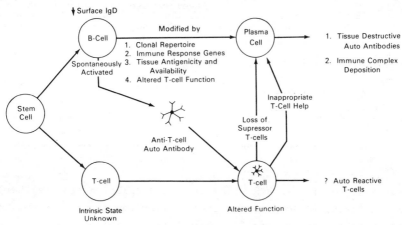

Fig. 13. Postulated mechanism of autoimmune disease in NZB mice. See text for details.

probably rapidly endocytosed anti-T-cell autoantibody. The immuno-globulin in NZB T-cells is present by four weeks of age (data not shown).

This result implies that NZB T-cells function in a sea of membrane-active autoantibody from at least four weeks of age. In order to determine whether NZB T-cells are intrinsically abnormal exposure to autoantibody must be prevented. Experiments performed to date have not satisfied this requirement.

DISCUSSION

The above results allow the construction of a model of the mechanism of autoimmune disease in NZB mice (Figure 13). B-cells are activated spontaneously. Whether T-cells are intrinsically abnormal, which would imply a defect of the common stem cell, is not known. The B-cell acti-vation is probably polyclonal at first. It is likely that secondary factors subsequently shape the spectrum of autoantibodies the B-cells produce. Such secondary modifying factors would include the clonal repertoire available, immune response genes, the specificity of anti-T-cell autoantibody produced and the antigenicity and accessibility of

TABLE 4

POSTULATED INTERACTION OF NZB AND NZW GENETIC TRAITS IN THE
PATHOGENESIS OF LUPUS-LIKE SYNDROME OF NZB x NZW $F_1$ HYBRID MICE

| Trait | Level of expression of trait in strain | | |
|---|---|---|---|
| | NZB | NZW | NZB x NZW $F_1$ |
| Primary (Spontaneous B-cell activation) | +++ | − | ++ |
| Secondary (*in vitro* anti-DNA pfc) | + | +++ | +++ |
| Clinical manifestation (DNA/anti-DNA immune complex nephritis) | + | − | +++ |

tissue antigens. In addition, altered T-cell function, either acquired
or intrinsic, would affect autoantibody production. Perhaps the most
important point in regard to human patients is that the clinical manifestations of autoimmune disease, tissue destructive autoantibodies and
the deposition of immune complexes, are several steps removed from the
primary immunologic lesion.

The interaction of NZB and NZW traits offers an illustrative example
of this point (Table 4). NZB B-cells have the primary defect of spontaneous, uncontrolled activation. NZW is normal. The NZB x NZW $F_1$
hybrid has an intermediate level of IgM secretion. The secondary
factor is the propensity to make anti-DNA antibodies. Bell, et al.,
have recently shown that in the artificially stimulatory conditions of
tissue culture NZW spleen cells spontaneously make many more anti-
single stranded DNA plaque-forming-cells than NZB mice[18]. The combination of B-cell activation from NZB and the potential to make anti-DNA
antibody from NZW combine in the $F_1$ to produce a mouse that, while less
immunologically activated, produces more anti-DNA antibodies and
succumbs to immune-complex glomerulonephritis considerably earlier than
does the NZB.

SUMMARY

B-cell surface IgD is decreased spontaneously in NZB mice and after
B-cell activation in normal mice. NZB splenic B-cells secrete more
than 40 times as much pentameric IgM as B-cells from normal strains.
NZB mice have a significant increase in the number of spleen cells
containing large amounts of cytoplasmic IgM by one week of age. The
coordinate increase of spontaneous TNP plaque-forming cells and plasma

cells suggests that the spontaneous B-cell activation is initially polyclonal. The IgM secreting cells are large and surface immuno-globulin positive but cannot be distinguished from non-secreting cells by their level of total surface immunoglobulin. The light scatter size distribution of T-cells is not altered in NZB spleen. B-cell activation in NZB mice occurs in the absencce of thymus-processed cells. IgM, presumably endocytosed anti-T-cell autoantibody, can be detected within peripheral NZB T-cells by four weeks of age. This may affect T-cell function. Spontaneous polyclonal activation of B-cells in NZB mice is a major factor in the pathogenesis of their autoimmune disease. The intrinsic state of NZB T-cells is unknown.

## ACKNOWLEDGEMENTS

We appreciate the technical assistance of Elinor Brown, Harry Goodman and Melvin Harding. We are particularly indebted to John Powell, Martin Miller and Robert Romanoff for the development of the computer software used to process the FACS data.

## REFERENCES

1. Allison, A.C., Denman, A.M. and Barnes, R.D. (1971) Lancet, 2, 135-140.

2. Krakauer, R.S., Waldmann, T.A. and Strober, W. (1976) J. Exp. Med., 144, 662-673.

3. Barthold, D.B., Kysela, S. and Steinberg, A.D. (1974) J. Immunol., 112, 9-16.

4. Chused, T.M., Moutsopoulos, H.M., Sharrow, S.O. and Mond, J.J. (1978) Arth. & Rheum., in press.

5. Vitetta, E.S. and Uhr, J.W. (1975) Science, 189, 964-969.

6. Cohen, P.L., Ligler, F.S., Ziff, M. and Vitetta, E.S. (1978) Arth. & Rheum., in press.

7. Bourgois, A., Kitajima, K., Hunter, I.R. and Askonas, B.A. (1977) Eur. J. Immunol., 7, 151-153.

8. Moutsopoulos, H.M., Boehm-Truitt, M., Kassan, S.S. and Chused, T.M. (1977) J. Immunol., 119, 1639-1644.

9. Izui, S. and Dixon, F.J. (1978) Fed. Proc. 37, 1373.

10. Morton, J.I. and Siegel, B.V. (1974) Proc. Natl. Acad. Sci. USA, 71, 2162-2165.

11. Hardin, J.A., Chused, T.M. and Steinberg, A.D. (1973) J. Immunol., 111, 650-651.

12. Cantor, H., McVay-Boudreau, L., Hugenberger, J., Naidorf, K., Shen, F.W. and Gershon, R.K. (1978) J. Exp. Med. 147, 1116-1125.

13. Botzenhardt, U., Klein, J. and Ziff, M. (1978) J. Exp. Med. 147, 1435-1448.

14. Eichmann, K. and Rajewsky, K. (1975) Eur. J. Immunol. 5, 661-666.

15. Shirai, T. and Mellors, R.C. (1974) Proc. Natl. Acad. Sci., 68,

190

1412-1415.

16. Shirai, T. and Mellors, R.C. (1972) Clin. Exp. Immunol. 12, 133-152.

17. Parker, L.M., Chused, T.M. and Steinberg, A.D. (1974) J. Immunol. 112, 285-292.

18. Shirai, T., Hayakawa, K., Okumura, K. and Tada, T. (1978) J. Immunol., 120, 1924-1929.

19. Bell, D.A., Roder, J.C. and Singhal, S.K. (1978) Arth. & Rheum., in press.

DISCUSSION

CANTOR:   The central point of your thesis is that NZB disease is due to a primary B cell abnormality.  This is based on elevated IgM but not IgG production of NZB B cells in T-depleted mice.  This is not surprising if the NZB environment combines "inappropriate" antigens which stimulate IgM (but not IgG) secretion.  Your view would be strengthened if NZB/nude mice or NZB "B" mice develop IgG autoantibodies and autoimmune disease.  Do they?

CHUSED:   The fact that NZB marrow transfers hypergammaglobulinemia to histocompatible mice while NZB given normal marrow do not exhibit hypergammaglobulinemia argues against "inappropriate antigens" in the envirnoment.
I feel that the role of T cells in the switch of NZB B cells from IgM to IgG needs to be examined as you suggest.  The rate of development of autoantibodies, particularly of IgM type such as anti-thymocyte antibody in "B" NZB mice, is currently being studied.

TALAL:   In addition to abnormalities of T and B cells, there are also defects in accessory cells.  Dr. Warner, and Dr. McCombs in my laboratory, demonstrated impaired function of spleen macrophage in NZB mice detected by an inability to support in vitro antibody synthesis.  It seems likely that a defect in a stem cell may be transmitted into three differentiated cell lines:  T cells, B cells, and macrophages.

COHEN:   Your conclusion of a primary T-independent defect in NZB B cells rests in part on two kinds of experiments using NZB mice depleted of mature T lymphocytes: adult, thymectomized, lethally-irradiated mice reconstituted with fetal liver cells, and NZB "nude" mice.  However, a case for T cell involvement in both of these cases can still be made.  For example, some types of T cells have been shown to function for a number of days following lethal irradiation.  Such radioresistant T cells could influence the abnormal activation of the B cells you noted.  Furthermore, some investigators believe that prethymic T stem cells can have regulatory functions as "NK cells" or possibly as "M cells".  An abnormality of such prethymic cells in NZB "nude" mice also could be related to the B cell abnormality.

CHUSED:   Although T cells that suppress B cell responses are generally considered radiosensitive and "NK" or "M" cells are not usually credited with suppression capabilities of thymic processed cells, I agree that if both of the two possibilities you suggest were true, our results could be explained by something other than a primary B cell abnormality.  However, when taken together with the fetal liver culture results of Datta and Schwartz, I feel the evidence is quite compelling.

DATTA:   In adoptive transfer systems where NZB cells or stem cells from marrow were transferred to thymectomized lethally irradiated histocompatible nonautoimmune strains (BALB/c or DBA/2), they went on to produce autoantibody (Deheer et al., J. Immunol.,1978; Siegel and Morton, Proc. Natl. Acad. Sci., 1974).

GERSHON:   What were your criteria for establishing that the cells with cytoplasmic Ig were T cells?

CHUSED:   Fixation does not remove surface immunoglobulin from the B cells which still stain more brightly than T cells; 40-50% of the cells are in the dull staining peaks on the left and by exclusion must be the T cells.

GENETIC, VIRAL AND IMMUNOLOGIC ASPECTS OF AUTOIMMUNE DISEASE IN NZB MICE

SYAMAL K. DATTA AND ROBERT S. SCHWARTZ

Department of Medicine and Cancer Research Center,
Tufts University School of Medicine, Boston, Massachusetts   02111

INTRODUCTION

Genetic, immunologic, viral and environmental factors are all implicated in the autoimmune disease and lymphoid malignancy that develop spontaneously in NZB mice[1,2]. However those factors which are fundamentally important in the etiology of NZB disease have not been clearly elucidated.  Genetic analyses of NZB mice employing improved virologic techniques, have revealed to us that viral genes segregate independently of "autoimmunity genes".  This has led to a clearer understanding of these factors and changed our approach to elucidate the fundamental nature of the defects in New Zealand mice.

Several studies have implicated xenotropic C-type RNA viruses in the etiology and pathogenesis of the autoimmune disease in NZB mice[2,3].  NZB mice express xenotropic viruses in high titers throughout life[4-6].  Viral antigens and antiviral antibodies are detected in the renal lesions of NZB and B/W mice along with antibodies to DNA[3].  It has been proposed that the xenotropic virus causes disturbed immunoregulation and autoimmunity in NZB mice, possibly by provoking an immune response against virus-producing thymocytes[2].

The presence of virus particles and viral antigens, however, do not prove that xenotropic viruses cause the disease of NZB mice.  Transmission of autoimmunity has not been achieved with cell-free filtrates from NZB mice, nor by transplanting NZB lymphomas which are known to produce large numbers of virus particles[7].  The explanation for these results came later.  Although expressed by mouse cells, the xenotropic viruses cannot enter and infect mouse cells, it can productively infect only cells of heterologous species[2,4].

We therefore attempted to ascertain the relationship between xenotropic virus and autoimmunity in NZB mice by genetic analyses.  This approach became feasible by the demonstration that the high grade expression of xenotropic virus in NZB mice is a genetically determined trait[5,6].

## I. IDENTIFICATION OF TWO XENOTROPIC VIRUS INDUCING LOCI (*Nzv-1 and Nzv-2*) IN NZB MICE

A highly sensitive, quantitative and rapid infectious center assay was developed to detect xenotropic viruses, the only kind of virus expressed by NZB mice. The assay reflected the *in vivo* virologic status of the animals and since it was done by biopsy of the spleens, the animals could be aged and time-dependent correlations between expression of virus and development of autoimmunity obtained[5,6].

Xenotropic virus was detected in high titers in all NZB mice tested (age range 1-12 months) by this assay[5,6]. Various non-autoimmune strains were tested and crossed with NZB mice. In all $F_1$ hybrids a similar picture emerged-dominant, highly penetrant expression of the NZB viral phenotype. Among the non-autoimmune strains, the SWR mice consistently failed to express infectious xenotropic and ecotropic viruses spontaneously[5,6,8,9]. Moreover, SWR mice never developed signs of autoimmunization. Thus a strain radically different from the NZB phenotype in virologic and autoimmune status was found. This provided the opportunity to dissect virus expression and autoimmunization in the progeny of crosses between NZB and SWR mice.

Quantitative xenotropic virus assays were done in the progeny of NZB and SWR crosses. Two independently segregating autosomal dominant loci (tentatively named, *Nzv-1 and Nzv-2*) determined expression of infectious xenotropic virus. The progeny mice segregated into high-virus, low-virus and virus-negative phenotypes. The segregation ratios suggested that one of the virus inducing loci, *Nzv-1* was responsible for high grade expression of xenotropic virus and the other, *Nzv-2* determined low titers of xenotropic virus. The high-virus phenotype occured in mice homozygous or heterozygous for the dominant allele of *Nzv-1*. When both *Nzv-1* and *Nzv-2* are present the low-virus phenotype as determined by *Nzv-2* would be masked by the high-titered virus expression caused by *Nzv-1*. Therefore the low-virus phenotype occured in animals that were either homozygous or heterozygous for the dominant allele of *Nzv-2* but had the recessive allele to *Nzv-1*. Virus negative mice had simultaneous homozygosity of the recessive alleles to *Nzv-1* and *Nzv-2*. Thus, the $F_1$ and

$F_1$ x NZB backcross mice were always high-virus expressors (titers similar to the NZB parent) whereas the $F_1$ x SWR backcross and $F_2$ progeny segregated into three phenotypes: high-virus, low-virus and virus-negative(undetectable xenotropic virus as in the SWR parent)[5]. Further, selective backcrosses and intercrosses using the low-virus and high-virus first backcross progeny and the virus-negative SWR parent have confirmed that *Nzv-1* and *Nzv-2* are independent autosomal dominant loci specifying high and low-virus phenotypes respectively[6].

We emphasize that these viral and genetic studies, in contrast to others, were not done on serially passaged[10] or chemically treated[11,12] fibroblasts but on fresh spleen cells. Therefore, we are dealing here not with induction of virus in tissue culture cells *in vitro* but with the actual virologic phenotype of the intact animal [5,6,13].

Repeated tests by splenic biopsy showed that the virologic phenotypes of the progeny mice were stable for many months[6].

In conclusion, NZB mice are homozygous for dominant alleles of two loci that determine spontaneous expression of xenotropic virus: *Nzv-1* and *Nzv-2*. By contrast, SWR mice are homozygous for recessive alleles at both loci. Since structural genes of xenotropic virus are integrated into the chromosomes of virtually all mice including the virus negative SWR strain[14,15] we favor the interpretation that *Nzv-1* and *Nzv-2* are regulatory genes that promote the continuous expression of endogenous viral genomes.

II.   INDEPENDENT SEGREGATION OF THE XENOTROPIC VIRUS INDUCING GENES FROM "AUTO-IMMUNITY GENES" IN NZB MICE

The relationship between expression of infectious xenotropic virus and the development of autoimmunization was studied in the progeny of the crosses between NZB and SWR mice[16]. Autoantibodies (anti-erythrocyte and anti-DNA antibodies), immune complex-deposit nephritis and lymphomas developed in the progeny mice. The virologic phenotype of the animal could be dissociated from the presence of either autoantibodies or nephritis. For example, mice that expressed titers of virus as high as the NZB parent failed to develop signs of autoimmunization even up to 24 months of age. By contrast, some ($F_1$ x SWR) backcross and $F_2$ mice that expressed low titers of virus developed autoimmune disease. Furthermore, a proportion of virus negative mice produced autoantibodies and were found to have typical immune-deposit nephritis. No viral antigens could be detected in the renal lesions of such virus-negative animals[16].

The incidence of nephritis in $F_1$ x SWR backcross mice was significantly higher than in the $F_1$ x NZB backcross or $F_1$ mice. This indicated that genes from the ostensibly normal SWR parent contributed to the development of nephritis in the backcross [16].

The dissociation between expression of xenotropic virus and production of auto-antibodies was not a peculiar feature of SWR x NZB crosses, similar results were found in crosses between NZB and B10·A, C57BL/6 and AKR mice[16].

III.   SEGREGATION OF GENES DETERMINING HIGH SERUM gp70 LEVELS FROM XENOTROPIC VIRUS
       INDUCING GENES AND IMMUNOPATHOLOGY IN NZB x SWR CROSSES

gp70 is the major glycoprotein component of the envelope of C-type RNA viruses. It is found in tissues and serum of virtually all mice[17,18]. The serum gp70 of all mice structurally resembles the envelope gp70 of the NZB xenotropic virus, but occurs independently from complete retrovirus particles[18]. In most strains including SWR the level of serum gp70 is low[17-19]. However, NZB mice express extremely high levels of this viral antigen in their serum due to an abnormally high synthetic rate[3,17].

In collaboration with Dr. Frank Dixon's Laboratory we determined the serum gp70 levels in NZB and SWR mice and the progeny of their crosses[19]. The serum gp70 values segregated to "NZB-like" and "SWR-like" levels in the progeny of the crosses. A complex genetic mechanism determined the inheritence of NZB-like serum gp70 levels. We found that the factors determining the expression of this retroviral protein were inherited independently of the genes (*Nzv-1* and *Nzv-2*) that determined the expression of infectious xenotropic virus. Autoimmune disease, including immune deposit nephritis could be dissociated from the levels of serum gp70 and from deposition of gp70 in the kidneys[19].

By contrast with the dissociation between expression of xenotropic virus, high serum gp70 levels and glomerulo-nephritis, the presence of antibodies to DNA and

increased levels of circulating immune complexes did correlate with the development
of renal lesions[16,19].

IV.  VIRUS EXPRESSION, SERUM gp70 LEVELS AND DEVELOPMENT OF MALIGNANT LYMPHOMAS IN
NZB x SWR CROSSES

A significant correlation was found between high-grade expression of serum gp70
and the presence of lymphomas in the progeny mice[19].  90% of these crosses with
lymphoid malignancies, also had the high-virus phenotype (i.e. expressed NZB-like
titers of xenotropic virus)[16].  But genotypes of the animals were important in determining the lymphoma  incidence.  $F_1$ and ($F_1$ x NZB) backcross progeny (these are all
high-virus) had a much higher incidence of lymphomas than the progeny of $F_2$ and
($F_1$ x SWR) backcross that had the high-virus phenotype[16].

Thus, we can conclude that the genes that determine the expression of infectious
xenotropic virus and the retroviral glycoprotein gp70 in NZB mice segregate independently from those that are involved in the autoimmune disease of these animals.  This
means that, although xenotropic virus and viral antigens may *participate secondarily*
in the formation of immunopathologic lesions in the NZB mouse or its crosses, they
are not a primary requirement.

V.  INHERITENCE OF AUTOIMMUNITY AND ABNORMAL LYMPHOCYTE FUNCTIONS IN NZB CROSSES

Identification and enumeration of "autoimmunity genes" in NZB mice have not been
achieved because the pattern of their inheritence is complex[20-24].  We and others
have shown that the expression of autoimmunity in NZB crosses is strongly influenced
by genes inherited from the "normal" parent[16,21,22].  This modification not only involves the time of onset and severity of the disease but also the nature of the
lesion.  Multiple genes that segregate independently appear to influence the development of each autoimmune disease marker[24].  This complexity is further exemplified
when cells of the immune system in NZB mice are analyzed.

Although attempts have been made to pinpoint a fundamental cellular defect that
is responsible for the autoimmune disease, the evidence is in favor of multiple
defects.

An age-dependent deficiency of different varieties of suppressor T cells in NZB
mice has been implicated to be a primary defect, causing deregulation of the immune

system and autoimmune disease[25-28]. However, this and other defects of T cell function may be secondary to the early degenerative changes in the thymic epithelium seen in these mice[29]. The appearance of natural thymocytotoxic autoantibodies which cause a loss of suppressor T cells, can also be dissociated from the production of anti-DNA antibodies in NZB mice[30]. Other authors, including our laboratory, failed to find a consistent decline of suppressor T cell function in old NZB mice[31-33]. Moreover, disturbance of suppressor T cell function may occur as a secondary phenomenon consequent to a primary B cell problem[34].

Certain abnormalities of B and T cells are detected in NZB mice much earlier than the loss of suppressor T cells[35-37]. We have also found evidence for hyperactive B-cells in NZB mice throughout life, beginning in the fetal stage of development. This B cell hyperactivity of NZB mice is not a secondary phenomenon due to any deficiency of suppressor T cells or excessive helperTcell activity and it is genetically transmitted in NZB x SWR crosses[33]. (Manny, N., Datta, S.K. and Schwartz, R.S. in preparation). Adoptive transfer studies by DeHeer et. al. also showed that B cells of NZB mice have the inherent potential of anti-erythrocyte autoantibody formation even in a normal host[38]. These studies imply a *primary genetically determined defect* at the *B cell level* that makes NZB mice refactory to immunoregulatory influences.

Others have demonstrated abnormalities in hematopoietic stem cells[39] and macrophages in NZB mice[40].

Thus widespread, apparently primary and independent defects in immunocytes some of which have been shown to be genetically transmitted[23,33] occur in NZB mice.

## INTERPRETATION AND SIGNIFICANCE OF THESE RESULTS

The xenotropic viruses are expressed in low titers by many mouse strains[2,5]. Indeed, most vertebrates appear to express this class of endogenous viruses while in the embryonic stage and they have been seen in placentas of many species[14]. Every mouse strain tested contains complete copies of the xenotropic provirus genome integrated in its chromosomes[14,15]. The genomic RNA of xenotropic class of viruses also called endogenous class I viruses) have been shown by molecular hybridization to share a high degree of homology to DNA from normal mouse cells[14,15]. This has been shown for inbred and wild mice, and even in mice that fail to express the complete virus. The expression of xenotropic viral genes occur in many cells undergoing normal differentiation and proliferation e.g. in thymocytes[41,42] lymphocytes[43,44], male genital tract and sperm[17], embryo, placenta[45], hematopoietic cells[46] and possibly regenerating liver[47]. The retroviral glycoprotein, gp70, structurally similar to the glycoprotein in the envelope of NZB xenotropic virus, exists as a normal serum protein in virtually all mice, irrespective of their virologic status[18]. Our studies confirmed that the expression of this retroviral gene product in the serum, as free gp70, is independent of the expression of complete retrovirus particles[19]. The normal SWR mouse although consistently xenotropic virus-negative when tested for complete infectious virus, does have the P12 of xenotropic virions in its cells[9] and expresses low levels of gp70 in its serum[19]. The endogenous xenotropic viral genes thus, have been conserved and have become part of the genetic makeup of all mice[14, 17,18]. They are therefore, under the control of similar regulatory mechanisms as normal cellular genes. NZB mice, which express these genes continously, may have abnormal regulatory genes.

Thus the xenotropic virus of NZB mice may be considered an "auto-virus" and gp70 a potential autoantigen. These viruses may play a secondary role in the autoimmune disease of NZB mice by supplying autoantigens, but in this regard they seem no different from erythrocyte autoantigens or DNA. The NZB model should therefore be distinguished from other models of immunologically mediated disease in which a foreign agent exogenous to the host incites an immune response (e.g., Aleutian disease of mink equine infectious anemia, lymphocytic choriomeningitis virus)[48]. Our results indicate that whether or not an autoimmune reaction occurs against *endogenous* (xenotropic) viral antigens or other antigens depends on genes ("autoimmunity" genes) that are distinct from viral genes. This interpretation sets the NZB model apart from the type of immune complex-mediated injury involving infection by *exogenous* viruses in which case the infectious agent participates in both the etiology and pathogenesis of immunological lesions.

Finally, these studies indicate that NZB mice have abnormalities in the expression of at least three independently segregating genetic traits: 1) those

regulating the endogenous proviral genes of the xenotropic virus  2) the genes regulating serum gp70 synthesis and  3) multiple genetically prescribed defects of B
cells, stem cells, thymic epithelium (plus T cells) and macrophages.  We have also
presented data indicating that there is no cause and effect relationship between
these genetic abnormalities but the fact that they are manifested simultaneously in
the NZB mice, is a clue to the fundamental nature of the defect in these animals
(Figure 1).

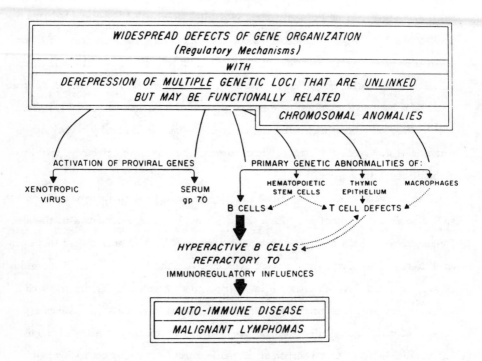

Figure 1.  *The NZB disease is a disorder of gene regulation.*

The NZB mice may have widespread defects in gene organization or genetic regulatory
mechanisms[49] with derepression of different cellular genes (including "proviral
genes" and "auto-antibody genes"), that are unlinked but may be functionally related.
The consistent chromosomal abnormalities that have been described in both adult
spleen cells and embryo cells of NZB mice may be also related to this fundamental

genetic disturbance[50,51]. A similar situation exists in humans with inherited diseases like Ataxia Telangiectasia[52] and Dysgammaglobulinemia[53], where chromosomal abnormalities are associated with lymphocyte dysfunction, immunodeficiency, auto-immune disease and lymphoid malignancy.

*ACKNOWLEDGEMENTS*

*This work was supported by U.S. National Institutes of Health Grant Nos. CA-19575 and CA-10018. Dr. Syamal K. Datta is the recepient of a Faculty Research Award from the American Cancer Society (FRA-#180). We thank Mrs. Bernice Kus for typing the manuscript.*

REFERENCES

1. Talal, N. (1976) Transplant. Rev., 31, 240-263.

2. Levy, J.A. (1976) Biomedicine, 24, (2), 84-93.

3. Yoshiki, T., Mellors, R.C., Strand, M. et. al. (1974) J. Exp. Med., 140, 1011-1027.

4. Levy, J.A., Kazan, P., Varnier, O., et. al. (1975) J. Virol., 16, 844-853.

5. Datta, S.K. and Schwartz, R.S. (1976) Nature, 263, 412-415.

6. Datta, S.K. and Schwartz, R.S. (1977) Viology, 83, 449-452.

7. East, J. (1970) Prog. Exp. Tumor Res., 13, 84-134.

8. Datta, S.K. and Schwartz, R.S. (1976) Eur. J. Cancer, 12, 977-988.

9. Stephenson, J.R., Reynolds, R.K., Tronick, S.R. et. al. (1975) Virology, 67, 404-414.

10. Levy, J.A. (1974) Am. J. Clin. Pathol., 62, 258-280.

11. Stephenson, J.R. and Aaronson, S.A. (1974) Proc. Nat. Acad. Sci., U.S.A., 71, 4925-4929.

12. Kozak, C. and Rowe, W.P. (1978) Science, 199, 1448-1449.

13. Melief, C.J.M., Datta, S.K., Louie, S. et. al. (1975) Proc. Soc. Exp. Biol. Med., 149, 1015-1018.

14. Gillespie, D. and Gallo, R.C. (1975) Science 188, 802-811.

15. Chattopadhyay, S.K., Hartley, J.W. and Rowe, W.P.: Personal Communication.

16. Datta, S.K., Manny, N., Andrzejewski, C., et. al. (1978) J. Exp. Med. 147, 854-870.

17.  Lerner, R.A., Wilson, C.B., Del Villano, B.C., et. al. (1976) J. Exp. Med. 143, 151-166.

18.  Elder, J.H., Jensen, F.C., Bryant, M.L. et. al. (1977) Nature, 267, 23-28.

19.  Datta, S.K., McConahey, P.J., Manny, N. et. al. (1978) J. Exp. Med. 147, 872-880.

20.  Burnet, F.M. and Holmes, M.C. (1965) Nature, 207, 368-370.

21.  Howie, J.B. and Helyer, B.J. (1968) Adv. Immunol., 9, 215-268.

22.  Ghaffar, A. and Playfair, J.H.L. (1971) Clin. Exp. Immunol., 8, 479-490.

23.  Warner, N.L. (1977) in Autoimmunity-Genetic, Immunologic, Virologic and Clinical Aspects, Talal, N. ed., Academic Press, New York, pp. 33-62.

24.  Braverman, I.M. (1968) J. Invest. Dermatol. 50, 483-499.

25.  Allison, A.C., Denman, A.M. and Barnes, R.D. (1971) Lancet, 2, 135-140.

26.  Barthold, D.R. Kysela, S.J. and Steinberg, A.D. (1974) J. Immunol., 112, 9-16.

27.  Krakauer, R.S., Waldman, T.A. and Strober, W. (1976) J. Exp. Med. 144, 662-673.

28.  Cantor, H., McVay-Bondreau, L., Hugenberger, J. et. al. (1978) J. Exp. Med., 147, 1116-1125.

29.  Gershwin, M.E., Ikeda, R.M., Kruse, W.L. et. al. (1978) J. Immunol., 120, 971-979.

30.  Raveche, E.S., Steinberg, A.D., Klassen, L.W. et. al. (1978) J. Exp. Med. 147, 1487-1501.

31.  Roder, J.C., Bell, D.A. and Singhal, S.K. (1975) in Suppressor Cells in Immunity, Singhal, S.K. and St. Sinclair, C.N.R. Eds., University of Western Ontario Press, London, Ontario, pp. 164-173.

32.  Fernandez, G., Friend, P., Yunis, E.J. et. al. (1978) Proc. Nat. Acad. Sci. U.S.A. 75, 1500-1504.

33.  Manny, N., Datta, S.K. and Schwartz, R.S. in preparation.

34.  Kermani-Arab, V. and Leslie, G.S. (1977) J. Immunol., 119, 530-536.

35.  Evans, M.M., Williamson, W.G. and Irvine, W.J. (1968) Clin. Exp. Immunol., 3, 375-383.

36.  Moutsopoulos, H.M., Boehmterwitt, M., Kassan, S.S. et. al. (1977) J. Immunol. 119, 1639-1644.

37.  Dauphinee, M.J. and Talal, N., (1973) Proc. Nat. Acad. Sci. U.S.A., 70, 3769-3772.

38.  DeHeer, D.H. and Edgington, T.S. (1977), J. Immunol., 118, 1858-1863.

39.  Warner, N.L. and Moore, M.A.S. (1971) J. Exp. Med., 134, 313-334.

40.  McCombs, C., Horn, J., Talal, N. et. al. (1975) J. Immunol., 115, 1695-1699.

41. DelVillano, B.C., Nave, B., Croker, B.P. et. al. (1975) J. Exp. Med., 141, 172-187.

42. Tung, J.S., Fleissner, E., Vitetta, E.S. et. al. (1975) J. Exp. Med., 142, 518-523.

43. Levy, J.A., Datta, S.K. and Schwartz, R.S. (1977) Clin. Immunol., Immunopathol. 7, 262-268.

44. Moroni, C. and Schumann, G. (1975) Nature 254, 60-61.

45. Benveniste, R.E., Lieber, M.M., Livingston, D.M. et. al. (1974) Nature, 248, 17-20.

46. Feldman, D. (1976) Proc. Nat. Acad. Sci. U.S.A. 73, 3710-3713.

47. Vincent, R.N., Mukherji, B.B., Mobry, P.M. et. al. (1976) J. Gen. Virol., 33, 411-419.

48. Oldstone, M.B.A. and Dixon, F.J. (1969) J. Exp. Med. 129, 483-505.

49. Britten, R.J. and Davidson, E.H., (1969) Science, 165, 349-357.

50. Emerit, I., Feingold, J., Levy, A. et. al. (1977) C.R. Acad. Sci. (Paris) 284, 249-253.

51. Fialkow, P.J., Bryant, J.I. and Friedman, J.M. (1978) Int. J. Cancer, 21, 505-510.

52. Hecht, F., McCaw, B. and Koler, R.D. (1973). N. Eng. J. Med., 289, 286-291.

53. Feingold, M., Schwartz, R.S., Atkins, L. et. al. (1969) Am. J. Dis. Child., 117, 129-136.

DISCUSSION

GERSHON:  How does your theory account for the finding of Steinberg that treatment of NZB or B/W mice with adoptive transfer of young thymocytes can completely ameliorate their disease?

DATTA:  The ameliorating effect of such therapeutic maneuvers as transplantation of young thymocytes or injection of thymic hormone into NZB mice have been transient.  Moreover, the therapeutic effect of the thymocyte transplants occurs only in young (three to four week) NZB recipients; after four months, they are refractory to such treatment.  (Ref. Gershwin, Steinberg et al., Arth. Rheumat.)

GERSHON:  I'm afraid either I didn't understand your answer or you didn't understand my question.  Your theory says that the cause of disease is resistance of B cells to regulation.  Since normal T cells prevent disease, B cell resistance to regulation cannot be the sole etiologic factor.

DATTA:  I am proposing that there are multiple genetically pre-scribed defects in NZB mice.  I agree with you that B cell resistance to regulation is not the sole etiologic factor in NZB disease.  On the other hand, injection of young T cells cannot prevent the disease completely and neither is the therapeutic effect long lasting.  In other words, the disease does not arise solely due to a regulatory T cell deficiency either.  Then we would expect severe autoimmune disease developing in nu/nu mice and not just a mild increase of ANA production that is found in nudes.  In NZB, we are dealing with a hyperactive population of B cells starting from fetal life (Manny, N., Datta, S.K., Schwartz, R.S., in preparation).  These B cells may be subdued partially and transiently with a pharmacologic dose of young thymocytes only at a young age but later they are refractory. Moreover the spontaneous hypersecretion of IgM that we find in the fetal liver cells of NZB and that Dr. Chused et al. and we have found in B cells from NZB spleens does not appear to be influenced by T cells (unmanipulated young T cells).  These increased numbers of IgM-producing  cells that also secrete IgM at an accelerated rate appear long before any suppressor T cell deficiency is detected in NZB mice.  These fetal, hyperactive B cells may be refractory to physiological signals (i.e., their threshold may be higher than normal) from not only T cells, but also idiotype-anti-idiotype net-work regulation (e.g., Jerne, Cosenza, and Kohler).
In normal strains, immature B cells expressing surface Ig and capable of secreting Ig that would react to self antigens is expected to be eliminated (i.e., clonal abortion of immature B cells on exposure to antigen).  In NZB mice, we do not know if the hyperactive B cells are making IgM with specificity to certain self-determinants or are simply polyclonally activated.  It is possible that there is a genetically programmed defect in elimination of such immature NZB B cells early in ontogeny.  These B cells then proliferate and be-come increasingly refractory to the finely tuned immunoregulatory signals that exist  under physiologic conditions.

WEKERLE:  In cultures of rat and mouse lymphocytes, one finds spontaneous blastogenesis, which probably reflects T lymphocyte responses to autochthonous non-T cells.  Did you, or someone else, find changes in spontaneous blastogenesis in NZB lymphocyte cultures?

DATTA:  This could be a response to proteins in the fetal calf serum used for the culture media or expression of embryonic antigens. We have not looked into this carefully.  Ziff and his associates have recently reported that NZB mice have an abnormally high T cell mediated cytotoxicity to histocompatible (H-2$^d$) targets from non-autoimmune strains like BALB/c and DBA/2 (J. Exp. Med., 1978).

DUPONT: The three human diseases, Bloom's syndrome, Fanconi's anemia, and ataxia telangiectasia, are monogenic diseases which all are characterized by chromosomal instability. The chromosomal abnormalities in at least Bloom's syndrome are multiple and consist of increased sister chromatid exchanges as shown by Dr. James German and co-workers. Are the chromosomal abnormalities in the NZB mouse characterized by increased sister chromatid exchanges?

DATTA: Your question is pertinent and we have mentioned these clinical examples in our manuscript, although I did not have the time to talk about them. In NZB, the consistent defects described by Fialkow et al., Lerner et al., and others are aneuploidy (especially involving chromosome no. 15) and increased breakage. Sister chromatid exchanges should be specifically looked for and studied.

CHUSED: The effect of administering NZB thymocytes on their disease has not been confirmed by Knight and Adams.
NZB plasma cells contain about four times as much IgM as those from normal mice. Since NZB cells secrete 40 times as much IgM but contain only 10 times as many plasma cells, this is consistent with the observation we have both made that each NZB plasma cell secretes about four times as much IgM.

TALAL: Your interesting model is compatible with virtually all of the data on NZB mice accumulated over the last 15 years. You postulate a hyperactive B cell that is refractory to immunoregulatory influences. Do you have any data that demonstrates this refractory characteristic of NZB B cells?

DATTA: At present, we have some preliminary results on the fetal liver cells of NZB mice which do not show any significant suppression of the IgM hyperproduction on co-culturing with thymocytes from fetal, neonatal, and/or young adult NZB or BALB/c mice.

MACKAY: As a long observer of human autoimmune disease, I have been impressed by B cell hyperactivity, expressed histologically by germinal centers and plasma cells and serologically by hyper-globulinemia and autoantibodies. Would you put primary emphasis on autonomous B cell overactivity or defective T cell regulation?
Also, with the fading emphasis on virus involvement in NZB disease, would anyone in the audience wish to support such involvement.

DATTA: We presented the evidence that B cell hyperactivity as evidenced by high IgM production is manifested before the T cell regulatory defects are detected. Chused et al. also presented similar evidence. I presented the data earlier that co-cultivation with thymocytes from normal strains does not suppress this hyper-activity of NZB fetal liver cells. This B cell overactivity could conceivably become more severe later from regulatory T cell deficiencies.

SCHWARTZ: I would like to try to reconcile the apparent discrepancy that arises from two lines of evidence on NZB mice. On the one hand, there are the findings that indicate abnormalities in suppressor T cells, and on the other hand, we now have the findings of an important abnormality of B cells. Indeed, Chused proposed to us that the primary abnormality in NZB mice is in the B cells. A way out of this

dilemma may be found by returning to Cantor's scheme of immunoregulation.  Can you conceive, Dr. Cantor, of the possibility that
signals from B cells (idiotypes?) could influence T cells?  If this
were so, could not a primary defect in B cells in turn produce an
imbalance in T cell function that might be expressed in the suppressor
T cell population?

CANTOR:  This is a formal possibility.  Alternatively, B cells of
NZB mice may secrete a species of Ig that does not activate feedback
suppression by T cells or directly "down-regulate" antibody formation
by B cells.

WARNER:  Dr. Cantor, could you clarify whether you are invoking a
specific subclass of IgG antibody that has this function on $T_s$ cells,
or whether another product from a subset of B cells is involved.
Earlier studies have shown that antibody-mediated suppression of
anti-sheep erythrocyte antibody responses are normal in NZB mice,
and hence the abnormality you are invoking would indeed need to be
probably isotype specific.  It is also perhaps worth noting that
the elevated Ig levels are only for IgM.  Further maturation of these
activated B cells does not seem to occur, perhaps compatible with
their refractoriness to regulation.

CANTOR:  It is likely that B cell products which activate $T_s$ cells
will represent a specialized set of Ig or Ig-like molecules.  There
is little or no direct evidence on this point.

GERSHON:  Bob, I don't think your model is supported by the available data.  First the reported NZB B cell defect is hypersecretion
of IgM.  This Ig isotype is not known to affect cells in the suppressor
network.  On the other hand, IgG antibodies can stimulate suppression
and do so by increasing the activity of the Ly l inducer of suppression.
Thus, the B cell defect in NZB mice could lead to autoimmunity only
in the absence of feedback suppression which would allow the Ly l Qa l+
inducer to work unabated.  Second, we have identified a mouse strain
which has a B cell defect in feeding back and activating the suppressor
system.  This is in the CBA/N and the male offspring of a CBA/N x X
$F_1$ cross.  These mice do not have the abnormal regulatory T cells
found in NZB mice.  Thus, to make your model feasible, several ad hoc
assumptions must be added.

CANTOR:  If the primary lesion in NZB were due to an intrinsic
"hyperactivity" of B cells, one might expect that polyclonal stimulation of B cells (e.g., with LPS) would induce anti-double stranded
DNA antibodies, a hallmark of the disease.  Is this the case?

DATTA:  I don't know; LPS treatment in vivo can provoke antibodies
against single-stranded DNA in "normal" mice and rabbits.

GERSHON:  I would like to add that in NZB and B/W $F_1$ mice, there
is an early (genetically programmed) degeneration of the thymic
epithelium (Devries and Hijmans, Gershwin E. et al.); this could
explain all the T cell differentiation defects.  Macrophage function
is also abnormal in NZB mice.  If macrophages play a role in T
lymphocyte differentiation (Beller and Unanue), that could also
contribute to the disorientation of the "T cell committee".

AUTOIMMUNITY AND LYMPHOPROLIFERATION: INDUCTION BY MUTANT GENE $lpr$, AND
ACCELERATION BY A MALE-ASSOCIATED FACTOR IN STRAIN BXSB MICE

EDWIN D. MURPHY AND JOHN B. ROTHS
The Jackson Laboratory, Bar Harbor, Maine 04609

ABSTRACT

Mutant gene $lpr$ (lymphoproliferation) is an autosomal recessive that appeared in
the 12th generation of brother x sister mating in the development of strain MRL
from a series of crosses involving strains LG/J, AKR/J, C3H/Di, and C57BL6/J. All
MRL/1 ($lpr/lpr$) mice develop massive generalized lymph node enlargement, starting
by 8 weeks of age and progressing to over 100 times control lymph node weights by 16
weeks of age. T cells constitute 88% of the lymphoproliferation. MRL/1 females die
at a mean age of $17.2\pm0.5$ weeks and males at $22.1\pm 0.5$ weeks, with immune complex
glomerulonephritis. Hypergammaglobulinemia develops with 2-fold increases in IgA,
IgM, and IgG2b and 10-fold increases in IgG1 and IgG2a. Antinuclear antibody titers
are detectable by 8 weeks of age and increase rapidly. Thymocytotoxic autoantibody
is demonstrable. Congenic inbred strain MRL/Mp-$lpr/lpr$, +/+ has been established.
Congenic inbred strains under development have reached the backcross generation
indicated in parentheses: A/HeJ (3), AKR/J (5), BALB/cJ (4), C3H/HeJ (4), C57BL/6J
(6), C57BL/10Sn (2), C58/J (3), NZB/B1NJ (4), SJL/J (7).

Recombinant inbred strain BXSB/Mp was developed from a cross between a C57BL/6J
female and an SB/Le male. The mean longevity of BXSB males is $22.1\pm1.9$ weeks, of
females $62.6\pm3.9$ weeks. Both males and females develop spontaneous autoimmune
disease characterized by moderate lymph node enlargement, splenomegaly, Coombs'
positive hemolytic anemia, hypergammaglobulinemia, immune complex glomerulonephritis
commonly with a nephrotic syndrome, and antinuclear, antierythrocytic, and thymocytotoxic autoantibodies. The autoimmune disease is greatly accelerated in males.
In $F_1$ hybrids the accelerating factor in males is provided only by a male BXSB
parent.

INTRODUCTION

Analysis of the inheritance of autoimmune manifestations in strain NZB and its
hybrids with other inbred strains has revealed complex patterns of multiple gene
interactions[1,2]. The genes that have been postulated have not been isolated on
other inbred strain backgrounds nor is their linkage known. The mechanisms of gene
action are poorly understood at the cellular level and unknown at the molecular
level. More rapid progress could be made with congenic inbred strain models differing at defined genetic loci.

This paper describes two different models for autoimmunity and lymphoproliferation developed in recent years in my research colony from inbred strains at The
Jackson Laboratory unrelated to strain NZB. The first involves an autosomal recessive mutant gene, $lpr$ (lymphoproliferation)[3-5], which produces massive T cell
lymphoproliferation and early onset autoimmune disease in all of 9 inbred strain
backgrounds at backcross generations ranging from N3 to N7. The second model,
recombinant inbred strain BXSB/Mp, develops a moderate predominantly B cell lymphoproliferation and autoimmune disease which is greatly accelerated in males[5,6]. In
reciprocal $F_1$ male hybrids the accelerating factor is provided only by a male BXSB
parent.

MATERIALS AND METHODS

Development of MRL substrains. The MRL substrains developed as a by-product of
a series of crosses involving strains AKR/J, C57BL/6J, C3H/Di, and LG/J (Fig. 1).

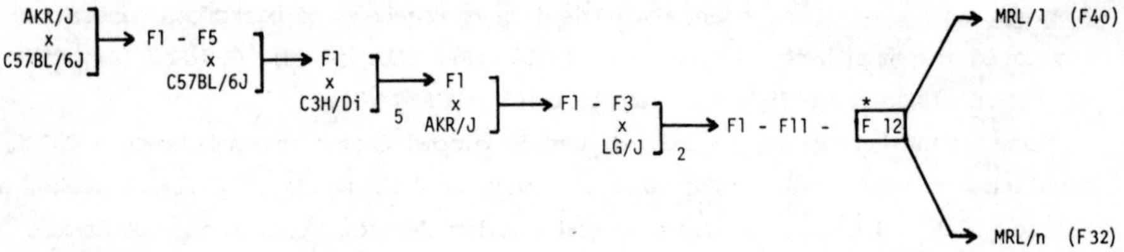

Fig. 1. Development of substrains MRL/l and MRL/n. F indicates generations of
brother x sister matings. Number at bottom of bracket indicates number of backcross generations. *The first case of early age lymphadenopathy occurred in the
female sibling of this breeding pair.

The series of crosses was begun in 1960 by Dr. M. M. Dickie in order to transfer
a mutation for achondroplasia ($cn$) from the high leukemic background of strain AKR
to a background without early incidence of leukemia. A new problem of dental
malocclusion developed and was not eliminated until the final backcrosses to strain
LG/J. From this point rigid inbreeding (brother x sister) was carried out. It is
estimated from the breeding history that the new composite genome is derived
75.0% from LG, 12.6% AKR, 12.1% C3H, and 0.3% C57BL/6. In the 12th generation of
inbreeding 5 of 12 offspring developed massive generalized lymph node enlargement
early in life. In the F12 mating the female sibling developed the lymphadenopathy,
the male did not. In the F13 matings an affected pair gave rise to the affected

substrain MRL/l. An unaffected pair after additional selection gave rise to the un-affected substrain MRL/n. The two substrains have at least 89% of their genomes in common. The original mutant gene $cn$ has been eliminated from these substrains and is maintained on an inbred strain related to the MRL substrain. Evidence will be presented showing that the massive lymphoproliferation and autoimmunity are con-trolled by a mutant autosomal recessive gene $lpr$, fixed in the homozygous state in substrain MRL/l. The gene has been transferred to substrain MRL/n (now designated MRL/Mp-+/+) by 5 cycles of cross-intercross matings reducing the estimated residual heterozygosity from 11% to 0.4% and producing the congenic inbred strain MRL/Mp--$lpr$/$lpr$ and +/+.

Development of recombinant inbred strain BXSB/Mp. Inbred strain SB/Le as origi-nally developed from a non-inbred stock was homozygous for the linked mutant genes satin ($sa$) and beige ($bg$) on Chromosome 13. Homozygous beige mice show pigment dilution and giant lysosomal granules homologous to those of the Chediak-Higashi syndrome in man. Lane and Murphy[7] reported the association of spontaneous pneumo-nitis with the homozygous beige genotypes of backcrosses of (C57BL/6J x SB/Le)$F_1$ hybrids to SB/Le. Atypical lymphoproliferations were observed in non-beige male (B6 x SB)$F_1$ hybrids and SB x (B6 x SB)$F_1$ backcrosses. In order to study the lympho-proliferative disease matings were set up to produce recombinants separating satin from beige, starting with an SB/Le-$sa$ $bg$/$sa$ $bg$ male and a C57BL/6J-+ +/+ + female:

1. (B6-+ +/+ + x SB-$sa$ $bg$/$sa$ $bg$)$F_1$ - + +/$sa$ $bg$
2. ($F_1$ x $F_1$)$F_2$-$sa$ +/$sa$ $bg$ (recombinant selected)
3. ($F_1$- + +/$sa$ $bg$ x $F_2$-$sa$ +/$sa$ $bg$)$F_3$-$sa$ +/+ + (selected)
4. ($F_3$-$sa$/+ x $F_3$-$sa$/+)$F_4$-$sa$/$sa$, $sa$/+, +/+
   Brother x sister matings continued with
   forced heterozygosity at $sa$ locus

Recombinant inbred strain BXSB/Mp is now at F27. The $sa$ gene has no detected influence on the expression of autoimmunity in this strain and is being eliminated from the expanded breeding colony.

Methods. The inbred and hybrid mice used in these studies were bred and main-tained in the author's research colony. Mice held for lifespan studies were fed a diet of Old Guilford #96 (21.3% protein) pelleted feed (Emory Morse Co., Guilford, Connecticut) and received chlorinated, acidified water *ad libitum*. All mice were examined weekly by palpation beginning at 6-8 weeks of age and were autopsied when near death. Prior to autopsy mice were exsanguinated via the retro-orbital sinus. Red and white cell numbers were determined with a Coulter cell counter and packed erythrocyte volumes were obtained. Serum was stored at -60C and later analyzed for serum protein alterations by cellulose acetate electrophoresis and for quantitation of specific immunoglobulin classes (IgA, IgG1, IgG2a, IgG2b, IgM) by agar gel radial immunodiffusion[8]. Total serum protein was quantitated spectrophotometrically using standard biuret procedures.

Mice of strains MRL/l, BXSB/Mp, and (NZB♀ x BXSB♂)$F_1$ male hybrids, and age--matched controls were evaluated for several autoantibodies at preclinical, early clinical, and clinical stages. Sera were tested for antinuclear antibodies by indirect immunofluorescence[9] using acetone-fixed rat liver sections as substrate. Fifty μl of diluted sera (1:1 to 1:200) was applied to the substrate, washed, and then overlayed with 100 μl of a 1:10 dilution of fluorescein isothiocyanate (FITC)--labelled goat anti-mouse (IgG fraction) gamma globulin (FITC-GAMG) that had been previously absorbed with mouse liver powder and ultracentrifuged. All immuno-fluorescence evaluations reported here were conducted using a Leitz Orthoplan micro-scope with darkfield illumination provided by an Osram HBO 200W high pressure Hg lamp. A KP490 (FITC) blue interference exciter and KP510 barrier filter combination was used throughout these studies. The presence and titer of thymocytotoxic auto-antibody was determined by the dye-exclusion complement-dependent microcytotoxicity method descrbied by Mittal et al.[10]. Thymocytes from 8-10 week-old male C57BL/6J mice were used as target cells. A single lot of rabbit serum was used as the source of complement (non-specific killing was 5% or less). Sera were defined as positive if there was a 50% or greater kill at dilutions of no less than 1:2. The direct antiglobulin (Coombs') test was performed to determine the presence of warm-type incomplete antibody on erythrocytes[11]. The titer of circulating erythrocyte auto-antibody was determined indirectly by incubation of 4-fold serial dilutions of selected sera with ficin-treated C57BL/6J erythrocytes followed by exposure to polyvalent goat anti-mouse gamma globulin[12].

The detection of renal immune complexes was accomplished following standard pro-cedures[13] by direct application of polyvalent FITC-GAMG to cryostat sections.

Enumeration of T and B cells at several stages in the pathogenesis of lympho-proliferation followed the methods described by Loor and Roelants[14]. One member of a pair of suspensions of mesenteric lymph node cells (prepared under conditions that inhibit capping) was exposed to a 1:2 dilution of AKR/J anti-C3H/HeJ (Thy-1.2) antisera; the duplicate suspension was incubated with normal mouse serum. Each pair of cell suspensions was washed and exposed to FITC-GAMG (1:10), rewashed, and fixed in cold EM-grade 2% formaldehyde solution. Three hundred cells were scored as either dead, fluorescence-positive, or negative. The specificity of the anti--Thy-1.2 antisera was tested by absorption with C3H/HeJ brain, AKR/J thymus and spleen, and MA/MyJ (Ly-1.1, Ly-3.2) spleen.

RESULTS

MRL substrains

Longevity. The mean life span of 28 MRL/1 females was 120±4 days, of 20 MRL/1 males 154±3 days (Fig. 2). The mean life span of 14 MRL/n females was 513±32 days, of 14 MRL/n males 646±33 days. In each substrain the females died at a significantly earlier age.

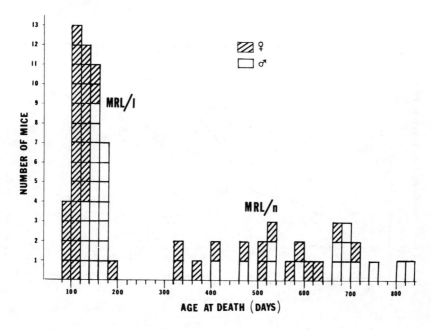

Fig. 2. Longevity of MRL/1 ($\ell p\hbar/\ell p\hbar$) and MRL/n (+/+) mice.

Lymphoproliferation. All MRL/1 mice develop massive generalized lymph node enlargement, starting by 8 weeks of age and progressing to over 100 times control lymph node weights by 16 weeks of age (Fig. 3). All lymph nodes are massively enlarged. Histologically there is blurring of nodal architecture (Fig. 4). The proliferation is predominantly lymphocytic with admixture of histiocytes and plasma cells (Fig. 5). There is 7-fold enlargement of the spleen and only slight enlargement of Peyer's patches. By 22 weeks of age there is doubling of thymic weight primarily restricted to the medulla since the cortex is atrophic. Perivascular infiltrates of lymphocytes, plasma cells, and histiocytes are common in lung, kidney, salivary gland, and liver. However, diffuse parenchymal infiltrates have not been observed. There is no leukemic blood picture. Some pathologists have raised the question of reticulum cell neoplasm, type B. Five attempts at transplanting enlarged lymph nodes of MRL/l to MRL/n, or $F_1$ hybrid recipients have been negative. During the last several weeks of life there is an average decrease of one-third of maximum lymph node size with extensive cellular necrosis and reappearance of nodal architectural features. Thus there is no clear evidence that the lymphoproliferative process is malignant. The possibility that it may be pre-lymphomatous has not been ruled out.

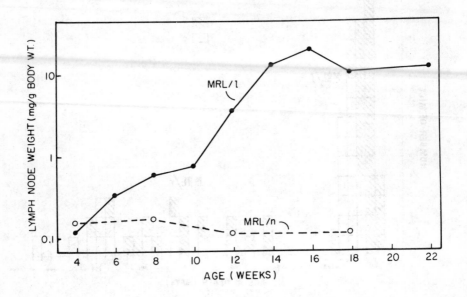

Fig. 3. Average weights of axillary, lumbar, and renal lymph nodes corrected for body weight in female MRL/l and MRL/n mice (n = 5-7 mice per age group).

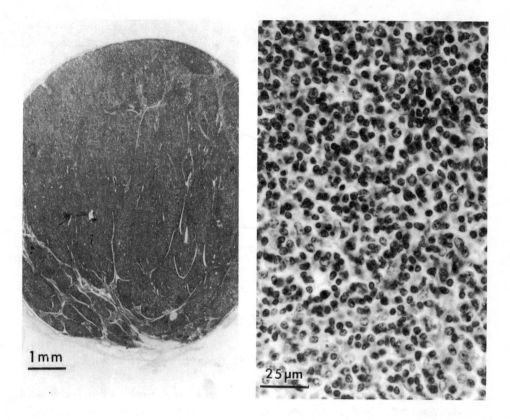

Fig. 4. Blurring of architecture in enlarged lymph node of MRL/1 mouse.

Fig. 5. Proliferation of lymphocytes with admixed histiocytes and plasma cells.

The proportions of thymic dependent (T) and thymic independent (B) lymphocytes in the mesenteric lymph node of MRL/1 and MRL/n mice were determined from ages 6 through 18 weeks by immunofluorescence using the Thy-1.2 alloantigen and surface immunoglobulin as T and B cell markers respectively. At 6 weeks of age the mesenteric nodes of both substrains weighed approximately 35 mg and contained similar percentages of T cells (MRL/1, 62.3; MRL/n, 66.1) and B cells (MRL/1, 32.1; MRL/n, 27.9). At 18 weeks of age the mesenteric node of MRL/n had increased in weight to 53 mg with no change in the proportion of the lymphocyte subpopulations. In contrast the mesenteric node of MRL/1 mice increased in weight to 2200 mg and the proportions of T and B lymphocytes had changed to 88.2 and 6.9% respectively (Fig. 6).

214

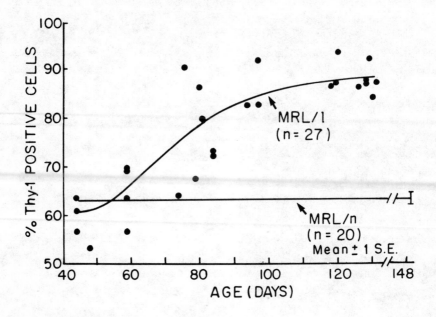

Fig. 6.  Frequency of T lymphocytes in MRL/l mesenteric lymph nodes.

Genetics.  Breeding tests involving $F_1$, $F_2$, and reciprocal backcrosses between MRL/l and MRL/n confirm the hypothesis of a single autosomal recessive gene controlling the massive lymphoproliferation.  None of 39 (MRl/l x MRL/n)$F_1$ hybrids and 0/29 $F_1$ x MRL/n backcross mice were positive.  However, 22/41 (53.7%) of $F_1$ x MRL/l backcross and 15/57 (26.3%) of $F_2$ mice had massive lymph node enlargement that developed by 4 months of age.  Similarly 10/50 (20%) of $F_2$ hybrids involving MRL/l and C57BL/6J and 15/55 (27%) $F_2$ hybrids involving MRL/l and SJL/J were positive.  This autosomal recessive mutant gene which first appeared in the 12th generation of inbreeding has been designated *lpr* (lymphoproliferation).  The results of linkage tests suggest a possible association between *lpr* and *Hm* (Hammer-toe, Chromosome 5) with 28% recombinants in the $F_1$ backcross generation.  Three-point crosses with *Hm* and *W* have been set up to test this linkage hypothesis.

The gene *lpr* has been transferred to MRL/n producing congenic inbred strain MRL/Mp-*lpr*/*lpr* and +/+ now at backcross generation N6.  Transfer of *lpr* to the following inbred strains is in progress (the number of cross-intercross cycles attained to date is indicated):  A/HeJ (3), AKR/J (5), BALB/cJ (4), C3H/HeJ (4), C57BL/6J (6), C57BL/10Sn (2), C58/J (3), NZB/BlNJ (4), and SJL/J (7).  The lymphoproliferation is expressed on all of the backgrounds tested thus far.  On C57BL/6J there is a decrease in the degree of lymph node enlargement and an increase in latent period.

MRL mice have the large body size of LG/J, are non-agouti (a/a) albinos (c/c), and have been typed for the following polymorphic loci: $Amy-1^a$, $Car-2^a$, $Dip-1^b$, $Es-1^b$, $Es-2^b$, $Es-3^c$, $Es-10^a$, $Got-1^a$, $Got-2^b$ $Gpi-1^a$, $Gpd-1^b$, $Gr-1^a$, $Hbb^d$, $Id-1^a$, $Ldr-1^a$, $Mod-1^a$, $Mod-2^b$, $Np-1^a$, $Pgm-1^a$, and $Pgm-2^a$. The phenotype of MRL cell surface allo-antigens so far determined include $H-2^k$, Ly-1.2, Ly-2.1, Ly-3.1, Thy-1.2, and $TL^-$.

Immunopathology. At autopsy the kidneys of MRL/1 mice show a subacute prolifera-tive glomerulonephritis. The glomerular lesions involve proliferation of both endo-thelial and mesangial cells and basement membrane thickening. Immunofluorescence studies reveal granular deposits of immunoglobulins in the capillary walls. Severe renal lesions show capsular cell proliferation, tubular damage, and casts. Six of 12 MRL/1 females had significant proteinuria, averaging a 9-fold increase over MRL/n. Necrotizing arteritis has been observed in lymph nodes, kidneys, spleen, and salivary glands. In lymph nodes the arteritis may be associated with hemorrhage and infarcts. The lungs show extensive perivascular and peribronchial lymphoproliferation frequent-ly accompanied by patches of atelectasis and of exudate containing macrophages. Skin lesions with hair loss and scab formation are common in the upper dorsal region. Erythematous lesions of the ears are very common and progress to necrosis of the tips of the ears. There is a swelling of the hind feet in approximately 15% of MRL/1 Mice.

The MRL/n mice do not develop the lymphoproliferative syndrome. They develop chronic glomerulonephritis. Necrotizing arteritis is very common. Approximately half of the animals die with malignant tumors, of which one-third are reticulum cell neoplasms.

Hypergammaglobulinemia is a consistent finding in MRL/1 mice with enlarged lymph nodes. Total serum protein determination and electrophoretic quantitation of serum protein classes were carried out for 22 MRL/1 and 15 MRL/n mice as con-trols, 4 months of age. Total serum proteins were increased from $6.0\pm0.15$ g/dl in controls to $7.0\pm0.2$ g/dl in MRL/1. Albumin was reduced from 3.9 to 3.2 g/dl. The alpha-globulins were unchanged. There was a 2-fold increase in Beta- and a 5-fold increase in gamma globulins. The major immunoglobulin classes were measured by radial immunodiffusion. There were 2-fold increases in IgA, IgM, and IgG2b and approximately 10-fold increases in IgG1 and IgG2a.

The sera of MRl/1 mice were negative for antinuclear antibody at 4 weeks of age. At a dilution of 1:10 the sera were positive in 3 of 9 MRL/1 at 8 weeks of age and in 9 of 9 by 20 weeks of age. In contrast, 2 of 10 MRL/n were positive by 20 weeks and 5 of 6 at 10 months of age. Thymocytotoxic autoantibodies were present in the sera of 4 of 11 MRL/1 at 17 weeks of age. Nine MRL/n were negative at this age. Incomplete erythrocyte autoantibodies were absent in 7 MRL/1 mice, 20 to 23 weeks of age.

Strain BXSB/Mp

Longevity. The mean longevity of BXSB males is 155±13 days, of females 438±27 days (Fig. 7).

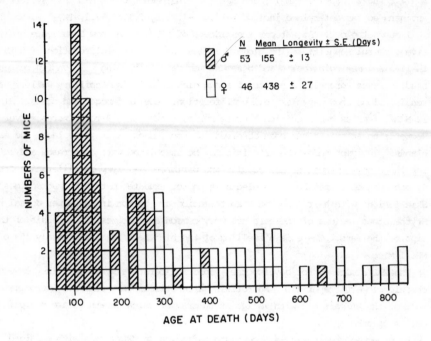

Fig. 7. Longevity of male and female BXSB mice.

Immunopathology. Both male and female strain BXSB/Mp mice develop spontaneous autoimmune disease characterized by moderate lymphadenopathy, splenomegaly, hemolytic anemia, leukocytosis, hypergammaglobulinemia, immune complex glomerulonephritis commonly with a nephrotic syndrome, and antinuclear, antierythrocytic, and thymocytotoxic autoantibodies. The disease is greatly accelerated and more severe in males than in females.

Fifteen clinically affected BXSB males, 21 to 29 weeks of age, had mean erythrocyte counts of $5.8\pm0.5 \times 10^{12}$/L and hematocrit values of 33±2%. Control erythrocyte counts on 6 strain C57BL/6J males, 20 weeks of age, were $10.4\pm0.2 \times 10^{12}$/L, and hematocrits 49±2%. Twenty-two clinically affected BXSB females 45 to 60 weeks of age had mean erythrocyte counts of $6.8\pm0.5 \times 10^{12}$/L and hematocrit values of 30±2%.

At autopsy the enlarged peripheral and abdominal lymph nodes in males retained their architecture but were infiltrated by a predominantly lymphocytic proliferation with admixed plasma cells and histiocytes. The lymphoproliferative process in females is similar. The enlarged spleens showed moderate hyperplasia of the white pulp and greatly increased erythropoiesis.

The kidneys of the BXSB males showed an acute to subacute exudative and proliferative glomerulonephritis. Tubular involvement was more striking than in the MRL/1 mice. Immunofluorescence studies revealed heavy granular deposits of immunoglobulins in the glomerular capillary walls and mesangium (Fig. 8). At autopsy the renal disease in females is a chronic glomerulonephritis.

The gamma globulin fraction comprised 17% of total serum proteins of clinically affected BXSB males, and 12% in clinically affected females, compared with 2% in 5-month old C57BL/6J males as controls. In the males there was approximately a 3-fold increase in IgM and IgG2b and a 9-fold increase in IgA and IgG1, compared with 4 month old C57BL/6 males as controls. IgG2a was undetectable in a population of 11 BXSB males. The values determined for 3 clinically affected females showed similar proportionate increases for IgM, IgG2b, IgA, and IgG1, with half the control value for IgG2a.

Antinuclear antibody was detected at a dilution of 1:10 in the serum of 1 of 5 BXSB males at 5 weeks of age, in 7 of 10 by 12 weeks, and in 15 of 16 by 25 weeks. Thymocytotoxic autoantibodies were detected in 1 of 4 BXSB males by 5 weeks of age, 6 of 10 by 12 weeks, and in 10 of 14 by 25 weeks. Eleven of 21 BXSB males are Coombs' positive by 25 weeks of age. Seven of 13 have circulating antierythrocyte antibody at similar ages.

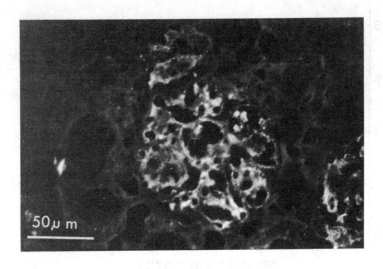

Fig. 8. Fluorescence photomicrograph of a glomerulus of a clinically affected BXSB male mouse. Stained with FITC-GAMG.

Genetics.  Recombinant inbred strain BXSB/Mp was derived from a cross between a C57BL/6J female and a SB/Le male as described in detail in Materials and Methods. The Y chromosome of strain BXSB can be traced back to the SB/Le male.  The auto-immune phenotype is transmitted as a dominant trait with expression in $F_1$ hybrids between BXSB and strains NZB/BlNJ, C57BL/6J, SJL/J and C3H/HeJ.  A possible exception involves the (AKR/J x BXSB)$F_1$ reciprocal hybrids in which clinical expression of autoimmune disease has not been observed by 11 months of age.  The hypothesis of a major dominant gene, as postulated for the autoimmune disease of strain NZB, has not been established in BXSB.  Recent observations of excess numbers of the auto-immune phenotype in $F_2$ and in $F_1$ backcross generations suggest that factors other than or in addition to a single dominant gene are operating.  The following isozyme and biochemical loci have been typed for strain BXSB: $Amy-1^a$. $Car-2^a$, $Dip-1^b$, $Es-1^a$, $Es-2^b$, $Es-3^c$, $Es-10^a$, $Got-1^a$, $Got-2^b$, $Gpi-1^a$, $Gpd-1^b$, $Gr-1^a$, $Hbb^s$, $Id-1^a$, $Mod-1^a$, $Mod-2^b$, $Np-1^a$, $Pgm-1^a$, and $Pgm-2^a$.  The phenotype of cell surface alloantigens is $H-2^b$, Ly-1.2, Ly-2.2, Ly-3.2, Thy-1.2, and TL$^-$.

A separate factor contributed by the BXSB male determines the time of appearance of lymph node enlargment and longevity in $F_1$ hybrid males.  Male (NZB♀ x BXSB♂)$F_1$ hybrids develop massive lymphoproliferation and have a markedly reduced life-span, 160±10 days, in comparison with the reciprocal male hybrids, 561±40 days (Fig. 9).

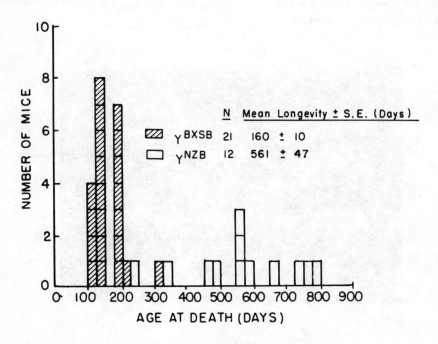

Fig. 9.  Longevity of reciprocal (NZB x BXSB)$F_1$ hybrid male mice.

DISCUSSION

The autosomal recessive mutant gene $lpr$ induces massive lymphoproliferation and early death with glomerulonephritis in all inbred strain backgrounds tested. In the case of strains MRL and NZB which develop spontaneous autoimmune disease there is marked acceleration of the development of autoimmunity. In the case of strains that do not appear to develop significant spontaneous autoimmune disease, such as C57BL/6 and C3H/He, one can postulate induction of autoimmunity.

In the $F_1$ hybrids with BXSB an accelerating factor is transmitted to male offspring by the male BXSB parent. If the factor is genetic it is linked to the Y chromosome.

A comparative study of the clinical and immunopathological findings in autoimmune models NZB, (NZB x NZW)$F_1$, MRL/1, MRL/n, and BXSB has been carried out at the Scripps Clinic and Research Foundation[15]. In this murine lupus "sweepstakes" strain MRL/1 developed the highest serum concentrations of IgG, ANA, anti-dsDNA, cryoglobulins, immune complexes, and rheumatoid factor. With the exception of one NZB x NZW)$F_1$ mouse, anti-Sm antibodies were found exclusively in MRL/1 and MRL/n[16]. Thus the congenic MRL/Mp-$lpr/lpr$ and +/+ provide the model of choice for the study of murine lupus.

The new congenic inbred strains and recombinant inbred strain BXSB, with early age of onset and uniform progression of spontaneous autoimmune disease, allow for precisely controlled studies of the interaction of genes, viruses, and sex-related factors in autoimmunity and related lymphoproliferation. They open the way for the determination of mechanisms of gene action in autoimmunity. The massive proliferation of T cells controlled by $lpr$ raises the question about what all these T cells are doing or not doing. Dr. Richard Gershon will present critical evidence for a disturbance in helper and suppressor cell interaction in MRL/1.

ACKNOWLEDGEMENTS

Supported in part by NIH Research Grants CA-22948 from the National Cancer Institute and AG-00250 from the National Institute on Aging and Contract CB-74174 from the National Cancer Institute. The Jackson Laboratory is fully accredited by the American Association for Accreditation of Laboratory Animal Care.

REFERENCES

1. Warner, N. L. (1977) in Autoimmunity: Genetic, Immunologic, Virologic, and Clinical Aspects, Talal, N. ed., Academic Press, New York, pp. 33-62.

2. Raveché, E. S., Steinberg, A. D., Klassen, L. W. and Tjio, J. H. (1978) J. Exp. Med., 147, 1487-1502.

3. Murphy, E. D. and Roths, J. B. (1977) Fed. Proc., 36, 1246.

4. Murphy, E. D. and Roths, J. B. (1978) Proc. 16th Internat. Cong. Hematol., Excerpta Medica, Amsterdam, pp. 69-72.

5. Murphy, E. D. and Roths, J. B. (1978) Mouse News Letter, 58, 51-52.

6. Murphy, E. D., Roths, J. B. and Lane, P. W. (1977) Proc. Amer. Assoc. Cancer Res., 18, 157.

7. Lane, P. W. and Murphy, E. D. (1972) Genetics, 72, 451-460.

8. Mancini, G., Carbonara, A. O. and Heremans, J. F. (1965) Immunochemistry, 2, 235-254.

9. Seligmann, M., Cannat, A. and Hamard, M. (1965) Annals N.Y. Acad. Sci., 124, 816-832.

10. Mittal, K. K., Mickey, M. R., Singal, D. P. and Terasaki, P. I. (1968) Transplantation 6, 913-927.

11. Cooke, A. and Playfair, J. H. L. (1977) Clin. Exp. Immunol., 27, 538-544.

12. Helyer, B. J. and Howie, J. B. (1963) Brit. J. Haemat., 9, 119-131.

13. Lambert, P. H. and Dixon, F. J. (1968) J. Exp. Med., 127, 507-522.

14. Loor, F. and Roelants, G. E. (1974) Nature, 251, 229-230.

15. Andrews, B. S., Eisenberg, R. A., Theofilopoulos, A. N., Izui, S., Wilson, C. B., Murphy, E. D., Roths, J. B. and Dixon, F. J. (1978) J. Exp. Med., (in press).

16. Eisenberg, R. A., Tan, E. M. and Dixon, F. J. (1978) J. Exp. Med. 147, 582-587.

DISCUSSION

SVEJGAARD: The human diseases mentioned by Dr. Dupont (ataxia telaniectasia and Fanconi's anemia) are characterized by increased DNA damage and/or defective DNA repair as described by Kidson and others. Has anyone looked at DNA damage and repair in any of the animal models of autoimmune disease?

MURPHY: Dr. Ronald Hart has expressed interest in looking at the lymphoproliferation in the MRL model.

WARNER: You mentioned that MRL/l mice have the highest levels of rheumatoid factor and cryoglobulin. This makes one think specifically of IgM and in view of our previous discussion on IgM in NZB mice, I am wondering whether the hypergammaglobulinemia you observed in the MRL/l mice is selective for any Ig isotype?

MURPHY: There is a twofold increase in IgA, IgM, and IgG2b, and more than a tenfold increase in IgG1 and IgG2a.

SCHWARTZ: Some aspects of the MRL model seem analogous with the human disease angioblastic lymphadenopathy, which is characterized by massive lymphoproliferation and autoimmunization.

CHUSED: Pursuing the point of whether these disorders are caused by T and/or B cells, have you had the opportunity to perform any immunologic manipulations such as thymectomy adoptive transfer of bone marrow?

MURPHY: Thymectomy between one and three weeks of age increases survival of MRL/l mice. So does 400 to 600R of whole body x-irradiation given between three and ten weeks of age.

THE CELLULAR SITE OF IMMUNOREGULATORY BREAKDOWN IN THE lpr MUTANT
MOUSE

RICHARD K. GERSHON, MARK HOROWITZ, JOHN D. KEMP, DONAL B. MURPHY &
EDWIN D. MURPHY

From the Laboratory of Cellular Immunology of the Howard Hughes
Medical Center (RKG, Director), The Dep't of Pathology Yale University
School of Medicine and the Jackson Laboratory, Bar Harbor, Maine.

ABSTRACT

Immune T helper cells from the MRL mouse substrain which carries
the lpr gene (MRL/1) mice can induce feedback suppressor cells but
respond poorly, to the suppressor signals they elicit.  As a result the
spleens of older MRL/1 mice contain extraordinarily high levels of T
cell mediated feedback suppression.  We discuss how these immunoregulatory defects can be related to each other and also responsible
for the lymphoproliferative and autoimmune disorders MRL/1 mice express.

INTRODUCTION

It has recently been shown that antigen stimulated Lyl cells can induce a non-immune set of T cells (surface phenotype Lyl23$^+$, Qal$^+$) to
participate in specific suppressor activity (1,2).  This immunoregulatory T-T interaction has been referred to as "feedback suppression"
(FBS) because a) the level of suppression exerted by a fixed number of
non-immune T cells increases in direct proportion to the numbers of
antigen stimulated Lyl cells added to them and b) a consequence of
the Lyl23 associated suppression is a reduction in the delivery of T
helper activity to B cells.

Since T cell immunoregulatory malfunction is currently thought to be
a major, perhaps dominant, factor in the complex etiology of clinical
and experimental autoimmune disease (3); and since NZB mice (a classical prototypical mouse strain for studies of naturally occurring autoimmune disease) have virtually undetectable levels of FBS (2), we have attempted to look at levels of FBS in other mice with naturally occurring
autoimmune diseases.  Towards this end we have examined the FBS circuit
in the MRL strain.  The autoimmune lessions which these mice express
are characterized in detail in this volume by Murphy and Roths. The
relavent point for this communication is that the MRL/1 substrain

develops a progressive T cell lymphoproliferative disorder, concomitant
with the production of numerous autoantibodies. This disease syndrome
seems to be controlled by a single recessive autosomal gene (lpr). A
second substrain, MRL/n, which shares at least 89% of the MRL/l genome,
but does not carry the lpr gene, serves as a reference strain for comparing cellular immunological defects.

RESULTS

   In the present study, we have compared the ability of these two
substrains to a) induce FBS b) deliver FBS and c) respond to FBS. The
typical experimental protocol used for asking these types of questions
consists of adding SRBC immune T cells to 2 types of assay cells; 1)
purified splenic B cells or 2) splenic B cells plus unfractionated non-
immune splenic T cells. (see refs. 1 & 2 for experimental details).
FBS was considered to be present when the immune T cells induced a
greater response in the first type of assay culture (pure B cells) than
in the second type (T & B cells). The results of 4 separate experiments using 3-4 month old MRL/l male (clinically affected) and MRL/n
mice are summarized in Table 1.

TABLE 1

| Group # | Source of Immune T cells | Source of Non-immune T & B cells | Amount of FBS |
|---------|--------------------------|----------------------------------|---------------|
| 1 | MRL/l | MRL/l | 0 |
| 2 | MRL/l | MRL/n | 0 |
| 3 | MRL/n | MRL/l | 4$^+$ |
| 4 | MRL/n | MRL/n | 1-2$^+$ |

   These results indicate that MRL/l mice have considerably more non-
immune splenic T cells which are capable of being activated to deliver
FBS than do age and sex-matched mice of the partner strain (compare
groups 3 & 4). The results also suggest that the immune T cells of the
MRL/l mouse are either resistant to FBS or fail to induce it, since
they are not suppressed by the non-immune T cells (compare groups 1 & 2
with groups 3 & 4). That resistance to FBS is the principle cause of
the failure to be suppressed is suggested by 2 lines of evidence.
First, adding immune MRL/l (but not MRL/n) T cells to group #3 in Table
1 leads to a significant decrease in FBS, demonstrating the ability of

MRL/l T cells to deliver a helper signal to B cells in the face of
MRL/n induced FBS. Second, adding small numbers of MRL/l immune T
cells to group 4 in Table 1 increases FBS indicating that MRL/l can
help induce FBS even at dose levels that are too low to induce B cells
to make antibody.

DISCUSSION

Thus, we conclude:

A.   The T cells from MRL/l mice are at least as good as those of
the MRL/n strain in inducing FBS.

B.   Nonimmune resting splenic MRL/l T cells deliver considerably
more FBS than do the cells of the partner strain.

C.   Immune T cells from MRL/l mice are relatively resistant to FBS.

These data can be explained if the mutation in the MRL/l mouse resulted in the production of Lyl inducer cells which are relatively
resistant to the FBS signal they induce. Thus, those cells which are
capable of recognizing and responding to self antigens start to proliferate and induce autoantibodies, as well as to induce FBS cells.
However, since the Lyl cells fail to respond to FBS signals, the circuit can not be completed, resulting in a) excess proliferation of Lyl
cells, b) increased autoantibodies induced by the proliferating Lyl
cells and, c) increased numbers of FBS cells also induced by the proliferating Lyl cell.

Preliminary results from ontogenetic studies are compatible with the
above hypothesis in that: a) relative resistance to FBS precedes the
development of the ability to deliver excess FBS activity in the spleen
b) the level of FBS activity that spleen cells can deliver increases
with age, in a fashion roughly parallel to the increase in lymph node
size, and c) Ly phenotyping studies have shown that FBS delivered by
the MRL/l spleen cells, even at the height of their activity, remains
$Lyl^{+}$ and is presumably dependent on cells in the Lyl23  pool, as are
all other examples of FBS so far studied.

It is interesting that the lesion in the immunoregulatory apparatus
of the MRL/l mouse is quite distinct from the lesion found in NZB mice.
This latter strain has a deficiency in the cells which deliver FBS.(2)
The end result however seems to be quite similar in both mouse strains.
Mapping the cellular sites in the immunoregulatory circuits where
genetic defects may occur in mutant mice should help to both increase
our understanding of these immunoregulatory circuits as well as to
allow us to think of directing specific therapy to the sites of the

lesions. At the very least our results show quite dramatically that autoimmune phenomena need not be mirrored by defective suppressor T cell activity, although the failure of the suppressor system to function properly may be the ultimate cause of disease.

ACKNOWLEDGEMENTS

Work supported in part by Grants CA-08593, AI-10497, CA-14216, CA-16359, CA-22948 and contract CB-74174 from the NIH, USPHS.

REFERENCES

1. Eardley, D.D., Hugenberger, J., McVay Boudreau, L., Shen, F.W., Gershon, R.K., and Cantor, H., (1978) Immunoregulatory circuits among T cell sets. 1. T helper cells induce other T cell sets to exert Feedback inhibition J. Exp. Med. 147:1106-1115.

2 Cantor, H., McVay-Boudreau, L. Hugenberger, J., Naidorf, K., Shen, F.W., and Gershon, R.K. (1978) Immunoregulatory circuits among T-cell sets: II Pysiologic role of feedback inhibition in vivo: absence in NZB mice. J. Exp. Med. 147:1116-1125.

3. Talal, N., 1977 Autoimmunity: Genetic, Immunologic, Virologic, and Clinical Aspects Academic Press, inc., New York.

DISCUSSION

TALAL: What T cell subpopulations are abnormal in the MRL/n strain which develops autoimmunity later in life? What differences do you find between MRL/1 spleen and lymph node?

GERSHON: We have not had the opportunity of testing old MRL/n mice. We have not found anything strikingly abnormal in their immuno-regulatory mechanisms at the age of 3-4 months.
No ability to induce B cells to make anti-SRBC antibodies can be found in the MRL/1 Ly 1 lymph node cells after two months of age while such activity remains in their splenic Ly 1 cells. We don't know if this is due to a loss of an ability or due to a dilution of anti-SRBC specific Ly 1 cells by the other lymph node cells involved in the lymphoproliferative disorder. By appropriate mixture experiments, we can show that the MRL/1 lymph node cells retain the ability to induce feedback suppression until the very end.

WARNER: If I followed your last point, is it possible that the specific defect is in the Qa-1$^+$ subset of Ly 1 T cells, which would, therefore, result in only poor T cell help and a failure of the feed-back loop? Could you also comment on whether there are Ly$^-$ T cells in the MRL/1 mice?

GERSHON: We don't have an antiserum which reacts with the Qa-1 allotype of the MRL mouse. However, our results strongly suggest that there is hyperactivity of feedback induction (and by analogy, Ly 1 Qa-1$^+$ activity) in MRL/1 mice. Thus, we suggest the predominant regulatory defect is a lack of response of the feedback inducer Ly 1,2,3 Qa-1$^+$ to the suppression it induces.
The lymph nodes are depleted of Ly 1,2,3 T cells. The splenic T cells show a normal distribution of the three major Ly defined sub-sets.

ROSE: Is it possible to transfer resistance to feedback suppression with cells or serum?

GERSHON: We have not yet tried this. It's a good idea. Perhaps Ed Murphy has tried this. Could you comment Ed?

MURPHY: We have not tried it yet.

"MOTHEATEN":  A SINGLE GENE MODEL FOR STEM CELL DYSFUNCTION AND EARLY ONSET
AUTOIMMUNITY

LEONARD D. SHULTZ
The Jackson Laboratory, Bar Harbor, Maine   04609

ROBERT B. ZURIER
Department of Medicine, Division of Rheumatic Diseases, University of Connecticut,
School of Medicine, Farmington, CT   06032

ABSTRACT

Mice homozygous for the recessive mutation called "motheaten" (me) develop
autoantibodies against thymocytes and double stranded DNA by three weeks of age.
Sera from littermate control mice were consistently negative for these autoantibodies.  The thymocytotoxic autoantibody was an IgM immunoglobulin as determined by its sensitivity to 2-mercaptoethanol.  This autoantibody selectively
killed immature thymocytes and reacted equally well against thymocytes of all
strains tested regardless of H-2 haplotype, Thy-1, or TLA specificity.  Lymphoid
cells from spleen, lymph node, Peyer's patch and peripheral blood were not sensitive
to killing by motheaten serum.  Motheaten mice appear to have increased numbers of
stem cells as determined by the spleen colony assay.  However, bone marrow from this
mutant does not save lethally-irradiated syngeneic recipients.  Sublethally-irradiated syngeneic recipients of motheaten bone marrow develop hyperimmunoglobulinemia
and antinuclear antibodies.  Transfer of bone marrow from motheaten mice into con-
genitally anemic $W/W^v$ recipients resulted in a cure of the anemia and the development of hyperimmunoglobulinemia and autoimmunity.

INTRODUCTION

Much of our understanding of the genetic control of autoimmunity has resulted
from experimentation with the NZB strain of mice and its $F_1$ hybrids.  Studies of
the genetic basis for autoimmune nephritis[1] and for spontaneous production of
autoantibodies to thymocytes[2], erythrocytes[3], and nucleic acids[4] in these mice
have indicated that the various manifestations of autoimmunity are controlled by
multiple loci including a single dominant gene that predisposes to generalized
autoimmunity and other genes that appear to have a modifying effect on the expression of specific autoantibodies[5].  Reports of elevated numbers of hematopoietic stem cells in NZB mice [6,7] suggest that development of autoimmunity is
associated with a defect in hematopoiesis.  However, experimentation on the role

of stem cell differentiation in NZB mice has been limited by the lack of a suitable coisogenic recipient for cell transfer experiments. In order to facilitate determination of the genetic control of autoimmunity, we have studied autoantibody production associated with single gene mutations. Recently, we have described immune dysfunction in mice homozygous for the recessive mutation called "motheaten" (me). Motheaten mice have a severe immunodeficiency as shown by depressed *in vivo* response to sheep red blood cells (SRBC) and poor reactivity of spleen cells in the graft-vs--host (GVH) assay[8], as well as impaired proliferative response to B cell and T cell mitogens and defective plaque-forming cell response *in vitro* to thymus-independent antigens. Paradoxically, polyclonal hyperimmunoglobulinemia is evident in motheaten mice by three weeks of age. Indication that the circulating immunoglobulins include autoreactive components was shown by the presence of immune complex nephritis and deposition of immunoblobulin in the thymus, skin, and lungs[8]. Motheaten mice are extremely shortlived. The mean lifespan is three to four weeks. None have survived beyond 8 weeks of age. The present study was carried out to test for specific autoantibodies and to determine the functional capabilities of stem cells in motheaten mice.

MATERIALS AND METHODS

Mice. The mice used in this investigation were obtained from the Animal Resources Department or from individual research colonies at The Jackson Laboratory. The husbandry of motheaten mice has been described[10]. The original mutation to motheaten occurred on the C57BL/6J strain of mice. Thus far it has also been backcrossed seven generations onto the C3HeB/FeJ-a/a background.

Thymocytotoxic autoantibody assay. Serial two-fold dilutions of sera were prepared in Hanks balanced salt solution supplemented with 2% heat-inactivated fetal bovine serum (Flow Laboratories). Two μl aliquots of each dilution were pipetted into Terasaki tissue culture plates (Falcon) and frozen at -60°C until use. Microcytotoxicity assays were carried out in duplicate according to a modification of the microtiter dye exclusion test[11]. Rabbit sera were selected from inbred strains of rabbits for low nonspecific cytotoxicity against mouse thymocytes and high cytolytic complement activity[12]. Viability counts were based on exclusion of 0.1% nigrosin dye. Two hundred cells per well were counted. The cytotoxic titer was defined as the highest dilution of serum that killed at least 50% of the target cells. Background level of % dead cells after incubation with littermate control sera was less than 15%. When the control data exceeded 15%, the experiments were repeated.

2-Mercaptoethanol (2-ME) treatment of serum. The method described by Hall et al was employed[13]. Equal volumes of sera from motheaten mice and 0.2 M 2-ME in phosphate buffered saline, pH 7.6 (PBS), were incubated for 15 min at 37°C followed by 4 hr at room temperature. The sera were than dialyzed in running tap water for 2 hr followed by overnight dialysis in PBS. An aliquot of pooled motheaten sera was diluted with PBS instead of 2-ME and similarly incubated.

Determination of autoantibodies to double stranded DNA (dsDNA). A previously described modified Farr assay was used[14]. Tritiated dsDNA was prepared from *Escherichia coli*: Column chromatography on hydroxyapatite and endonuclease treatment confirmed that the $^3$H DNA was not contaminated by single stranded DNA. Incubations were carried out with 10 μl serum samples.

Irradiation. Mice were irradiated in a GE 350 KVP Maxitron X-ray machine at 250 KV, 20 mA, with a 1 mm Al filtration and using a dose rate of 60-70 rads (R) per minute.

Spleen colony forming assay. Numbers of spleen colony forming units (CFU-S) in bone marrow of motheaten and littermate normal mice were determined according to the assay of Till and McCulloch[15]. Recipient mice (2-3 month old C57BL/6J males) were irradiated with 750 R and injected intravenously 18 hr later with $10^5$ C57BL/6J me/me or +/- bone marrow cells in 0.4 ml CMRL 1066 media (GIBCO). The recipient mice were killed 9 days later and spleens removed and placed in Bouin's fixative. Each experiment included a group of irradiated control mice receiving media alone.

RESULTS

Specificity of the thymocytotoxic autoantibody. Sera of all C57BL/6J me/me mice tested at more than 2 weeks of age were cytotoxic to thymus cells in the complement-dependent assay. In contrast, sera from littermate control mice were consistantly noncytotoxic. In order to examine the specificity of the anti-thymocyte autoantibody, we tested a pool of motheaten sera against thymocytes from C57BL/6J mice at various ages. Fig. 1 shows the sensitivity of thymus cells from 1 to 28-day old mice to motheaten sera. While maximum kill of 28-day old thymocytes was approximately 75%, nearly all of 1-day old thymocytes were killed by motheaten sera. These data suggest that the thymocytotoxic autoantibody selectively kills immature thymocytes. In order to test sera from motheathen mice against mature thymocytes, 4-week old C57BL/6J mice were injected with 5 mg hydrocortisone acetate intraperitoneally 48 hr before removal of their thymuses. This treatment depletes the thymus of immature cells and leaves functionally mature thymocytes[16]. As is shown in Fig. 1, thymus cells obtained from hydrocortisone--treated mice were resistant to lysis by motheaten sera.

232

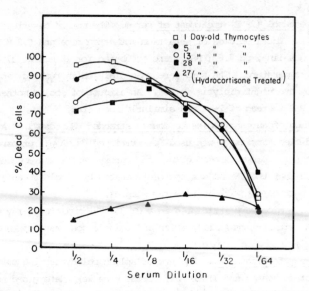

Fig. 1. Sensitivity of thymocytes from C57BL/6J mice at various ages and from hydrocortisone-treated 27-day old mice to the cytotoxic effect of C57BL/6J me/me sera. Each point represents the mean % dead cells from duplicate determinations at each dilution of sera pooled from 4-5 week old motheaten mice.

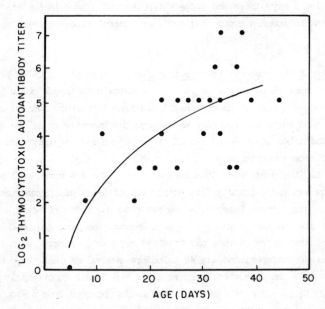

Fig. 2. Thymocytotoxic autoantibody titer of sera from individual C57BL/6J me/me mice at various ages. Sera were tested against thymocytes from 28-day-old C57BL/6J mice.

A small percentage of C57BL/6J mice have been reported to develop low levels of thymocytotoxic autoantibody after 6 months of age[17,18]. In contrast, C57BL/6J me/me mice have detectable levels of this autoantibody by 2 weeks of age. As is shown in Fig. 2, antithymocyte autoantibody appears to increase in titer over the short lifespan of this mutant. Treatment of serum from motheaten mice with 0.1 M 2-ME destroyed the cytotoxic activity against thymocytes. It thus appears that the thymocytotoxic activity resides in the IgM fraction of motheaten sera. Sera from motheaten mice were found to be cytotoxic for thymocytes of all mouse strains tested regardless of H-2 haplotype, Thy-1 or TLA specificity (TABLE 1). However, these sera were not cytotoxic for lymphoid cells from spleen, lymph nodes, Peyer's patches, or peripheral blood. Because of the presence of skin lesions in motheaten mice and the finding of immunoglobulin deposits in the skin[8], we examined cytotoxicity of motheaten sera against syngeneic epidermal cells. Motheaten sera were found to be highly cytotoxic for epidermal cells in the complement-dependent assay. Comparative cytotoxicity studies of motheaten sera against thymocytes, peripheral lymphoid cells, and epidermal cells will be published elsewhere.

TABLE 1

CYTOTOXICITY OF SERA FROM MOTHEATEN MICE AGAINST SYNGENEIC AND ALLOGENEIC THYMUS CELLS [a]

| Strain[b] | Thy-1 specificity | H-2 haplotype | TLA specificity | Mean $\log_2$ titer |
|---|---|---|---|---|
| C57BL/6J | 2 | b | −[c] | 5.7 |
| BALB/cJ | 2 | d | 2 | 5.3 |
| DBA/2J | 2 | d | 2 | 6.0 |
| NZB/BlNJ | 2 | d | 1,2,3 | 6.0 |
| C3H/DiSn | 2 | k | − | 5.7 |
| C58/J | 2 | k | 1,2,3 | 5.3 |
| A/WySn | 2 | a | 1,2,3 | 5.7 |
| 129/J | 2 | bc | 2 | 5.7 |
| SWR/J | 2 | q | 1,2,3 | 5.7 |
| PL/J | 1 | u | 1,2,3 | 6.0 |
| RF/J | 1 | k | − | 6.0 |
| AKR/J | 1 | k | − | 6.0 |
| B6.PL-Thy-1[a]/Cy | 1 | b | ?[d] | 6.0 |

[a] Sera was pooled from 3 C57BL/6J me/me mice at 5 weeks of age. Control sera pooled from littermates were not cytotoxic against these target cells.
[b] Thymus cells from 3 4-week old ♀ mice of each strain were individually tested for sensitivity to killing by motheaten sera.
[c] (−) = Negative for TLA specificities 1,2,3.
[d] (?) = TLA typing has not been reported for this strain.

Autoantibodies against dsDNA in sera from motheaten mice.  In addition to the development of thymocytotoxic autoantibody, motheaten mice at 3-5 weeks of age develop autoantibody against dsDNA.  Table 2 shows the level of anti-dsDNA in sera from 10 pairs of C3HeB/FeJ me/me mice and +/- littermates.  Serum from motheaten mice showed significantly higher binding of dsDNA compared with the littermate control sera (p < 0.01, Student's t-test).

TABLE 2

SPONTANEOUS PRODUCTION OF ANTIBODIES AGAINST ds-DNA IN C3HeB/FeJ-me/me MICE

| Age (days) | Sex | % DNA Binding me/me | +/- |
|---|---|---|---|
| 19 | ♀ | 20.7 | 4.6 |
| 20 | ♀ | 15.4 | 0.0 |
| 22 | ♂ | 17.1 | 3.7 |
| 25 | ♀ | 30.2 | 1.9 |
| 25 | ♀ | 16.4 | 5.5 |
| 25 | ♀ | 10.6 | 3.4 |
| 26 | ♂ | 18.1 | 1.3 |
| 33 | ♀ | 8.1 | 4.8 |
| 33 | ♂ | 20.3 | 5.6 |
| 38 | ♀ | 14.0 | 3.3 |
| Mean ± SE | | 17.1 ± 1.92 | 3.4 ± 0.59 |

Lifesparing capacity of marrow cells from motheaten mice.  Mice exposed to high doses (700-900 R) of whole body x-irradiation die after 1-2 weeks from hemhorrage and infection associated with leukocyte and platelet destruction.  Death can be prevented by repopulation with syngeneic bone marrow.  Bone marrow suspensions from 7 pairs of motheaten and littermate control mice were injected into syngeneic recipients 18 hr after irradiation with 750 R.  In each of two experiments, 5 lethally irradiated mice were injected with media alone.  Table 3 shows that while mice injected with littermate control marrow cells survived the effects of lethal irradiation, mice treated with bone marrow from motheatens survived no longer than mice receiving media alone.

TABLE 3

EFFECT OF BONE MARROW FROM MOTHEATEN MICE ON SURVIVAL OF LETHALLY IRRADIATED
RECIPIENTS

| Treatment of recipients [a] | No. of donors | Mean survival time (days) | p [b] |
|---|---|---|---|
| me/me marrow | 7 | 15.9 ± 0.89 | |
| media [c] | - | 14.0 ± 0.86 | NS |
| +/- marrow | 7 | > 60 | |

[a] Motheaten and littermate control donors were 16-27 days old.
[b] Each marrow suspension was injected into 3-4 recipients.
   Data analyzed by Student-Newman-Keuls multiple range test.
[c] In each experiment, 5 lethally irradiated mice were injected
   with 0.4 ml CMRL 1066 media.

Colony-forming units in bone marrow of motheaten mice. In order to determine
whether the inability of bone marrow from motheaten mice to save lethally-irradiated
recipients might be due to decreased numbers of hematopoietic stem cells, we compared the number of CFU-S in bone marrow from motheaten mice with that in marrow of
littermate controls. Marrow cell suspensions were prepared from 8 pairs of motheaten mice and normal littermates (7-31 days old). Each cell suspension ($10^5$ cells)
was injected intravenously into 750 R irradiated recipients. The number of CFU-S
was determined 9 days later from spleen colony counts. As is shown in Fig. 3,
motheaten mice do not have a deficiency in numbers of CFU-S. On the contrary, this
mutant has a significantly increased number of CFU-S/$10^5$ marrow cells compared with
littermate controls (p < 0.01, Student-Newman-Keuls multiple range test).

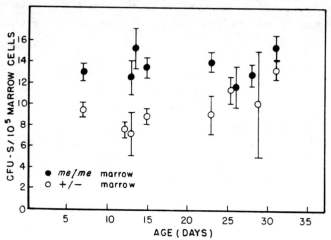

Fig. 3. Numbers of Colony-forming units in bone marrow from motheaten and littermate control mice. Each point represents the mean number of CFU-S from 4-5
marrow recipients.

<u>Transfer of autoimmunity with bone marrow from motheaten mice</u>. To establish a
hematopoietic microenvironment in which stem cells from motheaten mice would be able
to interact with normal stem cells, bone marrow from motheaten mice was transferred
into sublethally-irradiated syngeneic recipients. C57BL/6J mice were injected with
$2 \times 10^6$ bone marrow cells from C57BL/6J <u>me</u>/<u>me</u> or +/+ controls 18 hr after exposure
to 450 R. We found significant increases in IgM and anti-dsDNA in recipients of
motheaten marrow by 4 weeks after injection. By 20 weeks, recipients of motheaten
marrow had a 20 fold increase in serum IgM and showed a high level of anti-dsDNA
autoantibody. In addition, 5 of 10 recipients of motheaten marrow developed significant levels of thymocytotoxic autoantibody. In contrast, sublethally-irradiated recipients of normal marrow or media alone showed no significant change in
serum IgM levels and failed to develop thymocytotoxic or anti-dsDNA autoantibodies.

We also employed another experimental approach which enabled us to examine the
erythropoietic stem cell activity of motheaten mice as well as the transfer of autoimmunity. This approach utilized (WB/Re x C57BL/6J)$F_1$-$\underline{W}/\underline{W}^v$ mice. These mice have
a severe macrocytic anemia that can be cured with transplants of histocompatible
marrow cells[19]. They provide an unirradiated *in vivo* test system for assessing the
differentiation of erythropoietic stem cells. Groups of age-matched WBB6F$_1$ $\underline{W}/\underline{W}^v$
mice were injected intravenously with $2 \times 10^6$ bone marrow cells from C57BL/6J
<u>me</u>/<u>me</u> or littermate control mice. We found that marrow from motheaten mice cured
the anemia as quickly as marrow from littermate control mice. However, $\underline{W}/\underline{W}^v$
recipients of motheaten marrow developed hyperimmunoglobulinemia and high levels of
anti-dsDNA autoantibody by 100 days after injection. Although the anemia in recipients of motheaten marrow remained cured as shown by normal erythrocyte numbers and
mean cell volumes, $\underline{W}/\underline{W}^v$ recipients of motheaten marrow died by 20 weeks after injection. The proximate cause of death appeared to be pneumonitis and glomerulonephritis.

DISCUSSION

The most striking feature of autoimmunity in motheaten mice is its early onset.
By two to three weeks of age, there are significant levels of thymocytotoxic and
anti-dsDNA autoantibodies. The thymocytotoxic autoantibody appears to be specific
for immature thymocytes. As with the thymocytotoxic autoantibody described in NZB
mice [17,18], the sensitivity of thymocytes to killing by serum from motheaten mice
is not dependent on H-2 haplotype, Thy-1 or TLA specificity. Cocapping experiments
with anti-thy-1 antibody and thymocytotoxic autoantibody from NZB mice have indicated
that the NZB thymocytotoxic autoantibody reacts against a determinant common to both
the thy-1.1 and thy-1.2 alloantigens[20]. The increased cytotoxicity of motheaten
serum against immature thymocytes and lack of cytotoxicity against peripheral lymphoid cells might be associated with concentration of thy-1 antigen. It has been
shown that the concentration of thy-1 is greatest on immature thymocytes and

decreases on peripheral T cells[21,22]. A considerable accumulation of autoantibody
on the cell surface may be required to initiate complement-mediated lysis. Serum
from motheaten mice was also highly cytotoxic for epidermal cells. It is of
interest that epidermal cells express high levels of thy-1 antigen[23].

Immune complex glomerulorephritis and antinuclear autoantibodies in (NZB x NEW)$F_1$
mice resemble the autoimmune manifestations of systemic erythematosus in humans[24].
In (NZB x NZW)$F_1$ mice, low levels of anti-dsDNA are detectable at 4-6 weeks of age.
However, significantly increased levels are not seen until after 20 weeks of age[25].
In contrast, C3HeB/FeJ me/me have significantly elevated levels of anti-dsDNA by
three to four weeks of age.

The finding that bone marrow from motheaten mice has elevated numbers of hematopoietic stem cells as measured by CFU-S determinations but fails to save lethally
irradiated marrow transplanted recipients suggests either a decrease in the proliferative capacity of stem cells from these mice or an impairment in differentiative
capacity. The observation that marrow cells from motheaten mice cure macrocytic
anemia in $W/W^v$ recipients as rapidly as normal marrow indicates no impairment in
proliferative capacity. Since spleen colonies that arise in lethally-irradiated
recipients after transfer of syngeneic marrow are composed of erythroblasts, myelo-
cytes, and megakaryocytes, it is possible that the inability of marrow from motheaten mice to save lethally-irradiated recipients is due to a defect in differentiation to the lymphocyte or mononuclear phagocyte lineage. Alternatively, we
have not yet ruled out the possibility of an infective agent in marrow suspensions
from motheaten mice that would contribute to the death of irradiated recipients.
However, this possibility seems unlikely since comparison of marrow from motheaten
mice with littermate control marrow shows no increased expression of virus. In
addition, transfer of large numbers of marrow cells from motheaten mice into
sublethally-irradiated (450 R) recipients does not result in death.

Investigation of lymphocyte development in motheaten mice has revealed normal
numbers of thy-1-bearing cells (T cells) but only one-half to one-third the normal
numbers of immunoglobulin-bearing cells (B cells)[9]. In addition, the B cells present in motheaten mice have the characteristics of immature B cells. To reconcile
this finding with the increased numbers of plasma cells in lymphoid tissues of
motheaten mice[10] and with the increased level of serum immunoglobulins[8], we have
suggested that B cells in motheaten mice are subject to a nonspecific triggering to
plasma cell differentiation[9]. In addition to the defect in B cell differentiation,
macrophages from motheaten mice suppress both B cell and T cell proliferative
responses[9]. The finding that marrow cells from motheaten mice can transfer autoimmunity to sublethally-irradiated recipients suggests that the defect in lymphocyte
differentiation may be intrinsic to the stem cell.

238

ACKNOWLEDGEMENTS

This work was supported by National Institutes of Health Grants CA 20408 and AN 17309. The Jackson Laboratory is fully accredited by the American Association for Accreditation of Laboratory Animal Care. We thank Janis L. Wright , Page D. Keely and Patricia L. Miller for excellent technical assistance.

REFERENCES

1. Knight, J. G. and Adams, D. D. (1978) J. Exp. Med., 147, 1653-1660.
2. Raveché , E. S., Steinberg, A. D., Klassen, L. W. and Tjio, J. H. (1978) J. Exp. Med., 147, 1487-1502.
3. DeHeer, D. H. and Edingdon, T. S. (1976) Transplant. Rev., 31, 116-155.
4. Roubinian, J. R., Talal, N., Greenspan, J. S., Goodman, J. R. and Siiteri, P. K. (1978) J. Exp. Med., 147, 1568-1583.
5. Warner, N. L. (1977) in Autoimmunity, Genetic, Immunologic, Virologic, and Clinical Aspects, Talal, N. ed., Academic Press, New York, pp. 33-62.
6. Morton, J. I. and Siegel, B. V. (1971) Proc. Nat. Acad. Sci. U.S.A., 68, 124-126.
7. Warner, N. L. and Moore, M. A. S. (1971) J. Exp. Med., 134, 313-334.
8. Shultz, L. D. and Green, M. C. (1976) J. Immunol., 116, 936-943.
9. Sidman, C. L., Shultz, L. D. and Unanue, E. R. Manuscripts in preparation.
10. Green, M. C. and Shultz, L. D. (1975) J. Hered., 66, 250-258.
11. Mittal, K. K., Mickey, M. R., Singal, D. P. and Terasaki, P. I. (1968) Transplantation, 6, 913-927.
12. Fox, R. R., Cherry, M. and Shultz, K. L. (1978) J. Hered., 69, 107-112.
13. Hall, J. G., Smith, M. E., Edwards, P. A. and Shooter, K. V. (1969) Immunology, 16, 773.
14. Luciano, A. and Rothfield, N. F. (1974) Ann. Rheum. Dis., 32, 337-341.
15. Till, J. E. and McCulloch, E. A. (1961) Radiation Research, 14, 213-222.
16. Martin, J. W. and Martin, S. E. (1975) Nature, 254, 716-718.
17. Shirai, T. and Mellors, R. C. (1972) Clin. Exp. Immunol, 12, 133-152.
18. Shirai, T. and Mellors, R. C. (1971) Proc. Nat. Acad. Sci. U.S.A., 68, 1412-1415.
19. Russell, E. S. (1970) in Regulation of Hematopoiesis, Gordon, A. S. ed., Appleton-Century-Crofts, New York, 649-675.
20. Parker, L. M., Chused, T. M. and Steinberg, A. D. (1974) J. of Immunol., 112, 285-292.
21. Aoki, T., Hämmerling, U., deHarven, E., Boyse, E. A. and Old, L. J. (1969) J. Exp. Med., 130, 979-1001.
22. Owen, J. J. T. and Raff, M. C. (1970) J. Exp. Med., 132, 1216-1232.
23. Scheid, M., Boyse, E. A., Carswell, E. A. and Old, L. J. (1972) J. Exp. Med., 135, 938-954.
24. Howie, J. B. and Helyer, B. J. (1968) Advan. Immunol., 9, 215-266.
25. Steinberg, A. D., Pincus, T. and Talal, N. (1969) J. Immunol., 102, 788-790.

DISCUSSION

CANTOR:  Two questions:  First have you attempted to repopulate me/me neonates with C57BL/6 bone marrow?  Second, how do you account for the striking accumulations of granulocytes in the skin?

SHULTZ:  In answer to your questions:  First, these experiments are in progress.  Second, the granulocytic skin lesions may result from activation of complement at the site of antigen-antibody reaction. We postulate neonatal synthesis of autoantibody with high avidity for the skin.
An alternative hypothesis that the skin lesions reflect a cell-mediated autoimmunity against epidermal cells has not been ruled out.

CANTOR:  Do you agree that the normal capacity for me/me bone marrow cells to generate RBC and granulocytes combined with their defective capacity to generate normal lymphocytes points to a "split" after the pluripotential stem cell, separating a "pro-lymphocyte" from stem cells having granulocyte:RBC:monocyte differentiative options?

SHULTZ:  Our data is consistent with the hypothesis that motheaten mice have a defect in those stem cells in the lymphocyte differentiative pathway.

DUPONT:  There is an autosomal recessive inherited disease in man in which, in my opinion, has similarities with the disease you describe.  This is the Transcobalamin II deficiency which is lack of the B12 vitamin transporting serum protein.  In this disease, you see bone marrow stem cell abnormalities and severe T and B cell deficiency.  Have you looked for the presence of Transcobalamin in these mice?

SHULTZ:  We have not yet investigated the possibility of an association between immune dysfunction and vitamin B12 deficiency in motheaten mice.

GERSHON:  Do you have any evidence as to whether the skin and lung lesions are due to autoimmune reactions or to infectious agents?
Are the lung and skin lesions morphologically similar?
A point of information:  In contrast to other species, DTH lesions in mice are largely made up  of polymorphs.  Thus the presence of these cells does not necessarily imply an antibody or bacterial-mediated reaction.

SHULTZ:  Motheaten mice raised in specific pathogen-free environments do not show reduced incidence or severity of skin and lung lesions.  We postulate that these lesions are manifestations of autoimmunity.  Immunoglobulin deposits have been found in the skin and lungs of these mice.
While macrophages predominate in lung lesions of C57BL/6J me/me mice, neutrophils are the main cell type seen in skin lesions.
In a previous study of the genetic control of contact sensitivity in mice, we found a predominance of neutrophils at the site of reaction.

SCHWARTZ:  Do you have evidence of an autoantibody against stem cells?
Does the IgM in the motheaten mouse have restricted heterogeneity?

SHULTZ:  We have not investigated whether treatment of normal marrow cells with motheaten serum will affect stem cell function.

Immunoelectrophoresis of serum from motheaten mice has shown the polyclonal nature of the IgM. Both kappa and lambda light chains are evident in the broad IgM precipitin arcs.

WARNER: Your data on elevated IgM levels in young motheaten mice is reminiscent of findings in a subgroup of hormonally bursectomized chickens which have high levels of Ig. Recent studies in Lee Hood's laboratory have shown that the Ig in such birds is of restricted clonotypes. It thus might be of interest to more carefully analyze the motheaten IgM for similar restriction in Ig clonotypes.

INCIDENCE AND CHARACTERISTICS OF FUNCTIONAL B LYMPHOCYTES IN

MOTHEATEN MICE

PAUL W. KINCADE

Sloan-Kettering Institute for Cancer Research,
145 Boston Post Road, Rye, New York   10580

INTRODUCTION

The lymphoid, reticuloendothelial, and complement systems are
functionally intertwined and pictorial illustrations of this have
come to resemble metabolic pathways charts.  Defects in any one
part of the immune system might therefore result in inappropriate
and possibly pathological responses of the remaining normal effector
cells.  Autoimmune manifestations are often seen in patients with
immunodeficiency disease and lesions of both types result from the
recessive mutation of motheaten mice[1,2].  Homozygous defective
animals do not thrive, are deficient in B lymphocytes, develop hypergammaglobulinemia, have widely distributed Ig deposits, and usually
succumb to pneumonitis before 8 weeks of age.  This severe abnormality
could result from the homeostatic imbalance created by a partial
immune deficiency or might be explainable in terms of regulatory
defects intrinsic to the B cells.  We have used a simple cloning
technique to enumerate and partially characterize functional B cells
in the spleen, lymph nodes, and peripheral blood of these animals.
Numbers of these were markedly subnormal in all tissues except the
blood.  However, by several criteria the residual B cells seemed to
have attained a level of maturity comparable to those of their normal
littermates.  In contrast to the deficiency of B cells in these mice,
clonable B cells were present in normal or elevated numbers in tissues
of autoimmune (NZB x NZW)$F_1$ mice.

MATERIALS AND METHODS

Motheaten mice were either purchased as neonates or produced from matings of heterozygous breeders obtained from Jackson Laboratories. Most of these were from the fourth generation of C57 to C3H backcross matings. Preliminary evaluation of the defective animals confirmed reported descriptions[1,2] in that they were small relative to littermates, had characteristic skin lesions, and had immunoglobulin deposits in the thymus and kidneys. Sections of spleen and lymph nodes revealed a striking deficiency of lymphocytes in B areas and an abundance of immunoglobulin-containing plasma cells. In all experiments, heterozygous and apparently normal littermates were used as controls for comparison with the motheaten individuals at about three weeks of age. (NZB x NZW)$F_1$ mice were from a colony maintained at Sloan-Kettering Institute.

All reagents, culture techniques, and cell treatment protocols have been published in detail previously[3-5]. Cloning results shown were obtained by culturing cells in semisolid agar in the presence of 2-mercaptoethanol and lipopolysaccharide endotoxin (LPS). Most of the experiments were also performed in parallel with sheep red blood cells added to the cultures instead of LPS. Under either of these conditions colony numbers are directly proportional to numbers of B cells cultured and cloning efficiency is neither affected by macrophages or T cells.

RESULTS AND DISCUSSION

There are now a number of murine models of genetically controlled autoimmune disease and of these motheaten is the most severe. A major goal in studies of these animals must be to localize the cell type or types in which the mutant gene product is expressed. Recent developments in limiting dilution technology permit intrinsic differences in individual immunocytes to be easily discerned. This is in contrast to functional assays employing whole animals or mass cell cultures where the outcome is dependent on multiple cell types and regulatory events. In the system employed here, lymphocytes are dispersed in three dimensions in semisolid gel cultures containing mitogens. Cell-cell interactions are minimized by the gel matrix and each activated B cell proliferates without T cell or accessory cell dependence[4,5]. The relative incidence of functional B cells is revealed by this assay since numbers of proliferating foci are directly and linearly related to the number of cells cultured. All fetal and adult tissues of normal mice which contain surface Ig-bearing B cells also contain clonable B cells and these are heterogeneous with respect to immunoglobulin isotype, Ia antigen expression, and sensitivity to capping of their surface receptions[5]. This indicates that B cells representative of different maturational stages and experiences with antigen are capable of colony formation. One or more exceptional categories of B cells which do not clone under these conditions may be represented by mutant CBA/N mice[6,7]. As a preliminary assessment of motheaten mice, these same techniques have been used to determine the incidence and nature of their functional B cells.

Table 1

**INCIDENCE OF B LYMPHOCYTE COLONY-FORMING CELLS IN MOTHEATEN MICE**

|  |  | Experiment Number | | | | | |
|---|---|---|---|---|---|---|---|
|  |  | I | II | III | IV | V | VI |
| Motheaten (me/me) | Spleen | 1 ± 1* | 16 ± 5 | 0 | 3 ± 1 | 133 ± 26 | 103 ± 5 |
|  | Lymph nodes | 2 ± 1 | 8 ± 1 | 2 ± 1 | 5 ± 2 | 29 ± 5 | 32 ± 2 |
|  | Peripheral blood | 261 ± 9 |  | 90 ± 10 | 90 ± 19 | 107 ± 14 | 70 ± 10 |
| Normal Littermate (me/+ or +/+) | Spleen | 477 ± 15 | 459 ± 25 | 309 ± 23 |  | 832 ± 84 | 770 ± 85 |
|  | Lymph nodes | 361 ± 14 | 97 ± 9 | 177 ± 13 |  | 606 ± 26 | 734 ± 48 |
|  | Peripheral blood |  |  | 114 ± 6 |  | 625 ± 27 | 505 ± 19 |

* Mean number of colonies/$10^5$ cultured cells ± S.E.

The frequencies of colony-forming B cells in spleen and lymph nodes
of motheaten mice were very low relative to their littermate controls
(Table 1). There is wide variation in the size of individual colonies
in cultures of normal B cells and motheaten clones did not seem remarkable in this respect. The deficiency of lymphocytes in the thymus independent areas of these organs revealed by histological examination
and a diminished in vitro response to SRBC[1,2] correlate with this low
number of functional precursors. Lymphocyte numbers in peripheral
blood of motheatens have previously been found to be normal[1] and the
deficiency in clonable B cells was less dramatic and consistent than
for the solid tissues.

On a number of occasions, spleen cells from motheaten and littermate
mice were placed in conventional liquid cultures with mitogens and

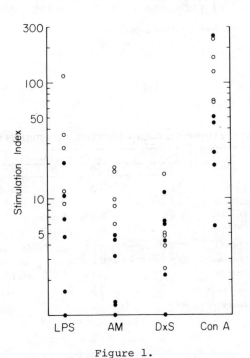

RESPONSES OF SPLEEN CELLS FROM
MOTHEATEN (•) OR NORMAL LITTERMATE (○)
MICE TO MITOGENS IN LIQUID CULTURES

Figure 1.

responsiveness assessed by $^{125}$IUdR uptake 72 hours after culture. As expected, responses of motheaten cells to the B-cell mitogens, lipo-polysaccharide (LPS), agar-derived mitogen (AM), and dextran sulfate (DxS) were low. Responses to the T-cell mitogen Con A were also usually lower than the controls (Fig. 1). The magnitude of stimulation in such liquid cultures is not directly proportional to numbers of functional immunocytes present and cooperation between as well as in-hibition by different cell types is known to occur. Interpretation of these findings is thus more difficult than with the results of cloning.

Immature B cells in fetal tissues of normal animals lack surface IgD and form colonies in the presence of rabbit anti-mouse δ serum. A variable proportion of the B cells in adult tissues express this iso-type and extremely small quantities of antiserum are required to

Table 2

**CHARACTERIZATION OF COLONY-FORMING B CELLS IN SPLEEN AND PERIPHERAL BLOOD OF MOTHEATEN MICE**

| SENSITIVITY TO ANTI-δ ADDITION | | Exp. I (Spl.) | Exp. II (P.B.) | |
|---|---|---|---|---|
| Motheaten (me/me) | Control | 21 ± 1* | 49 ± 11 | |
| | Anti-δ | 1 ± 2 | 2 ± 1 | |
| Normal littermate | Control | 837 ± 127 | 22 ± 3 | |
| (me/+ or +/+) | Anti-δ | 447 ± 87 | 10 ± 4 | |
| SENSITIVITY TO ANTI-Ig PRETREATMENT | | Exp. I (Spl.) | Exp. II (Spl.) | Exp. III (Spl.) |
| Motheaten (me/me) | Control | 24 ± 6 | 18 ± 2 | 6 ± 1 |
| | Anti-Ig | 7 ± 2 | 7 ± 3 | 7 ± 1 |
| Normal littermate | Control | 512 ± 70 | 546 ± 42 | 177 ± 20 |
| (me/+ or +/+) | Anti-Ig | 274 ± 12 | 338 ± 46 | 138 ± 19 |
| EXPRESSION OF Ia ANTIGENS | | Exp. I (Spl.) | Exp. II (P.B.) | |
| Motheaten (me/me) | C' alone | 2 ± 1 | 16 ± 5 | |
| | Anti-Ia + C' | 0 | 0 | |
| Normal littermate | C' alone | 50 ± 12 | 97 ± 9 | |
| (me/+ or +/+) | Anti-Ia + C' | 0 | 0 | |

* Mean number of colonies/10$^5$ initial cells ± S.E.

prevent their clonal proliferation[5]. Normal B cells from which this receptor is removed are tolerized rather than immunized by certain antigens[8,9]. In two experiments the sensitivity of motheaten B cells to anti-δ serum was comparable to that of their littermate controls (Table 2). This suggests that these receptors are displayed on the cell surface and that negative signals can be received through them.

Through a mechanism which is thought to be related to clonal abortion type tolerance susceptibility, young B cells are inactivated by preexposure to anti-Ig antibodies. By about six weeks of age the splenocytes of normal mice mature such that they can be incubated for 45 minutes in the presence of anti-Ig, anti-Ia, or anti-δ antibodies and washed without affecting their cloning efficiency[5]. Spleen cells from young motheaten mice were no more susceptible or resistant to this treatment than those of the controls (Table 2).

Another criteria of maturity of B cells is their expression of antigens encoded by the midportion of the major histocompatibility region. Acquisition of these Ia antigens by clonable B cells normally begins by the time of birth and such receptors could be particularly important in regulation of immune responsiveness[5]. In two experiments that were done with spleen and peripheral blood cells, motheaten B cells did not seem to be exceptional in their sensitivity to anti-Ia cytolysis in the presence of complement (Table 2).

We have just begun studies of the B cells in autoimmune (NZB x NZW)$F_1$ mice. In contrast to the deficiency in clonable B cells found in the solid tissues of motheaten mice, these were usually elevated in autoimmune B/W animals (Table 3). Studies are in progress to determine if there are intrinsic abnormalities in the B lymphocytes of these mice as has been suggested by some studies of New Zealand strain mice[10], or rather if the autoimmune phenomena are explainable solely in terms of regulatory abnormalities at the T-cell level[11,12].

Table 3

## INCIDENCE OF CLONABLE B CELLS IN AUTOIMMUNE B/W MICE

|  |  | Spleen | Lymph Nodes | Bone Marrow |
|---|---|---|---|---|
| 31-week old | (NZB × NZW)F$_1$ | 3180 ± 160* | 1872 ± 38 | 540 ± 22 |
| 40-week old | CBA/H | 865 ± 40 | 669 ± 40 | 222 ± 40 |

* Mean number of colonies/10$^5$ cultured cells ± S.E.

This initial assessment of motheaten mice does not reveal any intrinsic abnormalities in their B lymphocytes; just a deficiency in their numbers in lymphoid tissues. Mutations affecting the expression of cell surface Ia antigens and IgD would be expected to result in abnormal B-cell function since these receptors appear to be important for cellular interactions and susceptibility to tolerance. The reservation must be made that mutations could affect the function of receptors without this being obvious in the detection systems employed here. It is also conceivable that categories of B cells not capable of cloning are defective in motheaten mice and these would have escaped our evaluation. There is precedent for hypergammaglobulinemia associated with B-cell deficiency in chickens bursectomized late in embryonic life[13], and this could be explained in terms of deficient feedback antibody or an incomplete network of idiotypes and anti-idiotypes. However, the B-cell deficiency would not seem to account for the severity and early onset of pathology in these mice and could well be secondary to abnormalities elsewhere.

SUMMARY

A cloning technique was employed to enumerate and partially characterize the B cells of mutant motheaten mice. At three weeks of age, numbers of clonable cells were very low in spleen and lymph nodes relative to normal littermate controls. This difference was less impressive when peripheral blood cells were assayed. Conventional liquid cultures of motheaten spleen cells also reflected the diminished number of mitogen-responsive lymphocytes. Acquisition of IgD and Ia receptors appeared to occur normally and motheaten and littermate B cells were similarly sensitive to capping of surface immunoglobulin. By these criteria, B cells which are present in these immunologically compromised mice attain maturity and function normally. In contrast to the B-cell deficiency found in motheaten mice, autoimmune (NZB x NZW)$F_1$ mice had normal or elevated numbers of clonable B cells.

## ACKNOWLEDGMENTS

This work was supported by N.I.H. Grant CA 17404 and a Senior Investigatorship from the Arthritis Foundation. Ms. Grace Lee provided expert technical assistance.

## REFERENCES

1. Green, M.C. and Shultz, L.D. (1975) J. Hered., 66, 250-258.

2. Shultz, L.D. and Green, M.C. (1976) J. Immunol., 116, 936-943.

3. Kincade, P.W., Ralph, P. and Moore, M.A.S. (1976) J. Exp. Med. 143, 1265-1270.

4. Kurland, J.I., Kincade, P.W. and Moore, M.A.S. (1977) J. Exp. Med., 146, 1420-1435.

5. Kincade, P.W., Paige, C.J., Parkhouse, R.M.E. and Lee, G. (1978) J. Immunol., 120, 1289-1296.

6. Kincade, P.W. (1977) J. Exp. Med., 145, 249-263.

7. Kincade, P.W., Moore, M.A.S., Lee, G. and Paige, C.J. (1978) Cell Immunol. (in press).

8. Cambier, J.C., Vitetta, E.S., Kettman, J.R., Wetzel, G.M. and Uhr, J.W. (1977) J. Exp. Med., 146, 107-117.

9. Scott, D.W., Layton, J.E. and Nossal, G.J.V. (1977) J. Exp. Med., 146, 1473-1483.

10. DeHeer, D.H. and Edgington, T.S. (1977) J. Immunol., 118, 1858-1863.

11. Krakauer, R.S., Waldman, T.A. and Strober, W. (1976) J. Exp. Med., 144, 662-673.

12. Cantor, H., McVay-Boudreau, L., Hugenberger, J., Naidorf, K., Shen, F.W. and Gershon, R.K. (1978) J. Exp. Med., 147, 1116-1125.

13. Kincade, P.W., Self, K.S. and Cooper, M.D. (1973) Cell. Immunol., 8, 93-102.

DISCUSSION

WARNER:  Could you comment on whether the B cell colony assay
detects all B cells of a given differentiation stage, or only a
subset of them?

KINCADE:  We know that the assay detects immature B cells which
appear at 16.5 days in fetal liver and bear only IgM as well as B
cells which emerge later with the surface phenotypes $IgM^+$, $Ia^+$, and
$IgM^+$, $Ia^+$, $IgD^+$.  A comparable fraction of the B cells in newborn
spleen, adult bone marrow, spleen, lymph nodes, and fetal liver
form colonies.  These and other observations suggest that an entire
lineage of B cell development is detected.  On the other hand, CBA/N
mice never develop clonable B cells in any organ and their partial
deficiency seems to spare some functional B cells.  I have previously
proposed that these subsets belong to developmental pathways that
are separate from and parallel to the one detected by this assay.

WARNER:  Your data is suggesting an absence or deficiency of the
less mature B cells in the presence of excess numbers of mature plasma
cells.  How far back in the differentiation scheme from the mature
plasma cell do these mice have B cells?

KINCADE:  We have not determined the extent to which B cells can
differentiate towards a secreting plasma cell and still proliferate
in this cloning system.  In motheaten mice as in partially immunodeficient chickens, there is an abundance of Ig secreting cells and
very few B cells.  Perhaps in these situations the available B cell
precursors rapidly progress through the B cell stages to terminal
differentiation.  We have shown that the few functional B cells which
are present in motheaten mice and which fall within the window of
this assay are normal by several criteria.  The deficiency of
functional cells is much greater than that of total B cell numbers
however and there could be many phenotypically abnormal and possibly
non-functional B cells present.  Indeed Dr. Davidson and colleagues'
analysis with the FACS seems to reveal the presence of such abberent
cells.

# EVIDENCE FOR B CELL ACTIVATION OF MOTHEATEN MICE

WENDY F. DAVIDSON, SUSAN O. SHARROW, HERBERT C. MORSE, III AND THOMAS
M. CHUSED
National Institutes of Health, Bethesda, Maryland 20014 (USA)

## ABSTRACT

B lymphocytes from C57Bl/6J mice homozygous for the *me* gene exhibit
multiple phenotypic and functional abnormalities from an early age when
compared to normal littermates. These include a reduction in the
frequency of detectable sIg$^+$ cells in both spleen and lymph node,
alterations in the level of expression of surface IgM and IgD, increased cell size, increased expression of X-MuLV gp70, increased
plasma cell frequency and an increased frequency of TNP-specific plaque
forming cells. These observations provide strong evidence that *me/me* B
cells are polyclonally activated.

## INTRODUCTION

Mice homozygous for the recessive mutation motheaten (*me*) are short-
lived, have T and B cells which respond poorly to antigenic stimuli and
rapidly develop hypergammaglobulinemia and autoimmunity[1,2]. One interpretation of such a pattern of immunopathology is that *me/me* (me) B
cells are activated, possibly intrinsically. To address this point we
have examined spleen and lymph node cells from C57Bl/6J me and normal
littermates (nl),(*me/+* and *+/+*) mice one to four weeks of age for
abnormalities in the following parameters: frequency of surface
immunoglobulin positive (sIg$^+$) cells, cell size, surface immunoglobulin
isotype distribution, plasma cell frequency and expression of the major
coat protein (gp70) of xenotropic murine leukemia virus (X-MuLV). In
these studies, cells staining with affinity purified fluorescein-
conjugated F(ab')$_2$ fragments of immunoglobulin specific for mouse Fab
or μ chain or with fluorescein-conjugated F(ab')$_2$ fragments prepared
from antisera against X-MuLV gp70 were enumerated by flow microfluoro-
metery (FMF) performed on a FACS II[3,4].

## RESULTS

<u>Enumeration of surface immunoglobulin positive (sIg$^+$) cells.</u> Figure
1 shows the percentages of cells which stained with anti-Fab in spleens
from me and nl mice 8-33 days of age. At each age point, the me
population had a significantly lower frequency of sIg$^+$ cells than nls.

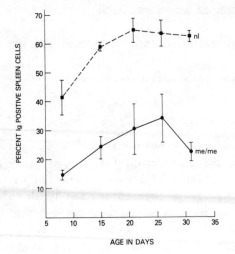

Fig. 1. Ontogeny of sIg$^+$ cells in me (solid) and nl (dashed) spleens. Me spleen cells have a decreased frequency of sIg$^+$ cells from one week of age.

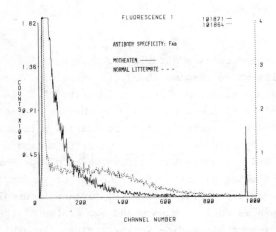

Fig. 2. Fluorescence distribution of total surface immunoglobulin of me (solid) and nl (dashed) spleen cells stained with anti-Fab. Me spleen contains more dull cells.

By the time adult levels of 50-60% sIg$^+$ cells were attained in normal animals, me spleens contained only 25-35% sIg$^+$ cells. A similar, though less marked, trend was also observed with lymph node cells. Less than 1% sIg$^+$ cells were detected in the thymuses of either me or nl mice of any age.

When analysed for cell size as measured by light scatter, me sIg$^+$ cells from both spleen and lymph node were significantly larger than the equivalent nl sIg$^+$ cells (data not shown).

B cell surface immunoglobulin isotypes. Figure 2 shows the fluorescence profiles obtained with anti-Fab stained spleen cells from one month old me and nl mice. At this age nl profiles are characterised by

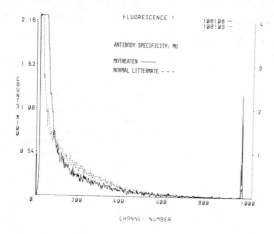

Fig. 3. Fluorescence distribution of surface IgM of me (solid) and nl (dashed) spleen cells stained with anti-IgM. The two profiles are quite similar.

the development of a peak of low to intermediate density sIg$^+$ cells (channels 200-500).

This peak, which has been shown to result from the acquisition of sIgD by sIgM$^+$ cells[5,6] is not seen with the me sIg$^+$ population. Instead, the majority (74%) of the me B cells are found between channels 50 and 200 and thus express very low levels of total surface Ig. Similar observations were also made with spleen and lymph node cells from me and nl mice 1-3 weeks of age.

Since anti-Fab detects both surface IgM and IgD it is possible that a lack of, or reduction in, the amount of either of these Ig isotypes could account for the low levels of sIg expressed on the majority of me B cells. Figure 3 shows the same populations of me and nl cells shown in Figure 2 but this time stained with anti-μ. With this reagent, the me and nl profiles were very similar, although me B cells generally had a higher proportion of cells with low surface Ig (42%) than control B cells (26%). It is also notable that the profiles obtained with me cells stained with anti-Fab or anti-μ were very similar in shape suggesting that me B cells express predominantly IgM. Together these data indicate that the IgD/IgM ratio for me B cells is significantly decreased. Similar decreases in the ratio of surface IgD/IgM have been reported following the stimulation of normal mice with mitogens[7] and SRBC[3].

Plasma cell content of me and nl spleens. The low density of surface immunoglobulin and the large size of B cells in me mice suggested that a significant number of these cells may be activated and

254

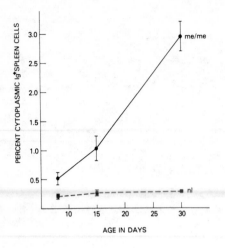

Fig. 4. Ontogeny of cells containing cytoplasmic immunoglobulin in the spleens of me and nl mice. Me spleen cells have a significantly higher frequency of positive cells by one week of age.

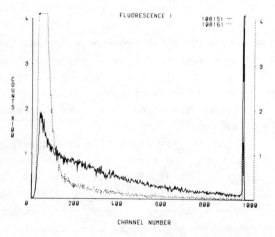

Fig. 5. Fluorescence distribution of X-MuLV gp70 of me (solid) and nl (dashed) spleen cells stained with anti-X-MuLV gp70 antibody. Me spleen contains significantly more bright cells.

possibly synthesizing Ig. Spleen cells were therefore fixed to render cytoplasmic Ig accessible, stained with either anti-Fab or anti-μ and then examined by FMF[3]. By this procedure, the percentage of cytoplasmic Ig$^+$ cells detected with either reagent was found to be significantly elevated in me spleens by as early as eight days of age and rose to 3%, or 10 times control levels, by 30 days of age (Figure 4).

Further evidence for polyclonal activation was obtained when spleen cells were examined for spontaneously occuring TNP-specific direct plaque forming cells (PFC). No significant differences in PFC counts were detected between me and nl mice until two weeks of age. There-

after, there was an exponential increase in TNP-specific PFC with age
in the me group (data not shown).

Membrane X-MuLV gp70 expresion. Lymphoid cells from other autoimmune strains such as NZB, BXSB and MRL/1 have been shown to express
high levels of X-MuLV gp70. This observation has also been made with
me mice. Figure 5 shows the fluorescence profiles obtained when spleen
cells from one month old me and nl mice were stained with anti-X-MuLV
gp70. Me cells expressed significantly more membrane gp70 than controls.

In contrast to the normal population, where gp70 expression on
spleen cells decreased over the first four weeks of life, spleen cells
from me mice 1-4 weeks of age all expressed similar levels of gp70. It
is also notable that in both the me and the nl populations large cells
expressed significantly more gp70 than small cells (data not shown).

DISCUSSION

Taken together, the increase in the size of $sIg^+$ cells, the combined
low levels of sIgM and apparently decreased sIgD/sIgM ratio, and the
increased frequency of plasma cells and spontaneous TNP-specific PFC in
me mice all provide strong evidence for polyclonal B cell activation.
The decreased frequency of $sIg^+$ cells may also reflect B cell activation, as antigen stimulated B cells have been reported to lose most of
their surface IgM and IgD[8]. Since me $sIg^-$ cells cannot be clearly differentiated from dull $sIg^+$ cells on fluorescence profiles our threshold
for me $sIg^+$ cells must be arbitrary. We may therefore have categorized
as negative some cells with an extremely low density of sIg. Data from
preliminary studies with mitogens suggests that blast cells have an
increased density of gp70. This, plus the additional observation that
large cells express more gp70 than small cells provides further support
for the suggestion that me B cells are activated.

The finding of phenotypic and functional changes in me mice as young
as one week of age suggests that me B cells are activated soon after
birth. The cause of this activation is unknown but may be the result
of an intrinsic defect of B cells as recently suggested for NZB mice[3]
or, alternatively, a secondary B cell abnormality caused by uncontrolled helper or dysfunctional suppressor T cells or an abnormally stimulating microenvironment. Although there is little evidence of such a
process, X-MuLV antigens may also play a role in the potentiation of
lymphocyte activation.

Extensive activation of the B cell population may provide an explanation for the poor responses of me spleen cells to SRBC[3] and LPS[9].

REFERENCES

1. Green, M.C. and Shultz, L.D. (1975) J. Hered. 66, 250-258.

2. Shultz, L.D. and Green, M.C. (1976) J. Immunol. 116, 936-943.

3. Chused, T.M., Moutsopoulos, H.M., Sharrow, S.O., Hansen, C.T. and Morse, H.C.,III, this volume.

4. Chused, T.M. and Morse, H.C., III (1978) in Origins of Inbred Mice, H.C. Morse, ed., Academic Press, New York, in press.

5. Scher, I., Sharrow, S.O., Wistar, R., Jr., Asofsky, R. and Paul, W.E. (1976 ) J. Exp. Med. 144, 494-506.

6. Vitetta, E.S., Melcher, U., McWilliams, M., Lamm, M.E., Phillips-Quagliata, J.M. and Uhr, J.W. (1975) J. Exp. Med. 141, 206-215.

7. Bourgois, A., Kitajima, K., Hunter, I.R. and Askonas, B.A. (1977) Eur. J. Immunol. 7, 151-153.

8. Abney, E.R., Cooper, M.D., Kearney, J.F., Lawton, A.R. and Park-house, R.M.E. (1978) J. Immunol 120, 2041-2049.

9. Davidson, W.F., Morse, H.C., III and Chused, T.M. (1978), in preparation.

DISCUSSION

WARNER: Could you comment on whether your data with motheaten B cells is comparable to similar studies with NZB derived B cells. Secondly, since the results suggest the specific absence of IgD (presumably M + D) bearing cells, is there any data to suggest whether the remaining B cells are on the less mature or more mature side of the absent M/D category. For example, is the frequency of IgG bearing cells increased?

DAVIDSON: The fluorescence profiles obtained with motheaten and NZB spleen cells stained with either fluoresceinated anti-Fab or fluoresceinated anti-$\mu$ are very similar in shape, suggesting that a high proportion of B cells in both strains have a similarly reduced IgD/IgM ratio, a property generally associated with B cell activation. Secondly, our data suggest a reduction in, rather than a total absence of surface IgD. When mean fluorescence intensity values of motheaten and normal litter mate cells stained with fluoresceinated anti-$\mu$ are compared, motheaten cells from mice as young as one week of age have surface IgM concentrations similar to four week old adult levels of surface IgM on motheaten B cells. Whether the subpopulation of cells with low surface IgM expression represents immature cells lacking surface IgD or very mature B cells has not been determined. Studies with fluoresceinated anti-$\gamma_1$ or $\gamma_2$ sera have shown no significant differences between cells from motheaten and normal litter mate mice of any age.

WARNER: Since your results on B cell differentiation are very similar between motheaten and NZB mice, it will be of considerable interest to ask whether the striking contrast between B stem cell colonies of young motheaten and old NZB/W mice as shown by Kincade, will also hold up when young NZB mice are studied. May I also ask whether the increased expression of xenotropic viral antigen is shown by host T cells and B cells or only by the B cells?

DAVIDSON: This will be an interesting aspect to follow up. My prejudice would be that young NZB mice will also have an increased B stem cell colony count. One trivial explanation for the apparent lack of B stem cell colonies in motheaten mice may be the extreme fragility of their cells to manipulation. There may be a selective loss of cells registering in this assay. In spleen of normal NZB, NZW, and DBA/2 mice, B cells express more surface xenotropic murine leukemia virus gp70 than T cells. In motheaten spleens light scatter analysis suggests that the large cells, which are predominantly B cells, express more gp70 than small cells. Furthermore, motheaten lymph node and thymus cell preparations express significantly less membrane gp70 than spleen cells.

DATTA: With regard to your finding of increased gp70 expression on the membrane of me/me B cells, did you find a similar increase in gp70 on the membrane of B cells of normal +/+ or me/+ litter mates that have been stimulated with polyclonal B cell mitogens like LPS?

DAVIDSON: Preliminary studies with LPS-stimulated cells have indicated a similar increase in gp70 expression to that observed with me/me spleen cells.

# GENETIC ASPECTS OF AUTOIMMUNE THYROIDITIS IN OS CHICKENS

LARRY D. BACON*, RANDALL K. COLE, CARL R. POLLEY, AND NOEL R. ROSE
Department of Immunology and Microbiology, Wayne State University
School of Medicine, Detroit, Michigan 48201 and Department of Poultry
Science, Cornell University, Ithaca, New York 14850

## ABSTRACT

Spontaneous autoimmune thyroiditis in OS chickens is polygenic, but
there are positive correlations between phenotype, cellular infiltration of the thyroids and antibody to thyroglobulin. The $B$-haplotype
has a strong influence on all parameters of disease, dependent on the
particular population studied.

## INTRODUCTION

Spontaneous autoimmune thyroiditis (SAT) in Obese strain (OS)
chickens is characterized by infiltration of the thyroids by mononuclear cells[1] and by production of serum antibodies to thyroglobulin
(TgAb), a storage product of thyroxine[2], starting at approximately 3
weeks of age. A correlation between TgAb and infiltration of the thyroids exists in early stages of disease[3]. However, delayed hypersensitivity to thyroglobulin has only been demonstrated later[4]. In the
current generation of 734 Obese strain (OS) chickens, 99% of the 12-
week-old birds have developed phenotypic symptoms of thyroiditis. The
severity of thyroiditis is greatly reduced[5,6], but not completely eliminated in bursectomized OS chicks, whereas neonatal thymectomy enhances thyroid pathology[7,8]. Although OS chickens have low levels of
circulating thyroid hormones, the phenotypic effects, including low
egg production, can be remedied by feeding thyroxine or a thyroxine
substitute[9]. Several recent reviews are available on the OS
chicken[10,11].

The OS strain of chickens was developed through genetic selection.
In the mid-1950's one of the authors (RKC) noted that a few females
in the Cornell C strain (CS), representing less than 1% of the flock,
developed symptoms characteristic of hypothyroidism. Breeders were
selected from affected families for the development of the hypothyroid
strain eventually named OS. Some inbreeding occurred in the strain in

*Present address: Regional Poultry Research Laboratory, U.S. Depart-
ment of Agriculture, SEA, East Lansing, Michigan 48823

the early years of selection, but once the strain was established in 1963, inbreeding was avoided to minimize genetic depression of reproductive traits. In this communication, we will describe several genetic factors that influence autoimmune thyroiditis in OS chickens and discuss the relevance of the $B$-haplotype to disease.

## THYROIDITIS IN OS CHICKENS IS POLYGENIC

The first report on the heredity of hypothyroidism in OS chickens established that the trait was polygenic and expressed with a degree of dominance[12]. Two mating experiments are relevant. In 1962 the incidence of hypothyroidism within the selected OS population was 50% for males and 81% for females. At that time four phenotypically obese OS males were mated with seven normal females from a White Leghorn strain unrelated to CS. Six percent of 68 $F_1$ males and 6% of 65 $F_1$ females were obese. During the following year, ten apparently normal $F_1$ females were backcrossed to five phenotypically obese OS males; 23% of 83 males and 42% of 79 females developed the typical OS syndrome.

Recently, W.E. Briles tested OS, CS and (OS x CS)$F_1$ birds for red blood cell antigens determined by each of the twelve known blood-group systems of chickens. The results confirmed earlier reports that the $B$-haplotype influenced the disease. The possibility of $L$, $C$, or $D$ locus effects required additional experiments. (OS x CS)$F_2$ experiments ruled out any influence by the $L$ or $C$ loci, but the $D$ locus requires additional work. Of the twelve known blood-group loci, only the $B$ haplotype, and possibly the $D$ locus, influence thyroiditis[13].

Data described below also show that one or more of the genetic loci determining severity of SAT is linked to the $B$-haplotype. However, the nominal influence of this gene complex on thyroiditis in non-inbred OS chickens that are segregating for $B$-haplotype[14], and the variability of occurrence of disease in birds of the same $B$-haplotype from substrain OSA or (OS x CS)$F_1$ hybrids supports the contention that thyroiditis is polygenic.

## DISEASE CORRELATIONS IN OS CHICKENS

The first hypothyroid CS females were slightly smaller than normal, developed long, silky body feathers, and accumulated large abdominal and subcutaneous fat deposits. The latter characteristic was responsible for naming the selected strain obese (OS). Individuals are assigned phenotypic scores of ++, +, $\pm$, or - for severe, moderate, suspect or nonaffected appearance, respectively. The selection scheme, described elsewhere[9], was based on phenotype alone, mating ++ birds at random with minimal inbreeding in the closed flock. We have previously determined that a high correlation ($r_s$ = 0.79)[3] exists between TgAb and thyroid cellular infiltration in (OS x CS)$F_1$ chicks, but no data have been presented on correlations of observed phenotype with TgAb or thyroid pathology. These values are of interest because breeders are selected from a phenotypically classified population at 8-10 weeks of age and 20 weeks of age.

The correlation of phenotype with TgAb and thyroid pathology has been calculated in (OS x CS)$F_1$ chickens during a recent experiment in which the influence of the $B$-haplotype on thyroiditis in $F_2$ chickens was under study. Briefly, $F_1$ chickens were produced by mating two normal CS $B^6B^6$ males with 16 $B^{13}B^{15}$ and nine $B^5B^{15}$ OS females at Cornell University, producing $B^5B^6$, $B^6B^{13}$, and $B^6B^{15}$ offspring. When the $F_1$ offspring were 12 weeks of age, phenotypic scores were assigned, serum antibody titers to thyroglobulin were determined by standard passive hemagglutination procedures employing chicken erythrocytes which had chick thyroglobulin attached by chromium chloride treatment[2,15], and thyroid infiltration was assessed histologically by LDB[16] for all $B^5B^6$ and $B^6B^{13}$ birds. The majority of the $B^6B^{15}$ birds were used to produce $F_2$ chickens. In Table 1 results of the non-parametric Spearman rank correlation coefficients are summarized. Correlations are presented separately for each $B$-haplotype group since $B$-haplotype influenced disease (see below). The following correlations are apparent.

a. TgAb and thyroid pathology. These traits showed a highly significant positive correlation in the susceptible $B^6B^{13}$ group, $r_s$ = 0.82; P $\leq$ 0.001, and in the less susceptible $B^5B^6$ group, $r_s$ = 0.59; P $\leq$ 0.001, of $F_1$ chickens.

b. Thyroid pathology and phenotype. These traits also had a highly significant positive correlation in both the susceptible $B^6B^{13}$ group ($r_s$ = 0.49; P $\leq$ 0.001) and the less susceptible $B^5B^6$ group ($r_s$ = 0.46; P $\leq$ 0.01) of $F_1$ chickens.

c. TgAb and phenotype. These two parameters also had positive correlation coefficients, although to a lesser degree. Thus, in the two susceptible groups of $B^6B^{13}$ and $B^6B^{15}$ chickens, the correlations were highly significant, $r_s$ = 0.40; P $\leq$ 0.01 and $r_s$ = 0.31; P $\leq$ 0.001, respectively. In less susceptible chickens the correlation coefficient, $r_s$ = 0.21, was non-significant.

In summary, highly significant positive correlations exist for thyroid pathology and TgAb and for thyroid pathology and phenotype in $F_1$ chickens regardless of their $B$-haplotype. A lower although highly

TABLE 1

CORRELATIONS OF PHENOTYPE, TgAb AND THYROID PATHOLOGY IN (CS x OS)$F_1$

| $B$-Haplotype | No. Birds | | TgAb | Pathology |
|---|---|---|---|---|
| $B^5B^6$ | 29 | Phenotype | $.21^a$ | .46** |
| | | TgAb | | .59*** |
| $B^6B^{13}$ | 66 | Phenotype | .40** | .49*** |
| | | TgAb | | .82*** |
| $B^6B^{15}$ | 87 | Phenotype | .31** | n.d. |
| | | TgAb | | n.d. |

$^a$Spearman rank correlation coefficient
** = P $\leq$ 0.01; *** = P $\leq$ 0.001
n.d. = not done

significant positive correlation coefficient exists between TgAb and
phenotype in $B$-haplotype susceptible birds.  There is no correlation
between these traits in less susceptible ($B^5B^6$) birds.

## $B$-HAPLOTYPE INFLUENCE ON AUTOIMMUNE THYROIDITIS

$B$-haplotypes in OS chickens were described in 1973[14], based on skin
graft rejection, graft-vs-host splenomegaly response and hemagglutination using alloantisera, all done within full-sib families of OS or CS.
These specific antisera were used to blood group the OS and CS populations at Cornell University.  We present both the standard $B$-nomencla-
ture, and our previous laboratory designation[13].  Two $B$-haplotypes
were identified in the OS population with approximately equal gene
frequency, $B^{13}$ (formerly $B^1$) and $B^5$ (formerly $B^4$), and a third, $B^{15}$
(formerly $B^3$) in low frequency, $\underline{f}$ = 0.04, in a heterozygous state.  In
CS the $B^{13}$, $B^{15}$, and another haplotype, $B^6$ (formerly $B^2$), were present
in moderate frequencies of 0.50, 0.30 and 0.20, respectively, whereas
the $B^5$ haplotype has not been observed.

OS inbred sublines.  The first evidence of a $B$-haplotype influence
on SAT was based on chicks produced from heterozygous $B^5B^{13}$ x $B^5B^{13}$
matings.  Most of the chicks were full-sibs in the OSA family[17], which
had an inbreeding coefficient of >0.40 at that time, but a few were
from the similarly inbred OSB and OSC families[16].  OSA, OSB and OSC
represent three separate inbred families developed from the original
OS by sib-matings of heterozygous parents[11].  We found that $B^{13}B^{13}$
and $B^5B^{13}$ birds have more severe symptoms of SAT than $B^5B^5$ siblings

at 3, 6, and 10 weeks of age. In that report we noted that numerous $B^5B^5$ birds produced by $B^5B^5$ homozygous matings in one inbred family, OSB, did have severe thyroiditis at 10 weeks of age. This finding suggested that the $B$-haplotype might influence disease differently in each inbred line.

Serum for TgAb titers of each bird produced from heterozygous matings in each inbred line was determined in 1974, 1975, and 1976. Sera of chicks 4-5 weeks old and juveniles 6-10 weeks old were analyzed separately using non-parametric statistics[14]. In 1975 and 1976 a number of OSB and OSC chicks were killed at 6 weeks of age and their thyroids rated histologically. The TgAb results are summarized in Figure 1.

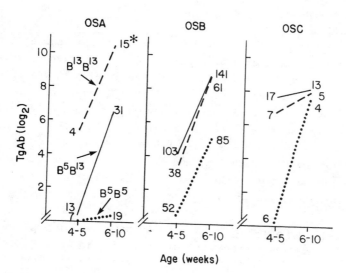

Fig. 1. Median titers of antibody to thyroglobulin in three sublines of OS.

*Numerals indicate number of animals.

We found that OSA ($B^5B^5$) chicks and juveniles had significantly less TgAb than did $B^5B^{13}$ or $B^{13}B^{13}$ OSA siblings. This finding corresponds well with earlier work where thyroid pathology was also observed[17]. Thyroid pathology was not determined in these birds because the offspring were needed for line reproduction. Large numbers of OSB birds were tested. The $B^5B^5$ chickens had significantly lower titers of TgAb than did $B^5B^{13}$ or $B^{13}B^{13}$ birds of the same age. However, as seen in Figure 1, there was a considerable increase in the median TgAb titer of $B^5B^5$ birds from 0.5 for chicks to 5.1 for juveniles. These results were supported by pathology scores of 6-week-old birds where $B^5B^5$ chickens had lower median pathology values, but no significant differences existed between them and their $B^5B^{13}$ or $B^{13}B^{13}$ siblings[13].

No OSC birds were produced from heterozygous matings in 1975 and limited TgAb data were collected in 1974 and 1976. TgAb titers of $B^5B^5$ chicks were significantly lower than those of $B^5B^{13}$ chicks, but not of $B^{13}B^{13}$ chicks (P = 0.07). TgAb titers of $B^{13}B^{13}$ and $B^5B^{13}$ chicks did not differ. At the juvenile age no significant differences for TgAb were found, as shown in Figure 1. The thyroid pathology scores of 6-week-old birds also showed no differences among $B$-haplotype groups[14].

Closed OS flock. In the closed flock of OS chickens at Cornell, $B^5$ and $B^{13}$ are segregating at about equal gene frequency, with $B^{15}$ at very low frequency. Essentially all birds develop thyroiditis. A study of 6-week-old chicks from OS $B^5B^5$ x $B^5B^{13}$ matings showed that a majority of OS $B^5B^5$ birds develop severe disease. Only 16 (37%) of 43 $B^5B^5$ chicks had a pathology score of $\leq 0.6$, the remainder having scores from 1.4 to 4.0. However, only two (5%) of 42 $B^5B^{13}$ siblings had this low level of pathology and others ranged from 2.0 to 4.0. Thus, while a majority of the OS $B^5B^5$ birds develop thyroiditis, $B^5B^{13}$ hatch-mates as a group have more severe SAT at 6 weeks of age (P $\leq$ 0.001)[18,11].

$F_1$ hybrids of OS and CS. Two studies have been carried out on $F_1$ chickens from matings between OS and CS to determine the role of the $B$-haplotype on SAT. In the first study, randomly selected CS hens with $B^{13}$, $B^{15}$ and $B^6$ haplotypes were mated with several $B^5B^{13}$ or $B^{13}B^{15}$ OS males in separate breeding pens at Cornell University. Fertile eggs were hatched at Wayne State University, the $B$-haplotype determined by hemagglutination, and the birds killed at 7 weeks of age. The Mann-Whitney U Test was used to determine the significance of differences among $B$-haplotypes on TgAb and thyroid pathology. A summary of the results is presented in Table 2. Whereas no $F_1$ birds of $B^5B^6$ and $B^6B^{13}$ genotypes had significant disease, significantly more had thyroiditis if they were $B^{15}B^{15}$, $B^{13}B^{15}$, $B^5B^{15}$, $B^{13}B^{13}$, $B^5B^{13}$, or $B^6B^{15}$.[3] It appears that the $B^{15}$ allele leads to more vigorous autoimmune response, and the $B^{13}$ allele moderates the response. $B^5B^6$ birds have milder disease.

TABLE 2

THYROIDITIS IN (OS x CS)$F_1$ CHICKENS

| Haplotype | Birds with thyroid pathology $\geq$ 0.4 or $\log_2$ antibody titer $\geq 4$[a] | |
|---|---|---|
| | Total | % |
| $B^5 B^6$ | 16 | 0 |
| $B^6 B^{13}$ | 9 | 0 |
| $B^{13} B^{13}$ | 18 | 28 |
| $B^5 B^{13}$ | 26 | 46 |
| $B^{15} B^{15}$ | 11 | 64 |
| $B^{13} B^{15}$ | 33 | 61 |
| $B^6 B^{15}$ | 17 | 76 |
| $B^5 B^{15}$ | 24 | 83 |

[a]The correlation value for antibody titer and pathology index is $r_s$ = 0.79, $P \leq 0.001$.

In a more recent $F_1$ experiment, from which the correlation data were collected, similar findings were obtained with 12-week-old $F_1$ chickens produced by $B^6 B^6$ CS males mated with $B^{13} B^{15}$ and $B^5 B^{15}$ OS females. The results are presented in Table 3. Only 14% of 29 $B^5 B^6$ birds developed significant thyroiditis in contrast to the $B^6 B^{13}$ and $B^6 B^{15}$ birds, where 57% or 58% of either group developed significant disease ($P \leq 0.001$). More $B^5 B^6$ and $B^6 B^{13}$ had significant SAT in this second study. Age may be an important factor. From these data on $F_1$ chickens, we suggest that $B^{13}$ and $B^{15}$ confer heightened susceptibility, whereas $B^6$ in the presence of $B^5$ is associated with resistance to SAT.

TABLE 3

THYROIDITIS IN (CS x OS)$F_1$ CHICKENS

| Haplotype | Birds with thyroid pathology $\geq$ 0.4 or $\log_2$ antibody titer $\geq 4$ | |
|---|---|---|
| | Total | % |
| $B^5 B^6$ | 29 | 14 |
| $B^6 B^{13}$ | 66 | 57 |
| $B^6 B^{15}$ | 88 | 58 |

DISCUSSION

We conclude from our genetic studies of the OS chicken that SAT is a polygenic trait, and that positive correlations exist among phenotypic symptoms of thyroiditis, cellular infiltration of the thyroids, and TgAb. The $B$-haplotype has a significant, time-related influence on SAT. The $B$ influence is barely discernable at early ages in non-inbred OS, inbred subline OSC and, to a lesser degree, inbred subline OSB. The $B$ influence is marked in OSA and $F_1$ hybrids of OS and the progenitor CS White Leghorns, but it is not a simple single determinant of disease. Some OSA $B^5B^5$ and some $F_1$ $B^5B^6$ non-susceptible chickens ultimately develop severe thyroiditis. This observation underlines the complexity involved in the $B$-haplotype influence.

The complexity of the $B$ influence may be in the number or magnitude of the other genetic determinants of disease present in chickens with a non-susceptible $B$-haplotype. In OSA some $B^5B^5$ did develop significant disease at a time when the inbreeding coefficient ranged from 0.4 to 0.7. It is easier to envision some segregation for loci not linked to the $B$-haplotype than two alternative forms of the $B^5$ alleles in these birds. Numerous mixed cultures of $B^5B^5$ OS lymphocytes have failed to respond to other $B^5B^5$ OS lymphocytes in culture, suggesting all $B^5B^5$ birds are identical in this assay[19].

Alternatively, the complexity of the $B$-haplotype influence may lie in undetected differences within the haplotype. Possibly more than one locus is involved within the $B$-haplotype, which is carried on a microchromosome[20]. Evidence exists that the $B$-haplotype is complex in chickens, although it appears crossing over is infrequent[21-23].

This conference attests to the fact the major histocompatibility locus is complex in mammals. Subloci might act through complementary gene action and more than one determinant may influence the immune response. Separate subloci may augment the immune response[24,25], while others may suppress[26]. In addition, there may be a non-specific effect on T-cell proliferation determined by genes at the $B$-haplotype[27].

The function of the $B$-haplotype genes in influencing autoimmune thyroiditis of OS chickens awaits elucidation. $F_2$ studies are underway and may demonstrate that genetic complementation is involved. The mechanism of $B$-haplotype control may ultimately be dependent upon the development of congenic lines of chickens. In addition to $B$-haplotype control, other genetic loci may be determining alterations in the target organ, or susceptibility to microbial or environmental factors, or abnormalities in the primary lymphoid organs[11].

ACKNOWLEDGMENTS
The skilled technical assistance of Ms. S. Kirchberger, Mr. Larry
Devroy, and Ms. Margaret Clark is greatly appreciated. This work was
supported by NIH research grant AM-20028.

REFERENCES
1. Kite, J.H., Jr., Wick, G., Twarog, B. and Witebsky, E. (1969)
   J. Immunol., 103, 1331-1341.
2. Witebsky, E., Kite, J.H., Jr., Wick, G. and Cole, R.K. (1969)
   J. Immunol., 103, 708-715.
3. Bacon, L.D., Sundick, R.S. and Rose, N.R. (1977) in Avian Immunology, Benedict, A.A. ed., Plenum Press, New York, pp. 309-315.
4. Welch, P.C. and Kite, J.H., Jr. (1971) Fed. Proc. (Abstract),
   30, 306.
5. Cole, R.K., Kite, J.H., Jr. and Witebsky, E. (1968) Science,
   160, 1357-1358.
6. Wick, G., Kite, J.H., Jr., Cole, R.K. and Witebsky, E. (1970)
   J. Immunol., 104, 45-53.
7. Wick, G., Kite, J.H., Jr. and Witebsky, E. (1970) J. Immunol.,
   104, 54-62.
8. Welch, P., Rose, N.R. and Kite, J.H., Jr. (1973) J. Immunol.,
   110, 575-577.
9. Cole, R.K., Kite, J.H., Jr., Wick, G. and Witebsky, E. (1970)
   Poultry Sci., 49, 839-848.
10. Wick, G., Sundick, R.S. and Albini, B. (1974) Clin. Immunol.
    and Immunopathol., 3, 272-300.
11. Rose, N. R., Bacon, L.D. and Sundick, R.S. (1976) Transplant.
    Rev., 31, 264-285.
12. Cole, R. K. (1966) Genetics, 53, 1021-1033.
13. Rose, N.R., Bacon, L.D., Sundick, R.S. and Briles, W.E. in
    Comparative and Developmental Aspects of Immunity and Disease,
    Gershwin, M.E. and Cohen, N. eds., Pergamon Press, New York,
    in press.
14. Bacon, L.D., Kite, J.H., Jr. and Rose, N.R. (1973)
    Transplantation, 16, 591-598.
15. Gold, E.R. and Fudenberg, H.H. (1967) J. Immunol., 99, 859-866.
16. Bacon, L.D., Kite, J.H., Jr. and Rose, N.R. (1974) Science,
    186, 274-275.
17. Bacon, L.D., Kite, J.H., Jr. and Rose, N.R. (1973) Fed. Proc.
    (Abstract), 32, 1026.
18. Bacon, L.D. (1975) Fed. Proc. (Abstract), 34, 980.
19. Jones, R.F. and Bacon, L.D. (1977) Fed. Proc. (Abstract), 36, 1191.
20. Bloom, S.E., Cole, R.R. and Bacon, L.D. Poultry Sci. (Abstract),
    57, in press.
21. Schierman, L.W. and McBride, R.A. (1969) Transplantation, 8,
    515-516.

22. Hala, K., Vilhemova, M. and Hartmanova, J. (1976) Immunogenetics, 3, 97-103.

23. Pink, J.R.L., Droge, W., Hala, K., Miggiano, V.C. and Ziegler, A. (1977) Immunogenetics, 5, 203-216.

24. Tomazic, V., Rose, N.R. and Shreffler, D.C. (1974) J. Immunol., 112, 965-969.

25. Benedict, A.A., Pollard, L.W., Morrow, P.R., Abplanalp, H.A., Maurer, P.H. and Briles, E.W. (1975) Immunogenetics, 2, 313-324.

26. Benacerraf, B., Kapp, J.A., Debre, P., Pierce, C.W. and de la Croix, F. (1975) Transplant. Rev., 26, 21-38.

27. Longenecker, B.M., Pazderka, F. and Law, G.R. (1972) Transplantation, 14, 424-431.

DISCUSSION

GASSER: Have you tested any B locus recombinants for susceptibility to thyroiditis?

BACON: Several hundred OS and CS chickens produced from B-haplotype heterozygous parents have been tested on alternate years since 1973. We never found evidence for a recombinant. However, some CS birds were found to be **trisomic** for the B-haplotype placing the locus (see ref. 20 of text) on a microchromosome.

ROITT: A standard antigenic challenge to well-inbred mice, i.e., with identical genome, reared under standard conditions produces a range of antibody responses. Is it necessary to postulate further genes influencing the variable responses in your $B^5B^5$ chickens?

BACON: No. However when one looks at the variability among inbred sublines and observes the early breeding data where several generations were needed before high proportions of birds developed hypothyroidism, and studies the $F_1$ and backcross data, it becomes evident that further genes are determining thyroiditis.

ROITT: Presumably then there is a question regarding the completeness of inbreeding of the $B^5B^5$ homozygotes.

BACON: OSA, OSB, and OSC had inbreeding coefficients ranging from 0.4 to 0.7 as described and were not fully inbred. While two $B^5$ haplotypes may have been present in these lines, it seems unlikely. It is more likely that residual heterozygosity at unlinked loci were determining the unexpected results. The OS line is not inbred, but a closed flock at Cornell (see ref. 9).

MACKAY: Does the sex of the bird have any influence on various expressions of thyroiditis in Obese chickens?

BACON: In the current OS strain, all symptoms are severe in both sexes. However in the first five years of selection, hypothyroidism was expressed more frequently and to a more pronounced degree in females than in males, and males were more apt to recover (see ref. 12).

LENNON: Have these birds been examined for the presence of auto-antibodies to TSH receptors?

BACON: Not to my knowledge.

VOLPE: The OS chicken thyroiditis is initially a B lymphocyte thyroiditis, but subsequently delayed hypersensitivity is detected. Is this sequence invariable? How long does it take?

BACON: It is variable and only occurs a month or more after severe thyroiditis has occurred (see ref. 8).

VOLPE: What is the significance of the change from a purely B cell thyroiditis to one associated with delayed hypersensitivity? Why does it happen?

BACON: Since delayed hypersensitivity is only observed long after severe thyroid infiltration and damage has occurred the significance must be minimal. I do not know, but possibly a different antigenic determinant(s) becomes involved.

SVEJGAARD: It may be of interest to note that Morten Simonsen has found evidence of an extremely strong linkage disequilibrium within the chicken MHC.

BACON: The $B^5$ allele may have come from a male other than CS, which was introduced in the early years of selection (see ref. 13), or $B^5$ may have been in CS in 1956 and subsequently lost by 1972 when blood was first tested for B antigens. Perhaps $B^5$ was maintained in OS during selection for fitness; that is, $B^{13}B^{13}$ birds may have had thyroiditis so severely that they produced few eggs or poorer quality semen and were selected against in favor of $B^5B^{13}$ or $B^5B^5$ birds.

WARNER: I think this might be an appropriate stage in today's session to raise a general point. In yesterday's session on human autoimmunity, we were particularly discussing the possible role of HLA linked genes. Yet until this present paper on thyroid auto-immunity in chickens, the subject of MHC relation to spontaneous autoimmunity in animals has not arisen. In none of our recent discussions on mouse models has there been consideration of the potential role of MHC linked genes, and it may also be of significance to note that in those discussions, we have been considering general-ized autoimmunity until in this present paper on chickens which concerns organ specific autoimmunity. I would like to hear some discussion on this general point.

SCHWARTZ: Because they are mutations of inbred strains!

MURPHY: Yes. But there are good ways of getting at the problem.

ROITT: The clustering of human organ-specific diseases on the one extreme and of non-organ specific disorders (e.g., with SLE) on the other pole of the autoimmune spectrum with only mild associa-tions between the two groups strongly indicates some fundamental differences in the mechanisms generating autoimmunity. The organ-specific group may show a greater MHC related component and if so, H-2 relationships in the murine lupus-like syndromes discussed might be less prominent.

GERSHON: It is **true** that if we study inbred mice, we will only notice autoimmune disease in those strains that either have or lack appropriate Ir genes. Thus, the way to test for MHC restriction would be to backcross the mutant locus onto mice of different H-2 haplotypes. Ed Murphy is doing this type of work and perhaps he might like to comment on this point.

MURPHY: We have reached backcross generations N4 to N6 with one pair in seven different inbred strains. The most striking strain background effect is a decrease in the degree of lymphoproliferation and an increase in survival time on strain C57BL/6J.

SPONTANEOUS AUTOIMMUNE DISEASE WITH ASSOCIATED ACQUIRED IMMUNE
DEFICIENCY IN CHICKENS

M.E. GERSHWIN, J. MONTERO, J.  EKLUND, H. ABPLANALP, R.M. IKEDA,
A.A. BENEDICT, L. TAM, AND K. ERICKSON
Sections of Rheumatology-Clinical Immunology, Departments of Medicine,
Pathology and Poultry Science, University of California, Davis and
University of Hawaii, Honolulu, Hawaii

ABSTRACT

   In 1974 sera from several chickens from two particular families of
hybrids, were found to have reduced levels of 7S Ig.  The expression
of this abnormality is a multifactorial inherited acquired dysgammaglobulinemia.  Affected birds, in addition to their Ig defects, have
an accelerated morbidity and mortality, including development of
Coombs' positive hemolytic anemia, cryoglobulins and rheumatoid
factor.  Furthermore, there is a significantly elevated risk of
thymomas and a statistically significant correlation between these
autoimmune manifestations and the acquired dysgammaglobulinemia.
The mechanism(s) for these defects are unclear, but an abnormality
in a T cell regulatory population is suggested.

INTRODUCTION

   Recently a line of chickens with an inherited immunodeficiency,
analogous to acquired agammaglobulinemia of humans has been described
(1,2).  This syndrome, similar to an earlier description of a
disorder by Lösch and coworkers is characterized by early   normal
Ig synthesis followed by dysgammaglobulinemia (3).  Furthermore, the
dysgammaglobulinemia has a variable expression including total absence
of serum IgG, marked elevation in serum IgG and generally elevated
serum IgM.  Of particular interest has been an increased premature
mortality and anemia in such birds.  Because of these later problems,
our laboratory has initiated a study to determine and quantitate
histologic and serologic alterations in this, and age matched control
lines of birds.  We report herein that these dysgammaglobulinemic
chickens have several associated immunologic disorders including
appearance of autoantibodies and development of lymphoproliferative
disease.

In 1973, three highly inbred lines were mated to produce $F_1$ progeny of two new lines (UCD 2 X UCD 3 $\longrightarrow F_1$-140 and UCD 7 X 3 $\longrightarrow F_1$-142). The $F_2$ and $F_3$ generations of these new lines as well as crosses among the $F_3$ progeny (140 X 142) were the subject of the original genetic studies of the dysgammaglobulinemia. The first abnormal birds, detected by sera immunoelectrophoresis, were found in the $F_2$ generations of both lines. Subsequent study of specific ancestor's serum indicated that the defect probably originated in a single line 3 male, who was a common parent to both line 140 and 142.

## MATERIALS AND METHODS

Chickens. All chickens were housed individually at the Hopkins Poultry Plant at the University of California, Davis or at the University of Hawaii. Matings and pedigree histories were maintained at Davis.

Genetics. To determine and evaluate the mode of inheritance of this disorder, preliminary crosses have been performed. These included the mating of abnormal cocks to normal and abnormal hens; mating of normal cocks to normal hens. All "normal" birds in these backcrosses were derived from line 140, but were shown to have normal Ig levels. The number of eggs set, the fertility of such eggs, the percentage of hatched eggs, and percentage of dysgammaglobulinemic offspring were quantitated.

Dysgammaglobulinemia. Line 140 and control chicks were bled from the jugular vein, the blood allowed to clot and the sera removed. Concentrations of sera Ig were performed by radial immunodiffusion using heavy chain specific anti-chicken 7S and 17S. Known reference standards were employed with each assay (4).

Hematology and Coombs' test. At 6 months of age and at serial intervals thereafter, birds were bled as above and complete blood counts including differentials performed. Additional red cells were removed by centrifugation and washed X 3 in phosphate buffer saline (PBS). A 5% red cell suspension was prepared and a monospecific micro direct Coombs' test using U-shaped microtiter wells prepared as previously described (5). Known positive and negative controls were included. In addition, the ability of serially aged chicken sera to agglutinate a 5% suspension of normal chicken RBC's was determined.

Chromoum survival studies. Autologous chicken red cells were prepared from birds with a positive and negative direct Coombs' test. These cells were labelled with sodium chromate-51, extensively washed and resuspended. The clearance of these labelled cells was determined by collection and monitoring of blood in a gamma scintillation counter (6).

Cryoprecipitates. Blood was collected from warmed chickens into pre-warmed sterile tubes and allowed to clot at 38-40°C. The serum was collected and placed at 4°C for 72 hours. The presence or absence of a cryoprecipitate was then determined (7). Select cryoprecipitates were further studied following a solubilization procedure, including 0.3 M NaCl for 1 hour at 38°C followed by an equal volume of distilled water for one hour at 38°C. This solubilization was performed only on extensively washed (X 5 and X 10) cryoprecipitates (8). The solubil-ized cryoprecipitates were studied by immunoelectrophoresis using anti-whole chicken sera, anti-chicken 7S and anti-chicken 17S. Furthermore, the ability of these solubilized cryoprecipitates to bind chicken 7S was determined by latex fixation as noted below. Finally, solubilized cryoprecipitate was homogenized in complete Freunds adjuvant and injected X 2 into New Zealand White female rabbits at 14 day intervals. Rabbits were bled and the antibody activity directed against normal chicken components studied by IEP.

Rheumatoid factor. Rheumatoid factor assays were performed by a modification of the Singer and Plotz latex fixation test using chicken 7S Ig (9). Briefly, the latex-globulin mixture was prepared by mixing 0.1 ml stock latex suspension (Dow Chemical 0.807 μm) with 0.5 ml 0.5 percent 7S Ig (DEAE cut) and 9.4 ml glycine-NaCl buffer pH 8.2. This mixture was stabilized by adding 0.4 percent bovine serum albumin (9). Serial dilution of test sera were performed in the buffer beginning at 1/20. Equal volumes of test reagent and diluted test sera were mixed and incubated at 56°C for 2 hours. Tubes were kept at RT over-night and centrifuged at 800 xg for about 3 minutes before reading.

Pokeweed mitogen induced immunoglobulin synthesis. Peripheral blood lymphocytes from normal and abnormal chickens were isolated. The ability of these cells to synthesize 7S Ig, in the presence or absence of pokeweed mitogen, was quantitated by radioimmunoassay (10). In addition, co-culture experiments were performed by mixing normal and abnormal lymphocytes in the presence or absence of PWM. The percentage suppression was calculated for each assay (11).

Histologic features. All birds were vigorously autopsied when morbund. Specific attention was placed on thymus, bursa, and spleen. Tissue from all parenchymal organs and bone marrow were prepared and preserved.  Sections were stained using hematoxylin and eosin and interpreted by an unbiased observer.

RESULTS

Genetics.  Backcross experiments suggest that the dysgammaglobulinemia is highly heritable, but not due to a single gene.  Moreover, the majority of the autoimmune phenomena discussed below are highly associated with the immunodeficiency, suggesting a common etiologic origin.  The fertility of abnormal cocks by abnormal hens was lower than of other crosses studied herein (Figure 1).  Moreover, there was a slight but not significant reduction in eggs hatched in crosses of abnormal parents.  However, and in particular, there were a significantly greater number of abnormal offspring when both parents had dysgammaglobulinemia.  On the other hand, there were a significant number of dysgammaglobulinemic offspring (20-30%) from matings of normal parents.

Serial immunoglobulin levels.  Until approximately 40 days of age, the sera levels of 7S and 17S Igs were indistinguishable from control birds.  Thereafter, study of the mean levels of birds revealed a progressive increase in 17S Ig and a depression for approximately 100 days of 7S (Figure 2).  However, these mean values are misleading and must be interpreted in view of individual data (Figure 3).  For example, analysis of four individual birds reveal animals with almost total absence of 7S Ig, marked elevation of 7S Ig, and either markedly elevated or normal levels of 17S Ig (Figure 3).

Co-culture experiments.  The production of 7S Ig by unstimulated lymphocytes from dysgammaglobulinemic birds is significantly lower than normal controls.  Furthermore, stimulation of these cultures with PWM increases the production of Ig for both normal and abnormal lymphocytes.  However, the magnitude of increment is significantly lower for abnormal cells (Figure 4).

The co-culture of abnormal and normal lymphocytes, in the presence of PWM, reveals a significant suppression of Ig synthesis by lymphocytes from dysgammaglobulinemic birds.  Preliminary evidence suggests that the cell responsible is a T cell (Figure 4).

Hematology. There was no significant differences in hematocrit, hemoglobin,  white cell count or differential in either 8 months or 15 month old birds when compared to controls (Table 1).  However, when the hematocrit of Coombs' positive birds were compared with Coombs' negative birds, a significant anemia was observed (Table 1).

Approximately 20% of birds 10 months of age had a positive Coombs'
test (Table 2); this figure progressively increased with age.    The
majority of such Coombs' reactions were IgM anti-erythrocyte antibodies.    There were only rare examples of detectable sera anti-
erythrocyte antibodies.

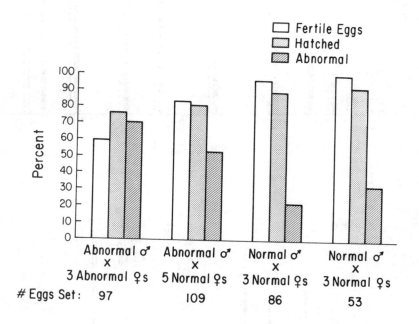

Fig. 1.    Percentage of birds with dysgammaglobulinemia in crosses of
abnormal and normal parents.

TABLE 1

COMPLETE BLOOD COUNTS

|                 | Line 140 | Coombs' positive | Controls |
|-----------------|----------|------------------|----------|
| HCT             | 28       | 21               | 30       |
| Hgb             | 9.3      | 7.0              | 11.1     |
| WBC (X $10^6$)  | 29       | 24               | 27       |
| RBC (X $10^6$)  | 2.97     | 2.33             | 3.04     |

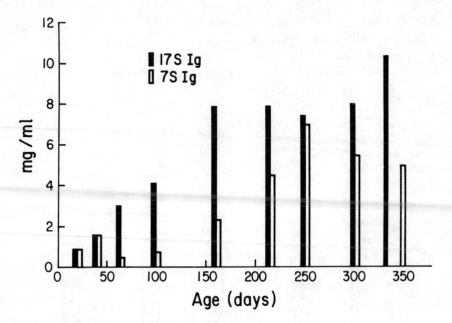

Fig. 2.   Mean levels of 7S Ig and 17S Ig in serially age line 140 birds.

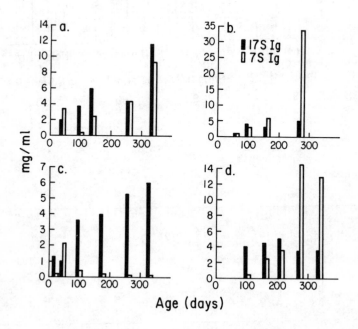

Fig. 3.   Levels of 7S Ig and 17S Ig in four individual birds with age.
Note the variable expression of the dysgammaglobulinemia.

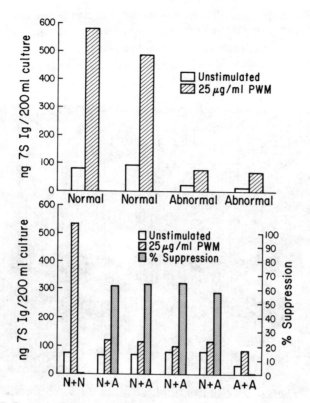

Fig. 4. 75 Ig synthesis by PBL from normal and abnormal chickens in the presence and absence of PWM(4A). Lymphocytes from abnormal birds suppressed Ig synthesis of normal birds (4B).

Chromium survival. There was a statistically significant increased decay of chromium labelled autologous red cells in Coombs' positive versus Coombs' negative birds (Table 3). For example, at ten days post-injection, there were 50% surviving chromium red cells compared to 20% in abnormal birds (p < .01).

Cryoprecipitates. Cryoprecipitates were detected in more than 50% of birds greater than 5 months of age (Table 4). Immunodiffusion of solubilized cryoprecipitates revealed the presence of IgM, occasionally IgG, and several other unidentified cryoproteins (Figure 5). Indirect analysis of such cryoproteins using rabbit anti-cryoprotein antibodies suggests the presence of IgM, IgG and IgA (Figure 5). Additionally, there appears to be rheumatoid factor-like activity in the cryoprotein. However, the specificity of this activity remains to be conclusively studied.

<u>Rheumatoid factor</u>.  Rheumatoid factor activity in the sera of abnormal birds can be detected as early as 6 months.  The frequency of this activity increases to 100% of birds at 18 months of age (Table 5). As with the cryoprecipitate, the specific activity of the rheumatoid factor remains to be assayed.

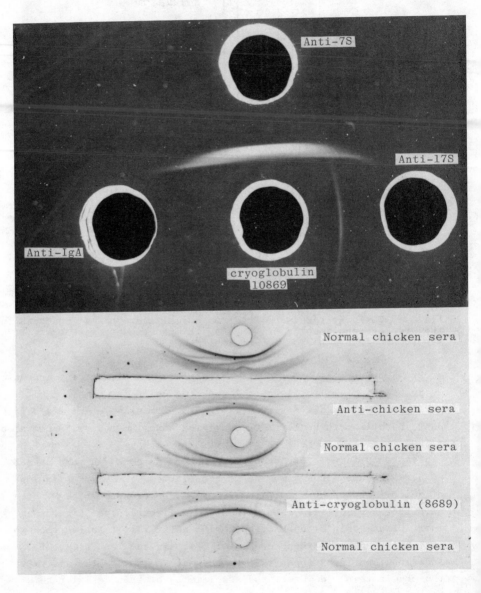

Fig. 5.  Immunodiffusion and immunoelectrophoresis of solubilized cryoglobulins.

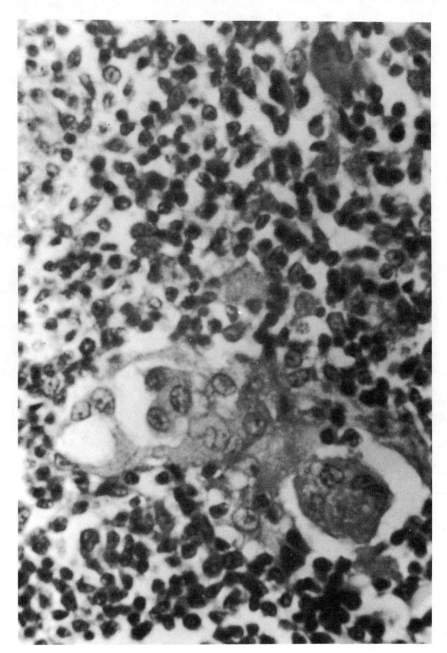

Fig. 6. Photograph (H&E X 150) of a thymoma from a 14 month old bird. Note the distinctly abnormal coalesced epithelial cells.

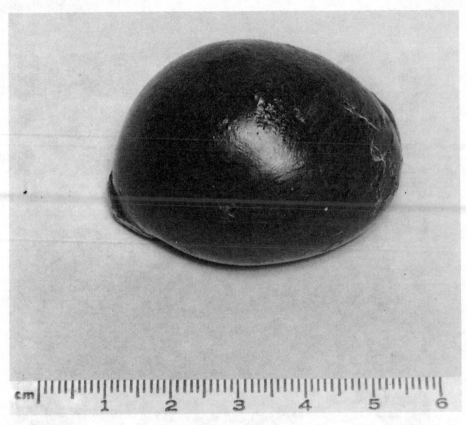

Fig. 7. Photograph of spleen from 15 month old bird with a Coombs' positive hemolytic anemia. Histologically the spleen is composed of an engorged reticulo-endothelial system.

TABLE 2

DIRECT COOMBS' (+) FREQUENCY WITH AGE

| Line | Hatch | Age(days) | # positive / # tested | % positive |
|------|-------|-----------|-----------------------|------------|
|  | Fall 1976 | 287-320 | 4/23 | 17.4 |
|  |  | 362-404 | 5/13 | 38.5 |
|  |  | 511-541 | 4/8 | 50 |
| 140 + Outbreds | Spring 1977 | 195-211 | 3/47 | 6.4 |
|  |  | 330-370 | 8/62 | 13.4 |
|  | Fall 1977 | 183-211 | 0/40 | 0 |
| 159 Control | Spring 1977 | 211 | 0/22 | 0 |
|  |  | 370 | 0/20 | 0 |

TABLE 3

AUTOLOGOUS CHICKEN RBC ($^{51}$Cr) SURVIVAL

| Group | % Remaining | |
|---|---|---|
| | Days | PM 10 |
| Coombs' Positive | $42 \pm 7$ | $20 \pm 7$ |
| Coombs' Negative | $76 \pm 5$ | $51 \pm 4^*$ |

* p < 0.01

TABLE 4

CRYOPRECIPITATES

| Line | Hatch | Age(days) | # positive / # tested | % |
|---|---|---|---|---|
| | Fall 1976 | 348–378 | 7/8 | 87.5 |
| | | 413–443 | 2/3 | 66.7 |
| 140 + Outbreds | Spring 1977 | 330–370 | 38/62 | 61.3 |
| | Fall 1977 | 180–211 | 46/63 | 73 |
| 159 Control | Spring 1977 | 211 | 1/16 | 6.25 |
| | | 370 | 3/20 | 15 |

Histology.   There was no uniform cause of death in birds.   None-theless, there were some examples of bacterial infection including pneumonia, and frequent thymomas (Figure 6).   The thymomas did not appear to be invasive.   In addition, both splenomegaly and hypo-

TABLE 5

RHEUMATOID FACTOR

| Line | Hatch | Age(days) | # positive / # tested | % |
|---|---|---|---|---|
| | | 348–378 | 6/8[2] | 75 |
| | Fall 1976 | 413–443 | 3/4[2] | 75 |
| | | 511–541 | 8/8[3] | 100 |
| 140 + Outbreds | Spring 1977 | 330–370 | 29/63[3] | 46 |
| | | 142–153 | 10/75[2] | 13.3 |
| | Fall 1977 | 180–211 | 13/58[3] | 22.4 |
| 159 Control | Spring 1977 | 211 | 1/16[2] | 6.25 |
| | | 370 | 0/20[3] | 0 |

[1] positive = 1/80 titer or greater

[2] tested with Hyland RF latex reagent

[3] tested with latex fixation using chicken 7S-Ig

splenia were noted (Figure 7). Finally, there were numerous examples of diffuse lymphocytic infiltrates throughout parenchymal organs, similar to that described for human pseudolymphoma. Thus far, there have been no major abnormalities observed in the bursa.

DISCUSSION

The usefulness of chickens as models for immunologic studies has long been recognized. Indeed, we believe that this unique line of mutant birds will prove to be of significant utility in our understanding of both autoimmunity and immunodeficiency. Although our data is thusfar preliminary and non-mechanistic, we suggest that the immunodeficiency is highly heritable with genetic control by one or more genes with incomplete penetrance. Furthermore, the statistical association of dysgammaglobulinemia and autoimmunity suggests a common etiology.

The birds described herein have, in addition to defects in immuno-globulin levels, appearances of a Coombs' positive hemolytic anemia, rheumatoid factor, cryoprecipitates, and thymoma. These features are strikingly similar to the New Zealand Black mouse. Indeed, the existence of isolated thymoma and pseudolymphoma is a very rare event in chickens and much unlike the several forms of avian lymphomas (12-14). Furthermore, the disease appears very similar to common variable immunodeficiency of humans because of its apparent multi-factorial inheritance, the acquired nature of the defect, variations among affected chickens, normal proliferative response to mitogens, failure of lymphocytes to synthesize 7S Ig, the presence of an apparent T cell suppression population, development of cryoprecipitates, and lastly, the occurrence of autoimmune disease (15).

The increased prevalence of rheumatoid arthritis and collagen vascular disease in human agammaglobulinemia is a well documented phenomena (16,17). The mechanism of these events has been suggested as due to an abnormal immunologic regulatory function. Indeed, we propose a similar defect herein. Furthermore, numerous examples at autopsy of pleuritis, pericarditis and synovitis have been found. Similarly, peripheral necrotic combs were occasionally in the birds during the winter months generally in association with cryoglobylobu-lins.

In humans, disorders of the immune system which lead to agammaglob-ulinemia have been considered to be primary defects in B cell matura-tion and terminal differentiation (18,19). The influence of other lymphoid subpopulations on this defect appears to be a critical determinant of this disease expression and related heterogeneous syndromes. There obviously remains a considerable number of experimental studies to be performed on these birds. Questions which must be answered include the genetic basis, the association between immunodeficiency and autoimmunity, the sequence of events resulting in thymomas, the influence of the later tumors on immune expression, and surface marker characteristics of lymphoid tissue. In addition, identification of a specific abnormality in a lymphoid subpopulations may lead itself to considerable therapeutic latitude. Such studies might lend considerable credence to the use of newer therapeutic modalities in humans.

## ACKNOWLEDGEMENTS

The authors wish to thank David Milich, Jasmin Shen, and Leona Schmidt for their technical and secretarial assistance.

## REFERENCE

1. Benedict, A.A., Abplanalp, H.A., Pollard, L.W., and Tam, L.Q. (1977) in Avian Immunology, Benedict, A.A. ed., Plenum Publishing, New York, pp. 197-205.

2. Benedict, A.A., Chanh, T.C., Tam, L.Q., Pollard, L.W., Kubo, R.T. and Abplanalp, H.A. (1978) in Comparative and Developmental Aspects of Immunity & Disease (eds. Gershwin, M.E. and Cooper, E.L.) in press Pergamon 1978.

3. Lösch, U. and Hoffman-Fezer, G. (1973) Zbl. Vet. Med. A. 20, 596-605.

4. Mancini, G., Carbonara, A.O. and Heremans, J.F. (1965) Immunochemistry, 2, 235-254.

5. Gershwin, M.E. and Steinberg, A.D. (1974) Clin. Immunol. Immunopath. 4, 38-45.

6. Wright, R.H. and Kreier, J.P. (1969) Experimental Parasitology, 25, 339-352.

7. Grey, H.M. and Kohler, P.F. (1973) Seminars in Hematology, 10, 87-112.

8. Weisman, M. and Zvaifler, N. (1975) J. Clin. Invest., 56, 725-739.

9. Singer, J.M. and Plotz, C.M. (1956) Am. J. Med., 21, 888-892.

10. Wu, L.Y.F., Lawton, A.R. and Cooper, M.D. (1973) J. Clin. Invest., 52, 3180-3189.

11. Waldmann, T.A., Broder, S., Blaese, R.M., Durm, M., Blackman, M. and Strober, W. (1974) Lancet, 2, 609-613.

12. Hans-Christoph, L. (1964) in Avian Tumor Viruses, National Cancer Institute Monograph 17. Beard, J. ed., U.S. Department of Health, Education and Welfare, Public Health Service, National Cancer Institute, Bethesda, Maryland. pp. 37-62.

13. Biggs, P.M. (1976) in Differential Diagnosis of Avian Lymphoid Leukosis and Marek's Disease. Payne, L.N. ed. Commission of the European Communities, Luxemborg. pp. 67-86.

14. Campbell, J.G. (1969) Tumours of the Fowl. J.B. Lippincott Company, Philadelphia.

15. Gershwin, M.E. and Steinberg, A.D. (1976) Semin. Arthritis Rheum., 6, 125-164.

16. Fudenberg, H. and Solomon, A. (1961) Vox. Sang. (Basel) 6, 68-78.

17. Good, R.A., Kelly, W.P. Rotstein, J., and Varco, R.L. (1962) Progr. Allerg., 6, 187-319.

18. Gajl-Peczalska, K.J., Park, B.H., Biggar, W.D. and Good, R.A. (1973) J. Clin. Invest., 52, 919-928.

19. Waldmann, T.A. (moderator) (1978) Ann. Int. Med., 88, 226-238.

DISCUSSION

LENNON:  Was there any histological evidence of skeletal muscle cellular infiltration?

GERSHWIN:  There was no histologic evidence of skeletal muscle cellular infiltration.

CHUSED:  Is the phenotypic variation in disease manifestation due to residual genetic variation or to nongenetic influences?

GERSHWIN:  We believe that the immunopathology is transmitted by one major gene with other associated minor modifying genes.  At present Drs. Abplanalp and Eklund are performing selective out-breeding experiments to better define these relationships.  Further-more since these birds are not completely inbred, the issue of genetic contributions to phenotypic variations in disease expression becomes difficult to define.  However, we have recently, by retrospective analysis, defined a similar syndrome in a family of line 3 chickens. This line, you will recall, is where we believe the original muta-tion occurred.  Study of these inbred birds should shed further light on this phenomenon.  Similarly we hope to address the issue of infectious agents on disease expression and to identify possible associates with B alleles.

WARNER:  For the record, could you comment on whether the presence of Gumboro's disease (infectious bursal disease) has been determined in this line of birds?

GERSHWIN:  To my knowledge, we have not determined whether our birds are infected with Gumboro's disease.  We certainly plan to look into this possibility.

CANINE SYSTEMIC LUPUS ERYTHEMATOSUS:
PHENOTYPIC EXPRESSION OF AUTOIMMUNITY IN A CLOSED COLONY*

ROBERT S. SCHWARTZ, FRED QUIMBY AND JANINE ANDRÉ-SCHWARTZ

Department of Medicine and Cancer Research Center,
Tufts University School of Medicine, Boston, Massachusetts 02111

Systemic lupus erythematosis (SLE) in the canine species occurs in a form that is remarkably similar to SLE in human beings.[1] About 75% of dogs with SLE are females. The disease is characterized clinically by arthritis, autoimmune hemolytic anemia, thrombocytopenia and glomerulo nephritis. Facial rashes with a histology typical of that of lupus dermatitis have been found in several animals. Some dogs may also develop seizures, presumably on the basis of involvement of the central nervous system. Dogs with SLE have positive LE cell tests, antinuclear antibody and antibodies to both single-stranded and double-stranded DNA. Circulating immune complexes can be detected in their serum by means of a Clq binding assay. The renal lesion is that of membranoproliferative glomerulo-nephritis and is characterized by the presence of nodular deposits of immunoglobulin along the glomerular basement membranes. In some dogs with lupus it has been possible to elute from the diseased kidneys anti DNA antibodies.

Over the past 10 years we have developed a colony of dogs derived from an original mating pair in which the dam had SLE and the male was normal.[2] This colony (the QS colony) is now in the 5th inbreeding generation. After a latent period of approximately 7 years clinical and pathological signs of SLE as well as other autoimmune diseases have begun to appear in dogs of this colony. The 10 years required to develop this colony was due to several reasons, including the sexual physiology of the dog, the long incubation time before the appearance of clinical signs of disease, reduced fertility that accompanied inbreeding, and a marked increase in fetal wastage with successively inbred generations. Despite these restraints the QS colony has been established as a unique group of dogs with a remarkable array of immunological abnormalities.

The original dam, a miniature poodle, had SLE with arthritis as the predominant feature. This animal also had autoimmune hemolytic anemia and immunothrombocytopenic purpura, both of which were successfully treated. The sire was a normal miniature poodle. Brother to sister and cousin matings have been carried out for 5 generations. The diseases that developed in this colony include SLE, rheumatoid arthritis, autoimmune thyroiditis, Sjögren's syndrome, and celiac disease. One dog

*Supported by NIH Grant AM 09350

had both Sjögren's syndrome and autoimmune thyroiditis. In 2 dogs SLE was complicated by typical immune deposit glomerulo-nephritis.

Many of the dogs in this colony have at least one serological abnormality.
Most dogs have multiple autoantibodies, including antinuclear antibodies, the LE
cell phenomenon, antibodies to double stranded DNA, antimicrosomal antibodies,
rheumatoid factor, antithyroid antibodies, antibodies to colloid antigen of the
thyroid, and circulating immune complexes as detected by the Clq binding assay.
Two dogs with Sjögren's syndrome had antibodies that bound specifically to the
nictitans (this structure is the major tear forming organ of the canine eye). Many
dogs also have cyroprecipitates and in some animals there are cyroglobulins with
binding affinity for DNA. About 70% of diseased dogs in this colony have lymphocytotoxic antibodies in their serum.

Thus far Sjögren's syndrome has been described only in human beings. Pathological findings suggestive of this disorder have been found in NZB mice, but in that
species the clinical manifestations of Sjögren's syndrome were absent. Canine
Sjögren's syndrome is characterized by severe conjunctivitis, corneal ulcerations,
gingivitis, and in one animal, lymphocytic thyroiditis·. The Schirmer test demonstrates a pronounced reduction in the formation of tears. Pathological changes
include chronic inflammation of the nictitans, which contains massive infiltration
by lymphocytes, and lymphocytic infiltration and destruction of the lacrimal and
salivary glands.

Autoimmune thyroiditis in dogs of the QS colony is characterized by clinical
signs of hypothyroidism, decreased levels of thyroxin in serum, antibodies to
thyroid colloid antigen and lymphocytic infiltration of the thyroid gland. Some
animals of this colony have antibodies to thyroid colloid antigen and normal levels
of thyroxin. It is of interest that the incidence of autoimmune thyroiditis in
this colony has increased with successive generations of inbreeding and in the
fourth generation 57% of the animals have autoantibodies to thyroid colloid antigen.

During the development of the QS colony 2 other colonies were developed down
to the 3rd generation of inbreeding and then discarded. The A colony was developed
from the mating of a pure bred German Shepherd female with SLE and a normal German
Shepherd male. The colony did not survive because of the unexpected development of
multiple neoplasms in many of the descendents. Other animals in this colony develop
a severe periportal fibrosis of the liver that resembled histologically chronic
biliary cirrhosis in human beings. All descendents of the original breeding pair
had antinuclear antibody and positive LE cell tests. Several animals had antibodies
against parietal cells but in no case were we able to detect antibody against smooth
muscle or mitochondria. Malignancies of the mammary glands, uterus, kidney, and
lymphoid system were found in dogs of the A colony. There was a relatively high
incidence of primary nephrofibroma. This canine neoplasm has features found in

Wilm's tumor of children.

The incidence of multiple neoplasms in dogs of the A colony was compared with that in several other groups. Figures were obtained from review of all necropsies conducted in dogs at the Angell Memorial Animal Hospital in 1975. A second group was supplied by Dr. Harold Casey of the Armed Forces Institute of Pathology and contains information on over 1000 necropsies of German Shepherd dogs. When the A colony was compared to these other sources, the incidence of multiple neoplasms was significantly higher. For instance, 83% of the dogs of the A colony had multiple tumors, whereas 24% of the Angell Memorial Animal Hospital cases and 8% of the Armed Forces Institute of Pathology cases had multiple tumors. Thus, the A colony was characterized by the development of autoantibodies, a peculiar hepatic lesion having some histopathological features of chronic biliary chirrhosis, and a very high incidence of multiple primary neoplasms.

One dog of this colony is of particular interest. This animal, Casper, developed a macrocytic anemia and extremely low levels of vitamin $B_{12}$ in serum (40 picograms per ml; normal, 160 to 240 picograms per ml.) The serum folate level was normal. The thyroxin level was 0.44 ng/ml (normal 1.2 - 2.7 ng/ml). The serum gastrin level was 280 pg per ml (normal 50-80 pg per ml). The serum contained autoantibodies to glomerular basement membranes, gastric perietal cells, canine testicle, baby rat liver nuclei, thyroid colloid antigen and human intrinsic factor. There was also radiological evidence of severe ankylosing spondylitis. Biopsies of the stomach revealed infiltration by lymphocytes in the tunica propria and mucosa. A Schilling test demonstrated that vitamin $B_{12}$ was not absorbed even after the administration of hog intrinsic factor.

Clinically the animal had signs of anemia, hypothroidism, arthritis of the spine and there was a marked reduction in the sperm count. This dog represents the first example of pernicious anemia in a non-human species. It is remarkable that the animal had autoantibodies to human intrinsic factor and gastric parietal cells along with autoimmune thyroiditis, autoimmune orchitis and ankylosing spondylitis. Antibodies against parietal cells were found in two of Caspers cousins.

The B colony was developed through the mating of 2 German Shepherds with SLE. This colony was used in various attempts to trigger the development of SLE in dogs. Animals of this colony were chronically exposed to ultraviolet light. Each animal received 34.5 microwatts/$cm^2$ of energy for 6 hours daily, as estimated by combining all UV spectro bands between 290-339 nanometer wavelengths. Each dog was given this treatment for 5 consecutive days a week for 16 consecutive months. Each week the dorsum of each animal was shaved. Nine of 10 dogs survived the 6 month period of this investigation. 8 of 9 animals died of natural causes 2 years after the UV radiation. One animal developed generalized amyloidosis and immune complex nephritis. One dog had chronic aggressive hepatitis and another had periportal fibrosis. In no case was there evidence of SLE, either clinically or serologically.

Six animals of this colony were treated with hydralazine, a drug known to trigger antinuclear antibodies and a lupus like syndrome in human beings. The drug was given in a dose of 10 mg/kg daily for 8 months. During the administration of hydralazine 3 of the dogs developed positive antinuclear and LE cell tests. No dog developed signs of SLE during the 8 months the drug was administered. Control dogs from the same colony were also monitored for serologic abnormalities, which developed in the same incidence as the animals treated with hydralazine. Therefore, we were unable to demonstrate that hydralazine can induce systemic lupus or seriologic abnormalities in the dog.

Several dogs were immunized with native or denatured DNA. The nucleic acids were emulsified in Freund's adjuvant and administered in multiple subcutaneous sites in 2 doses. None of the animals given native DNA or a methylated BSA-denatured DNA complex developed seriologic abnormalities during the period of the study. Thus, despite a genetic background that would favor the production of antinucleic acid antibodies, none of the animals in this group of experiments reacted to deliberate immunization with DNA in several forms.

While this work was proceeding, 169 dogs with a variety of autoimmune diseases were identified at the Angell Memorial Animal Hospital. The male:female ratios for rheumatoid arthritis, autoimmune hemolytic anemia and SLE were 1:1, 1:3, and 1:1, respectively. However, out of 19 cases of SLE only one was found in a castrated female. This finding is statistically significant and suggests that castration may deter the development of SLE in dogs.

Results in the QS colony favor a genetic interpretation of SLE in dogs; however, a pseudo-genetic environmental factor (infectious agent?) cannot be excluded. Our experience in breeding dogs with SLE has shown that the kind of disease in the progeny is highly variable. In the A colony we saw a high incidence of multiple primary neoplasms and chronic hepatitis in the progeny. Most of the offspring had antinuclear and other autoantibodies in their serum. In the B colony all progeny had multiple serologic abnormalities but autoimmune disease did not develop. In the QS colony the offspring have developed SLE, Sjögren's syndrome, thyroiditis, celiac disease and multiple serologic abnormalities.

NZB mice have a high incidence of autoimmune hemolitic anemia and a low incidence of nephritis. When they are mated with "normal" NZW or SWR mice the $F_1$ progeny have a high incidence of nephritis and a low incidence of autoimmune hemolitic anemia.[3,4] We believe that multiple genetic factors influence the type of autoimmune disease that develops in both mice and dogs. Some of these factors may be inherited from the normal parent.

Based on studies of murine, canine and human SLE[5] we postulate the existence of 2 classes of genes: (a) those that permit a general disposition to autoimmunity

(Class I genes) and (b) those that determine the phenotype of the disease (Class II genes). Class I genes would be related to lymphocyte function and immunoregulation. They could include Ir genes and other genes that control the activities of B cells and T cells. Permissive alleles of these genes would be associated with the production of autoantibodies. Genes of Class II in our formulation determine the manifestations of disease, e.g. nephritis, rash and autoimmune hemolytic anemia. Alleles of Class II genes would, for example, specify the size of immune complexes, determine functions of the complement system, or specify the nature of the cellular reactions to inflammatory stimuli. Both classes of genes are complex and consist of multiple unlinked dominant and recessive components. Both structural and modifier genes could be involved. The classification we propose is functional and does not imply linkage or mode of inheritance.

We speculate that all individuals with SLE (mice, dogs, humans) have both classes of genes. They produce multiple autoantibodies (Class I genes) and develop lesions of varying severity and type (Class 2). Thus, in SLE the 2 classes of genes are interacting or complimentary. A normal individual could have either class of genes, but not both. Genes of Class II cannot be expressed in the absence of complementary alleles of Class I.

Support for this concept comes from the observation that the $F_1$ hybrid of 2 normal strains of mice can be susceptible to autoimmune disease. Obata et al[6] found that neither C57Bl/6 nor 129 mice spontaneously formed autoantibodies against the antigen $G_{IX}$ (gp70). However, all $F_1$ hybrids of these 2 strains produced substantial amounts of the autoantibody and had histologic evidence of autoimmune disease. The conclusion of these authors was that "the autoimmune response required 2 dominant genes, each parent contributing a high response allele to the hybrid". This effect may have been due to the kind of gene complimentation that is known to affect immunologic responsiveness and suppressor T cells[7]. The net operational effect was that of a system of 2 or more genes inherited from both normal parents.

We recognize that some of the data may be explained by one or more dominant or recessive gene of reduced or variable penetrance and that such genes need be present in only one parent. For example, the inheritance of polydactyly in humans is an example of a dominant gene of reduced penetrance in which phenotypically normal parents have children with extra digits. However, we believe that the finding of autoantibodies in healthy members of the QS colony (and in first degree relatives of humans with SLE[5]) the findings in hybrids in NZB[3,4] mice and the results of Obata et al[6] are more consistant with the dual gene hypothesis.

Genes of Class I are of particular interest because their permissive alleles can set the stage for autoimmunization. The phenotypes of these alleles might be identifiable in normal individuals by in vitro tests of immune function. Our hypothesis predicts that abnormal immunoregulation will be found in healthy consaguineous relatives of patients with SLE. Genes of Class II could at present be identified only by mating tests.

In summary, the uniform and spontaneous development of autoimmune disease in highly inbred mice is *ipso facto* evidence that the disorder has a genetic basis. The data in canine and human SLE are consistent with this interpretation. We have tried to codify these complex genetic influences by proposing a hypothesis of 2 classes of genes, one that relates to immunoregulation and another that specifies immunopathologic lesions. An experimentally testable prediction of this hypothesis is that permissive alleles of one of these 2 classes of genes are detectable in the relatives of individuals with SLE. This formulation also suggests the hypothesis that structural genes for autoantibodies are encoded within the germ line. This may explain why normal mice that are treated with endotoxin produce antibodies to nucleic acids.[8] We believe the concept of two classes of interacting genes provides a dynamic basis for further analysis of SLE and other autoimmune diseases.

REFERENCES:

1. Lewis, R.M., Schwartz, R. and Henry, W.B., Jr. Canine systemic lupus erythematosus. Blood 25:143, 1965.

2. Lewis, R.M. and Schwartz, R.S. Canine systemic lupus erythematosus. Genetic analysis of an established breeding colony. J. Exp. Med. 134:417, 1971.

3. Knight, J.G. and Adams, D.D. Three genes for lumpus nephritis in NZB x NZW mice. J. Exp. Med. 147:1653, 1978.

4. Datta, S.K., Manny, N., Andrzejewski, C., André-Schwartz, J. and Schwartz, R.S. Genetic studies on autoimmunity and retrovirus expression in crosses of New Zealand black mice. J. Exp. Med. 147:854, 1978.

5. Block, S.R., Winfield, J.B., Lockshin, M.D., D'Angelo, W.A. and Christian, C.L. Studies of twins with systemic lupus erythematosus. Amer. J. Med. 59:533, 1975.

6. Obata, Y., Stockert, E., Boyse, E.A., Tung, J.S. and Litman, G.W. Spontaneous autoimmunization to $G_{IX}$ cell surface antigen in hybrid mice. J. Exp. Med. 144:533, 1976.

7. Dorf, M.E., Stimpfling, J.H., Cheung, N.K. and Benacerraf, B. Coupled complementation of Ir genes. "Ir Genes and Ia Antigens" (H.O. McDevitt, editor). Academic Press (New York), 1978, p. 55.

8. Fischbach, M., Roubinian, J.R. and Tatal, N. Lipipolysaccharide induction of IgM antibodies to polyadenylic acid in normal mice. J. Exp. Med. 120:1856, 1978.

DISCUSSION

ROSE: The two-gene concept is very useful, but can it explain the occurrence in one family of two such different autoimmune diseases as thyroiditis and lupus?

SCHWARTZ: The most likely explanation for this is that the dogs are not inbred and what we are observing is the sorting out of independently segregating genes.

GERSHWIN: New Zealand mice raised under germ-free conditions have no major alteration of immunopathology. Would you compare this observation with your own experience with germ-free studies in dogs.

SCHWARTZ: It is important to distinguish between the endogenous xenotropic virus of NZB mice and exogenous type C viruses. Since the former agents are products of cellular genes, their expression would not be affected by a germ-free environment. By contrast, a germ-free environment could influence a disease in which an exogenous, horizontally transmitted agent is involved.

GERSHWIN: Considering your concept of horizontal transmission of SLE in dogs, can you expand further on the anecdotal associations of SLE transmission in households with dogs.

SCHWARTZ: All that is known now is in the form of anecdotes, of which there are several. Epidemiologic studies may shed more light on this problem.

LENNON: In family studies of dogs with lupus where the father was normal did he ever develop autoantibody at a later date?

SCHWARTZ: No.

VOLPE: In those animals who are seropositive, but who have no clinical evidence of systemic lupus erythematosus, have you biopsied the kidneys or other organs at that stage?
Have you castrated any young female dogs to determine whether you can change the sex distribution during the subsequent development of SLE?

SCHWARTZ: The answer to both questions is "No".

CHUSED: The high frequency of serologic abnormalities in the dog colony and Casper's experience upon returning to it suggest a horizontally transmitted agent. Have you looked for serologic abnormalities in animals injected with material from lupus dogs?

SCHWARTZ: Yes, we have. Antinuclear antibody and positive LE cell tests developed in both mice and puppies injected at birth with cell-free filtrates prepared from the spleens of dogs from the SLE colony.

KÖHLER: I am afraid, I cannot agree completely with you in your definition of class 1 genes. As far as the immune system is concerned, self-recognition and auto-anti-idiotypic antibodies are part of the normal function. By this definition, class 1 genes are required for a normally functioning immune response and its absence would actually lead to disease.

SCHWARTZ: I should have been more precise. What I meant to refer to was alleles of particular genes that would permit the appropriate abnormality in central immune function to be expressed.

MACKAY: I would agree with the "class 1 gene" as I raised the same concept yesterday in relation to autoantibodies in relatives of chronic hepatitis patients. Class 2 genes determining inflammatory responsiveness could be so numerous that such might not be inherited "en bloc". I am slightly sceptical about the infectious agent. There seems no anecdotal evidence of owner-pet associations of lupus in Melbourne but I agree that epidemiologic data with serologic studies are needed. The occurrence of lupus in the one dog of the germ-free litter returned to the colony could be related to the B cell activation occurring when the dog experienced conventional conditions. It would have been of interest if the other germ-free littermates had been returned to conventional conditions elsewhere than in your colony and observed for a year or so for features of lupus.

SCHWARTZ: The two gene concept does not imply in any way that the two classes of genes are transmitted "en bloc". The concept classifies these genes functionally, but makes no statement about their number, location, or mode of inheritance.
Everyone is skeptical about an infectious agent in lupus; nevertheless, evidence in support of this view is accumulating in several different laboratories. All workers in this field have thus far been very cautious -- as I have tried to be -- in interpreting their results. As for the absence of anecdotes in Melbourne, one has to seek them out.

ROITT: If an infectious agent has to interact with individuals possessing both type 1 and 2 genes, are you not a little surprised at the high frequency of success you obtained on transfer of cell free extracts to mice and puppies?

SCHWARTZ: Genetic susceptibility to the induction of antinuclear antibodies by Type C viruses has in fact been well documented. Therefore, there is no essential conflict between a genetic theory of lupus and the participation of a virus.

ROITT: Regarding thyroiditis, is it classical in serology and histology? Is the thyroiditis observed atypical and are thyroid autoantibodies produced?

SCHWARTZ: The thyroiditis is not typical of Hashimoto's disease in that there is severe atrophy of the gland. Antithyroid antibodies are present.

WARNER: Since Dr. Mackay has raised the question of the association of B8 with lupus in man, could Dr. Svejgaard tell us the current status of this rather unsettled issue?

SVEJGAARD: Combined calculations on the data available at the HLA and Disease Registry in Copenhagen show a significant increase of B8 in Caucasian patients with SLE.

# MYASTHENIA GRAVIS IN DOGS:  ACETYLCHOLINE RECEPTOR DEFICIENCY WITH AND WITHOUT ANTI-RECEPTOR AUTOANTIBODIES

VANDA A. LENNON

Department of Neurology, Mayo Clinic, Rochester, MN 55901, U.S.A.

ANTHONY C. PALMER

School of Veterinary Medicine, the University of Cambridge, England.

CHRISTINA PFLUGFELDER and R. J. INDRIERI

School of Veterinary Medicine, the University of California, Davis, CA 95616, U.S.A.

## INTRODUCTION

The defect in neuromuscular transmission which is characteristic of myasthenia gravis (MG) is now known to be due to a deficiency of acetylcholine receptors (AChR) in the postsynaptic membrane[1-3]. Auto-antibodies to AChR are detectable in the sera of 87% of MG patients[4] and reduction of AChR in muscle has been found in association with *in situ* binding of antibodies [3,5] and complement[3].

MG can be induced experimentally in animals by immunization with AChR[6]. However, MG also occurs spontaneously in dogs[7-10] and, as in man, both congenital and acquired forms of canine MG have been described[9]. We have now studied serum from 7 dogs with spontaneously occurring MG (5 acquired and 2 congenital) and muscle samples from one dog with acquired and one with congenital MG to ascertain whether the abnormality underlying defective neuromuscular transmission is similar to MG in man. Our initial findings have led us to believe that further studies of spontaneous models of MG in animals will be of value in identifying etiologic factors triggering the autoimmune response in acquired MG as well as in defining the pathogenesis of congenital and other forms of MG which do not appear to have an autoimmune basis.

## MATERIALS AND METHODS

Dogs. Clinical details are summarized in Table 1.  Full details of several of the dogs have been published elsewhere[9,10]. The history of the Akita (#4) is noteworthy because the initial onset of weakness occurred approximately 10 days after the owner's yard was sprayed extensively with an insecticide containing four organophosphate compounds (diazinon-4-2; 0,0 diethyl 0-47590; 0,0 dimethylphosphorothioate and *Dursban*), a chlorinated hydrocarbon (*Toxaphene*) and a fungicide (dodecylguanidine).  The dog recovered spontaneously during hospitalization

and required no medication on discharge but relapsed in less than 2
months after returning home. She again recovered spontaneously during
a further 11 days of hospitalization. Approximately 2 months after
discharge a third relapse occurred which subsided after 3 weeks treatment with pyridostigmine.

TABLE 1

CLINICAL DETAILS OF MYASTHENIC[a] DOGS

| Breed | Sex | Age at Onset | EMG Decrement | Dilated Esophagus | Course |
|---|---|---|---|---|---|
| 1. German Shepherd[b] | F | 7 yr. | n.d.[c] | + | died |
| 2. Collie[8] | M | 2 yr. | + | + | recovered[d] |
| 3. St. Bernard[9] | F | 8 yr. | n.d. | + | died |
| 4. Akita | F | 5 yr. | + | + | recovered[d] |
| 5. German Shepherd | F | 3 yr. | + | + | died |
| 6. J. Russell Terrier[8] | M | 2 mo. | + | − | died |
| 7. J. Russell Terrier | F | 2 mo. | n.d. | − | died |

[a] Diagnosis was confirmed by clinical improvement after administration
of neostigmine or edrophonium.

[b] At autopsy an anterior mediastinal tumour was found, black in colour
with the histologic appearance of melanoma.

[c] n.d. = not done.

[d] Both dogs still clinically in remission requiring no medication,
July, 1978.

Diagnosis in all cases was confirmed by either or both clinical improvement after administration of neostigmine or edrophonium, and a
decrement in muscle action potential during repetitive motor nerve
stimulation (Fig. 1). All cases of acquired MG occurred in large
breeds while the 2 dogs with congenital MG were Jack Russell terriers,
a small breed of dog. At age 2 months the Jack Russell terriers developed muscle weakness with fatigue typical of MG which was temporarily
relieved by administration of neostigmine. Serum was obtained from the
mother and an unaffected littermate of one terrier. In all cases of
acquired MG onset was at age 2 years or older and esophageal involvement was prominent. Unlike acquired MG, the esophagus was not clinically affected in the congenital form of MG.

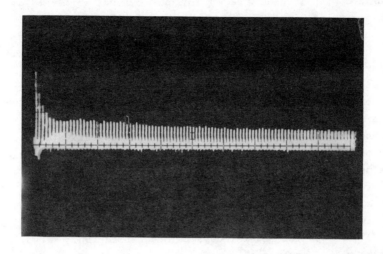

Fig. 1. Decrement in muscle action potential during repetitive motor
nerve stimulation (50/sec for 2 seconds). German Shepherd #5, with
acquired myasthenia gravis.

Antibody Assays. Sera were tested for antibody to AChR by an immunoprecipitation radioimmunoassay[11] using polyspecific rabbit anti-dog
immunoglobulin. The IgG class of the antibody was confirmed by re-
testing sera positive for antibody using rabbit anti-dog IgG. To provide antigen, AChR from muscle of a near-term fetal dog was solubilized
with 2% Triton X-100 and complexed with an excess of [125]I-labelled
α-bungarotoxin. Sera were also tested for antibodies reactive with
muscle striations by indirect immunofluorescence using unfixed mouse
muscle as substrate and fluoresceinated IgG of rabbit anti-dog IgG
(Cappel Laboratories, Downington, PA). Dilutions were commenced at
1/60 because more concentrated dog serum induced a diffuse apparently
non-specific fluorescence of muscle fibers.

Determination of AChR Content of Muscle. The amount of AChR in
muscle was determined on homogenized muscle samples of known weight[5].
Washed pelleted membranes were extracted with 2% Triton X-100 and aliquots of each extract were incubated overnight with [125]I α-bungarotoxin

to label AChR. Radiolabelled AChR was precipitated by subsequent sequential addition of 5 μl anti-AChR antibodies (from rats hyper-immunized with Torpedo AChR[12]) followed by goat anti-rat Ig. The specificity of binding of $^{125}$I α-bungarotoxin to AChR was demonstrated by inhibition with benzoquinonium or non-radioactive α-bungarotoxin. AChR concentration was expressed in terms of moles of $^{125}$I α-bungaro-toxin in the precipitate after subtraction of non-specific binding. The proportion of AChR which was already complexed with antibody when extracted from muscle was estimated as the ratio of cpm of $^{125}$I α-bungarotoxin-AChR complexes precipitated by rabbit anti-dog Ig in the presence of 5 μl of normal dog serum to cpm precipitated in the presence of an excess of dog anti-AChR autoantibodies and was expressed as a percentage.

To investigate the effect of freezing on the yield of AChR from muscle, carcasses of female Lewis rats were prepared as described previously[11] and either extracted immediately, or slowly frozen and stored at -20° or -40° for several days before extraction. Fresh carcasses of rats with experimental autoimmune myasthenia gravis (EAMG), induced by immunization with Torpedo AChR[12], were similarly extracted to provide additional controls for the samples of dog muscle.

RESULTS

Antibodies reactive with AChR solubilized from homologous muscle were detected by radioimmunoassay in the serum of 4 out of 5 dogs with acquired MG, but were not detected in the St. Bernard or in either of 2 dogs with congenital MG or in any of 8 non-myasthenic dogs (Table 2). The lowest antibody titer ($0.24 \times 10^{-9}$M) was found in a Collie, which subsequently had a clinical remission, and the highest titer ($49.00 \times 10^{-9}$M) was found in a female German Shepherd (case #1) which died and was found at autopsy to have an anterior mediastinal tumour. Repeating the assays with class specific rabbit anti-dog IgG confirmed the IgG nature of the anti-AChR autoantibodies.

The sera were tested also by indirect immunofluorescence on sections of skeletal muscle for the presence of anti-striational antibodies (Table 2, Fig. 2a). Anti-striational antibodies were detected only in the serum of German Shepherd #1, which was positive at a dilution of 1/240. The pattern of fluorescence, consisting of delicate fine striations on a light background (Fig. 2a), was quite different from the pattern obtained with serum from a human patient with MG which consisted of sharply demarcated striations on a contrasting dark interstriational zone (Fig. 2b).

TABLE 2

ANTI-ACH RECEPTOR AND ANTI-STRIATIONAL AUTOANTIBODIES IN DOG SERA

| Diagnosis | Moles $(\times 10^{-9})$ AChR Bound/Liter Serum | Anti-striational Antibody Titer |
|---|---|---|
| Non-myasthenic | | |
| normal | 0.00 | 0 |
| " | 0.00 | n.t.[a] |
| " | 0.00 | 0 |
| " | 0.00 | 0 |
| " | 0.00 | 0 |
| unaffected dam (MG#7) | 0.00 | 0 |
| unaffected littermate (MG#7) | 0.00 | 0 |
| congenital nystagmus | 0.00 | 0 |
| Myasthenia Gravis | | |
| Acquired | | |
| 1. German Shepherd | 49.00 | 1/240 |
| 2. Collie | 0.24 | 0 |
| 3. St. Bernard | 0.00 | 0 |
| 4. Akita | 3.12 | 0 |
| 5. German Shepherd | 3.09 | n.t. |
| Congenital | | |
| 6. Jack Russell Terrier | 0.00 | 0 |
| 7. "      "      " | 0.00 | 0 |

[a] n.t. = not tested

In order to establish the validity of comparing yields of AChR extracted from fresh samples of muscle with yields from samples which had been shipped frozen on dry ice, preliminary extractions were performed with rat muscle. Yields from non-myasthenic rat carcasses did not differ significantly whether extracted without delay or after rapid or slow freezing and storage at $-20°$ or $-40°$ for at least a week (Table 3). By contrast, the mean yield of AChR from muscles of a group of 8 rats immunized with AChR was reduced to 36% of the non-myasthenic value. Furthermore, on the average, 52% of AChR extracted from the muscle of rats with experimentally induced MG were found to be complexed with immunoglobulin.

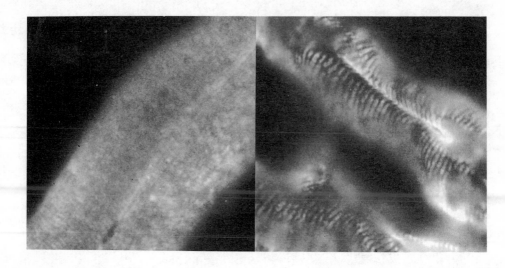

a                                               b

Fig. 2.   Sera tested by indirect immunofluorescence on unfixed frozen
sections of normal mouse muscle.   a) German Shepherd #1 (x1000).   Note
delicate, fine striational pattern on a light background.   b) Human MG
patient (x630).   Note sharply demarcated striations on dark interstriational zone.

The yields of AChR in extracts from non-myasthenic dog muscles
ranged from 0.19 to 1.29 x $10^{-12}$ moles per gram (Table 4).   The average yield from the normal dog muscles (0.37 x $10^{-12}$ moles/g) closely
approximated the yields from non-myasthenic rat muscle (Table 2).
Leg muscle from one normal dog (#3, a cardiovascular research subject)
yielded 2½ - 3 times more AChR than the three other normal muscles
tested (0.19, 0.22 and 0.30 x $10^{-12}$ moles/g) which included an inter-
costal muscle biopsy from dog #3.   The high normal value (0.77 x $10^{-12}$
moles/g) was in the range of values obtained from fetal muscle (#4)
and denervated adult muscle (3 samples from #5), both of which would
be expected to contain more AChR than normal, but there was no history
of nerve injury in the leg of dog #3.   The yields of AChR from multiple
specimens of 2 myasthenic dogs (mean = 0.05 x $10^{-12}$ moles/g) were consistently and significantly less than those of the non-myasthenic dogs
(14% of average for the normals).   The myasthenic process appeared
uniformly to involve muscles selected randomly for sampling.   A striking difference between AChR extracted from muscle samples of dogs with

congenital and acquired forms of MG was that only in the latter did
AChR in muscle exist complexed with antibody.  The first muscle biopsy
of the Akita, obtained at the peak of the dog's clinical weakness,
yielded 0.08 x $10^{-12}$ moles AChR/g, 100% of which was complexed with
antibody.  Although the total amount of AChR was slightly less (0.06 x
$10^{-12}$ moles/g) in a second biopsy taken 4 months later, 40% or 0.02 x
$10^{-12}$ moles AChR/g of muscle was free of antibody at that time.  The
presence in muscle membranes of AChR free of antibody was accompanied
by clinical improvement and a 33% reduction in the serum concentration
of anti-AChR antibodies (Table 4).

TABLE 3

YIELDS OF ACHR AND IMMUNE COMPLEXES FROM FRESH AND FROZEN RAT MUSCLE

| Immunogen with Adjuvants | Low Temperature Storage | Moles AChR $(x10^{-12})$/gram | % Complexed with Antibody |
|---|---|---|---|
| none | < 3 hours at 4° | 0.44 | 0 |
| " | < 3 hours at 4° | 0.49 | 0 |
| " | 7 days at -40° | 0.44 | 0 |
| " | 10 hours at 4° then 6 days at -20° | 0.41 | 0 |
| Torpedo AChR[a] | < 3 hours at 4° | 0.14 | 65 |
| " | " | 0.12 | 78 |
| " | " | 0.19 | 35 |
| " | " | 0.21 | 39 |
| " | " | 0.12 | 84 |
| " | " | 0.16 | 37 |
| " | " | 0.19 | 14 |
| " | " | 0.15 | 61 |

[a] 7.5 µg dose, prepared and injected as described previously[12].

DISCUSSION

Spontaneously acquired MG in dogs was associated with demonstrable
autoantibodies to AChR in serum in 4 out of 5 cases.  Anti-AChR antibodies have not previously been reported in canine MG, but are characteristic of acquired MG in man[4] and are strongly implicated in the
disease pathogenesis[1-3].  Lack of anti-AChR antibodies in the fifth
dog, a St. Bernard with a clinical history characteristic of acquired
MG, cannot presently be explained with certainty.  In preliminary
experiments, using an immunofluorescence technique described

TABLE 4

YIELDS OF ACHR AND IMMUNE COMPLEXES FROM DOG MUSCLE

| Diagnosis | Muscle | Moles AChR $(\times 10^{-12})$/gram | % Complexed with Antibody |
|---|---|---|---|
| Non-myasthenic | | | |
| 1.   normal | sartorius[a] | 0.22 | 0 |
| 2.   normal | pectineus | 0.30 | 0 |
| 3.   normal | intercostal[a] | 0.19 | 0 |
|      " | leg muscle[a] | 0.77 | 0 |
| 4.   term fetus | pooled muscle | 0.73 | 0 |
| 5.   neurogenic atrophy[b] | thigh muscle I | 1.16 | 0 |
|      " | thigh muscle II | 0.74 | 0 |
|      " | biceps femoris | 1.29 | 0 |
| Myasthenia Gravis | | | |
| 1.   Acquired (#3) | sartorius | 0.08 | 100 |
|      "          " | pectineus[c] | 0.06 | 60 |
| 2.   Congenital (#7) | leg muscle I | 0.03 | 0 |
|      "          " | leg muscle II | 0.04 | 0 |
|      "          " | leg muscle III | 0.04 | 0 |

[a] Muscle stored at 4° < 48 hours; all others were stored at -40° for periods ranging from 1 week to 7 months. A control non-myasthenic sample of muscle was extracted at the same time as each myasthenic sample.

[b] Prolapsed L1-2 intervertebral disc one month.

[c] Second biopsy taken 4 months after first. Dog clinically improved. Serum antibody titer at time of first biopsy was $3.12 \times 10^{-9}$ M and at time of second biopsy was $2.06 \times 10^{-9}$ M.

elsewhere[12], antibodies which bind to the surface of living cultured muscle have been detected in this dog's serum and in sera of both German Shepherds and the Akita but not in serum of the Collie (now in remission) or in the dogs with congenital MG (V.A. Lennon and G. Jones, in preparation). This suggests that the St. Bernard's serum contained antibodies directed to an antigenic determinant of the AChR exposed on the outside of the plasma membrane. If the determinants were close to or actually a part of the ACh binding site, antibodies would not be detected in the radioimmunoassay because the solubilized antigen employed was complexed with α-bungarotoxin, which binds essentially irreversibly to the ACh-binding site. Antibodies reactive with the toxin-binding region of AChR have been described in both human patients[13,14] and in serum of a rabbit immunized with AChR[15].

The highest titer of anti-AChR antibodies was found in a German Shepherd (#1) in association with anti-striational antibodies. Thymoma has been reported in association with MG in dogs[8] and although the histopathologic description (melanoma) of the anterior mediastinal tumour of German Shepherd #1 was not consistent with its being of thymic origin, it is noteworthy that in humans, higher titers of anti-AChR antibodies occur in MG patients with thymoma[4], and anti-striational antibodies occur with very high incidence with thymoma with and without MG[16]. The specificity of the anti-striational reactivity of the dog's serum was not examined in detail but the pattern of reactivity was quite different from that of the human subject with which it was compared. Considerable heterogeneity of anti-striational specificities has been described in the serum of human subjects with MG[16].

Lack of anti-AChR antibodies in the 2 dogs with congenital MG is consistent with the reported lack of anti-AChR antibodies in a human subject with congenital MG[17]. In the human patient, non-autoimmune pathogenesis of congenital MG was further substantiated by the patient's lack of beneficial response to treatment by plasmapheresis. This contrasted with the response of 7 patients with acquired MG in whom removal of anti-AChR antibodies by plasmapheresis was associated with progressive improvement in strength after a lag of at least 2 days.

The reduced amount of AChR extractable from muscle of dogs with congenital and acquired MG suggests that decreased sensitivity of the postsynaptic membrane to ACh is the basis of neuromuscular transmission failure in both conditions. The reduction in AChR content of the muscle could reflect either an absolute reduction in receptor protein or it may result from a reduced binding capacity or binding affinity of the AChR for $^{125}I$ $\alpha$-bungarotoxin (or ACh). The degree of reduction in AChR in myasthenic dogs (approximately 14% of the normal content) was greater than in the rats immunized with AChR (36% of normal content). Clinical signs of weakness (EAMG) in the rats were minimal. Lindstrom and Lambert[5] in a similar study of muscle from rats with EAMG reported that the minimum total AChR content of rat muscle was not much less than 40% of normal regardless of clinical severity of weakness. However increased severity of disease was paralleled by an increase in the proportion of remaining AChR which was complexed with antibody. In their study 99% of the receptors were complexed with antibodies in rats with profound weakness. In a combined biochemical and electrophysiological study of biopsied muscle from human patients with MG, the same authors found a linear relationship between the amount of AChR free of antibody and the sensitivity of the postsynaptic membrane to ACh (i.e. the amplitude of MEPPS). The average content of AChR free of antibody reported for their 8 patients with severe weakness, who had more than 60% reduction in MEPP amplitude, was $0.04 \times 10^{-12}$ moles/g. It is noteworthy that $0.04 \times 10^{-12}$ moles/g was the average AChR content of 3 separate but simultaneous muscle biopsies from a severely affected dog with congenital MG (#7, at age 5 months). The dog had no AChR complexed with antibody. In the Akita with acquired MG, 100% of the muscle AChR $(0.08 \times 10^{-12}$ moles/g) were complexed with antibodies at the peak of disease severity. During subsequent clinical improvement only 60% were complexed with antibodies and the serum antibody titer was reduced. It is of interest that the onset of clinical signs in this dog occurred 10 days after the beginning of exposure to an insecticide mixture which contained several ingredients known to affect neuromuscular transmission. It is possible that these drugs further impaired neuromuscular transmission at junctions in which the safety factor was reduced by subclinical MG, or that a drug might in some way have induced autoimmunity to AChR.

The results of this study indicate that acquired and at least one form of congenital MG share a common pathophysiological basis, namely failure of neuromuscular transmission due to a deficiency of AChR in the postsynaptic membrane. However the mechanisms responsible for AChR deficiency are quite distinct. Autoimmunity, although of unknown etiology, is undoubtedly the major if not sole cause of acquired MG. The basis of the congenital form of MG, whether failure of synthesis of AChR, accelerated degradation, faulty membrane insertion, low binding affinity for ACh or structural abnormality of the neuromuscular junction, remains to be ascertained. Experimental models of autoimmune myasthenia gravis in rodents[1,2,6] have provided valuable insights into the pathogenesis of the most common form of MG. The study of spontaneously occurring MG in animals should be equally valuable in identifying etiologic factors triggering the autoimmune response in acquired MG as well as in defining the pathogenesis of forms of MG which do not appear to have an autoimmune basis.

SUMMARY

Study of serum from 7 dogs with spontaneously occurring MG (5 acquired and 2 congenital) and muscle samples from 2 (one acquired and one congenital) revealed that:

(a) acquired MG in the dog resembles the human disease in that both anti-AChR and anti-striational antibodies are demonstrable;

(b) autoantibodies were not found in the congenital form of MG;

(c) there was a deficiency of AChR in both acquired and congenital MG, but only in the acquired form did AChR in muscle exist complexed with antibodies.

ACKNOWLEDGEMENTS

We thank Millie Thompson and Gregory Jones for excellent technical assistance, Dr. R. E. W. Halliwell for a gift of rabbit anti-dog Ig and Dr. P. E. Zollman for specimens of normal dog muscle and serum. We are grateful to the veterinarians (N.W. Johnston, T.F. Tunney and J.V. Goodyear) who kindly referred cases #3, 6 and 7 to A.C.P. This work was supported by grants form the Muscular Dystrophy Association (V.A.L.), the Muscular Dystrophy Group of Great Britain (A.C.P.), the National Institutes of Health (NS 11719-04 [V.A.L.] and NIAMDS AM 16717-03 [C.P.]) and the Los Angeles Chapter of the MG Foundation (V.A.L.).

REFERENCES

1.  Lennon, V.A. (1978) Human Pathology, (in press, September).
2.  Drachman, D.B. (1978) New Engl. J. Med., 298, 136-142; 186-193.
3.  Engel, A.G. et al., (1977) Mayo Clinic Proceedings, 52, 267-280.
4.  Lindstrom, J., et al., (1976) Neurology, 26, 1054-1059.
5.  Lindstrom, J., and Lambert, E.H. (1978) Neurology (Minneap.), 28, 130-138.
6.  Lennon, V.A. (1976) Immunol. Commun. 5, 323-344.
7.  Ormrod, A.N. (1961) Vet. Rec., 73, 489.
8.  Hall, G.A. et al., (1972) J. Path., 108, 177-180.
9.  Palmer, A.C. and Barker, J. (1974) Vet. Rec. 95, 452-454.
10. Johnston, N.W. and McDonald, I.F. (1977) Vet. Rec., 101, 216.
11. Lindstrom, J., et al., (1976) J. Exp. Med., 144, 726-738.
12. Lennon, V.A. (1977) In Cholinergic Mechanisms & Psychopharmacology Ed. D.E. Jenden, 24, 77-92, Plenum Press.
13. Almon, R.R. et al, (1974) Science, 186, 55-57.
14. Vincent A. (1978) First Int. Conf. on Plasmapheresis & Immunobiol. of Myasthenia Gravis (in press).

15. Zurn, A.D. and Fulpius, B.S. (1977) Eur. J. Immunol., 8, 529-532.
16. Peers, J., et al., (1977) Clin. Exp. Immunol., 27, 66-73.
17. Newsom-Davis, J., et al., (1978) Neurology, 28, 266-272.

DISCUSSION

GASSER: Have you attempted to induce myasthenia gravis in rat strains other than Lewis?

LENNON: We have tested Brown Norwegian rats (which differ from Lewis rats at the MHC) and have found that they are susceptible to induction of experimental autoimmune myasthenia gravis using either eel or Torpedo ACh receptors as antigen in the same doses and adjuvants as used for Lewis rats.

GERSHWIN: Do you have an explanation for the observation that all seven of your dogs were pure breeds rather than our all-American favorite, the mutt?
Do myasthenic dogs have any of the immunopathology seen in humans with MG, i.e., other autoantibodies?

LENNON: Spontaneous MG has also been reported in Labradors, and we presently have a Siamese cat with MG. Perhaps the reason MG has been reported only in pure breed animals is because of the expense involved in referring a pet to a veterinary neurologist.
The only other autoantibody we assayed was the antistriational antibody.

HARRISON: It is interesting that you stress the significant role of the acetylcholine receptor antibody in myasthenia gravis yet find no correlation between antibody titer and severity of disease. Would you care to comment?

LENNON: No correlation is seen in groups of patients between the titer of anti-AChR antibodies (as measured by radioimmunoassay with $^{125}I$-$\alpha$BT-AChR complexes) and the grade of disease severity. However within individual patients, lowering of antibody titers, e.g., upon withdrawal of penicillamine or immediately after plasmapheresis, is associated with improvement of clinical status.

MACKAY: Would you comment on the St. Bernard dog which did not have anti-receptor antibody. Was histology done on muscle or thymus in the dogs with autoimmune myasthenia, as the thymic lymphoid change in human myasthenia remains as an unsolved mystery of the disease.

LENNON: The 3 myasthenic dog sera without detectable anti-AChR antibodies were additionally tested by indirect immunofluorescence using living cultures of fetal dog muscle as substrate. IgG from the St. Bernard's serum bound to the cultured muscle cells, but sera from the dogs with congenital MG did not bind. These preliminary data suggest that the St. Bernard had antibodies reactive with an antigen exposed on the surface of fetal muscle membranes. If this proves to be an antigenic determinant of the AChR, it would put it close to the ACh binding site, which could explain why antibodies were not detected in the radioimmunoassay. The solubilized antigen used in that assay had $\alpha$- bungarotoxin (a molecule of 8000 daltons) bound to the ACh-binding site. In answer to your second question, no muscle or thymic histology reports are available at this time, except for the note of an anterior mediastinal tumor (melanoma) in the first german shepherd.

VOLPE: Could you tell us about any studies of delayed hypersensitivity in either human or spontaneous animal autoimmune myasthenia gravis?

LENNON: Abramsky and his colleagues in Israel and Richman and colleagues in this country have both reported the presence of antigen reactive peripheral blood lymphocytes in human patients with MG (using electric organ AChR as antigen in lymphocyte transformation assays). We do not yet have any information about cellular immune responses in dogs with spontaneously acquired MG. Although delayed-type hypersensitivity and lymphocyte transformation responses to AChR are demonstrable in rats with EAMG, there is as yet no evidence that cell-mediated immunity plays an effector role in the disease pathogenesis.

WEKERLE: One comment to the possible role of the thymus in the pathogenesis of myasthenia gravis. We have been culturing rat and mouse thymic reticulum as monolayers, and found that, under certain culture conditions, classical striated muscle colonies appear, which express remarkable amounts of acetylcholine receptors (AChR) on their surface. The development of these thymic muscle clones is genetically controlled by sex factors, and by the major histocompatibility gene complex.

This finding prompted us to postulate a two-step pathogenesis of MG, with double genetic control (Lancet, March, 1977). In a first step thymic muscle progenitor cells, which are genetically susceptible to induction are induced to myogenic differentiation and thus to AChR expression. This would be a necessary, but not sufficient, condition for step II to occur. In animals with concomitant genetically determined high reactivity to AChR determinants, thymic lymphocytes would recognize and react against the AChR neoantigens. This mechanism fits well into Dr. Schwartz's concept of class 1 and class 2 genes required for autoimmunity to occur: in our case, class 2 genes would determine the responsiveness of lymphocytes to AChR, and class 1 genes would control myegenic inducibility of thymic stem cells.

MORSE: To follow up on Dr. Wekerle's comment, it is noteworthy that pathologic studies of thymi from patients with myasthenia gravis have shown persistence of myoid elements which normally disappear early in life, and that such cells have been shown to have receptors for alpha-bungarotoxin. These findings may be of major importance in terms of abnormal sensitization of thymus-processed cells to acetylcholine receptors.

VOLPE: It is of interest that in Graves' disease, there is a reasonably good correlation between the titer of thyroid stimulating immunoglobulin (an antibody to the TSH receptor) and the severity of the hyperthyroidism, unlike the state of affairs in myasthenia gravis. This is undoubtedly because the thyroid stimulating immunoglobulin does not reduce the TSH receptors on the thyroid cells, unlike the situation in myasthenia. Thus, in Graves' disease, there is only one variable, i.e., the titer of the antibody and the reduced number of receptors.

# GENETIC ASPECTS OF ALEUTIAN DISEASE OF MINK

DAVID D. PORTER

Department of Pathology, UCLA School of Medicine, Los Angeles, California 90024

ABSTRACT

Aleutian disease is a chronic or persistent viral infection of mink, and is caused by a parvovirus. Highly virulent virus strains cause serious disease in all types of mink, while less virulent virus strains cause serious disease and death primarily in mink of the Aleutian genotype. Mink with progressive disease develop marked hypergammaglobulinemia and extremely high levels of viral-specific antibody. Deposition of circulating immune complexes causes severe glomerulonephritis and arteritis. The tissue lesions may be decreased by immunosuppressive drug therapy or infection in utero, or increased by inoculation of a killed virus vaccine prior to challenge by live virus. While it is apparent that both viral and host genetic factors play major roles in the development and extent of the lesions of Aleutian disease, additional work is needed for a complete understanding of the role the genetic factors play in this disease.

INTRODUCTION

In 1941 a blue-gray coat color mutation arose in standard dark wild-type mink. These mink were bred and the new coat color was found to be an autosomal recessive trait. The animals were called Aleutian mink because the coat color resembled that of the Aleutian blue fox. Breeding Aleutian mink with other recessive coat color mutants resulted in wide dissemination of the Aleutian gene in the mink-ranching areas of the world. Mink carrying this gene were noted to be difficult to raise due to a fatal progressive disease, described by Hartsough and Gorham in 1956[1]. Aleutian disease was originally thought to be a genetic disease, but was shown to be transmissible and probably viral in 1962[2,3,4]. The pathogenesis of Aleutian disease has been studied extensively, and a number of reviews are available [5,6,7,8]. More recently, progress has been made in characterizing the virus[9]. In this review I will emphasize genetic factors which contribute to Aleutian disease.

## MINK

Mink (Mustela vison) are members of the order Carnivora and family Mustelidae which includes the wolverine, ferret, skunk and weasel[10]. Mink are commercially raised in large numbers for their pelts, and Aleutian disease represents the major health problem faced by mink ranchers. The reproductive physiology of mink is well described[11,12], and the coat color genetics of mink is well understood[13]. In most areas, mink breed once annually in early spring and an average of 4 young are born after a 50 day gestation period[14]. Mink are usually bred twice on commerical ranches, and while most offspring result from the second mating, a few result from the first mating[12,15].

Brother-sister and other close family matings are usually avoided by ranchers, therefore the mink are genetically heterogeneous except for coat color genes.

There is a tendency for delayed implantation of fertilized ova, so that mink which breed early in the season have a longer gestation period than those bred late in the season. This tendency for delayed implantation is less than that seen in a number of other species in this family[10]. Mink which carry the Aleutian gene have a lysosomal abnormality similar to that of humans with the Chediak-Higashi syndrome[16]. The lysosomal abnormality is probably responsible for the dilution of the coat color, and is thought to make such mink more susceptible to a number of infectious diseases, including Aleutian disease.

## ALEUTIAN DISEASE VIRUS

The transmissible nature of Aleutian disease was easily established[2,3,4], and early characterization studies used the organs of infected mink as a source of virus[17]. Aleutian disease virus has been shown to be transmitted vertically and horizontally on mink ranches[7]. The relative importance of the two modes of transmission is the subject of some controversy. It proved extremely difficult to isolate the virus in cell culture systems. We eventually found that the virus would replicate in feline renal cells if the original isolations were attempted at temperatures markedly lower than the body temperature of mink (39°C)[18]. The temperature required for viral isolation varies with various viral strains, with the highly virulent Utah-1 isolate having a higher optimum temperature than less virulent strains. It has been postulated that temperature-sensitive viruses may lead to persistent infections, based principally upon experiments in cell cultures[19]. The virus is markedly cell-associated. Cultured virus produces Aleutian disease in mink, has equal infectivity for cell cultures and mink, and can be reisolated from mink at all times after infection. Cultured virus has the same properties as virus isolated from in vivo sources[20], and no difference in the immunologic characteristics of several strains of virus from the two sources has been noted.

The characteristics of Aleutian disease virus have been tabulated in detail[9]. The purified virions are naked icosahedra with a diameter of 23-25 nm and have a buoyant density of 1.41 to 1.43 gm/ml in CsCl [18,20]. The virus principally replicates in cell nuclei, although some antigen is observed in the cytoplasm[18]. The virion contains 1.2 x $10^6$ daltons of single-stranded DNA, and 3 or 4 structural polypeptides [21]. These characteristics are typical of members of the autonomous parvovirus group[22], and the marked stability of the virion to heat and chemical treatment is shared with other members of this viral group.

## VIRAL GROWTH AND THE HOST IMMUNE RESPONSE

In vivo growth of Aleutian disease virus is initially rapid and peak titers of $10^8$ to $10^9$ infectious particles per gram of spleen and liver are observed 10 days after infection[23]. At later times, viral titers slowly fall and eventually stabilize at approximately $10^5$ infectious units per gram of spleen and $10^4$ units per ml of serum[6]. We did not find a substantial difference between the viral titers in Aleutian and non-Aleutian mink during the early phase of viral growth using the Utah-1 strain of virus[23]. The equally virulent Guelph strain produced similar antigen titers in both types of mink 10-12 days after infection[24].

Immunofluorescence[23], and electron microscopic studies[25], have shown viral antigen and viral particles in the cytoplasm and nucleus of macrophages in the spleen, lymph node and liver (Kupffer cells). It appears that the virus probably replicates in macrophages since antigen and virions are found in nuclei. It is possible that some of the antigen found in the cytoplasm of macrophages could be immune complexes which were phagocytosed.

Immunofluorescence, complement fixation and counterelectrophoresis tests have been used to demonstrate viral antibody in mink with Aleutian disease, and all the tests appear to have the same specificity[26]. The antibody appears to be directed against virion capsid antigens[26], and agglutination of virions by antibody can be shown by electron microscopy[18,20]. Antibody first appears 7-10 days after infection and as early as 30 days post-infection the titers are extraordinarily high as compared with usual viral infections[23]. Progressively infected mink develop a marked hypergammaglobulinemia. We found 570 normal mink to have a gamma globulin level of 0.74 g/100ml, while 683 naturally infected mink had 3.5 g/100 ml[27], with a maximal level of 11 g/100 ml. The elevated gamma globulin is nearly all 6.4S IgG and the half-life of the IgG is decreased about one-third in hypergammaglobulinemic mink[28]. The IgG may show restricted electro-phoretic mobility by 40 days after infection[29], and about 10% of mink surviving infection for one year develop a monoclonal gammopathy[30], which may eventually disappear.

Antibody to the virus of the IgM and IgA classes may be found in low titer and a transient increase in IgM and lasting increase in IgA are found after infection[31]. IgA antibody may be selectively deposited in the glomerular lesions[32], but another study found IgM in the glomerular deposits[33].

Virus in the serum of chronically infected mink exists as com-plexes with antibody, but the complexes themselves are infectious[34]. Attempts to neutralize the virus by mink antibody result in agglu-tination of the viral particles, but the infectivity is not reduced in vivo or in the cell cultures[18,23]. Complexes smaller than virion size are present in such large amounts in the serum of chronically infected mink that they may be observed directly by analytical ultra-centrifugation of the serum[28].

The responses of normal and infected mink to a large variety of an-tigens other than Aleutian disease virus have been examined in detail. If mink are immunized with various antigens prior to infection with Aleutian disease virus, the antibody titers to these antigens usually fall despite a marked increase in gamma globulin levels[28,35,36]. Mink which are infected with Aleutian disease virus show about one-quarter the response to several antigens as compared with normal mink [28,37,38]. Mink which have been infected with Aleutian disease virus for 6 months or more may show normal immune responsiveness[38]. Simul-taneous exposure to Aleutian disease virus and an unrelated antigen results in a normal initial immune response to the unrelated antigen which rapidly falls during the period of increasing gamma globulin[39]. These results may be explained on the basis of antigenic competition with Aleutian disease viral antigen. It has been shown that peripher-al blood lymphocytes from Aleutian diease virus-infected mink are

markedly less responsive to T-cell mitogens, and indirect evidence
suggests that an absolute decrease in T-cell numbers occurs[40]. No
information on viral specific T-cell immunity is presently available.

PATHOLOGY

The principal lesion of Aleutian disease is a systemic plasmacytosis involving the bone marrow, lymphoid tissues, liver and kidneys
[8,41,42]. The time course of appearance of the lesions has been
throughly studied[43]. The degree of plasmacytosis is easiest to evaluate in the kidneys, since normally no such cells are present and
the diffuse interstitial plasma cell infiltrates can usually be distinguished from the occasional examples of chronic pyelonephritis.
The extent of plasmacytosis is directly proportional to the serum
gamma globulin level and to the degree of glomerulonephritis and
arteritis[27].

Glomerulonephritis is nearly always present in severe Aleutian
disease and the resulting renal failure is the usual cause of death
in infected mink[8]. Light and electron microscopic studies have shown
a typical glomerulonephritis with especially marked proliferation of
mesangial matrix material[44,45]. Immunofluorescence studies have shown
that moderate deposits of immunoglobulin and very marked deposits
of complement occur along capillary walls in mesangial regions[23,33,46,
47]. Viral-specific antibody can be eluted from the kidneys[23,48], but
viral antigen can be demonstrated in the immune deposits only after
elution of immunoglobulin except in mink infected transplacentally.

Arteritis affecting large muscular arteries in the heart, liver,
brain and kidneys is seen in severe Aleutian disease, especially in
mink of the Aleutian genotype. Immunoglobulin, complement and viral
antigen may be demonstrated in the lesions[49]. Cyclic variations in
the concentrations of several proteins of the coagulation system have
been observed in Aleutian disease[50], but the role of intravascular
coagulation in promoting the tissue lesions is not clear. Infected
mink may become anemic and immunoglobulin can be shown on the surface
of erythrocytes. Erythrocyte eluates contain viral specific antibody
[48].

During the early phase of intense replication of Aleutian disease
virus no clinical illness nor tissue lesions can be appreciated[23].
If mink antibody to the virus is passively administered to animals at
the time of peak viral replication, cytolysis of infected cells associated with acute and chronic inflammation occurs[51].

If mink are immunized with a killed Aleutian disease viral vaccine
and subsequently challenged with live virus, they develop markedly
enhanced tissue lesions[51]. When cyclophosphamide is continuously
given in immunosuppressive amounts to infected mink, tissue lesions
can be completely suppressed but viral titers are not affected[52].
Mink infected transplacentally develop much milder tissue lesions
than mink infected as adults, and such animals have higher viral
titers and lower antibody levels than the adult mink[53]. The presently available evidence suggests that the viral infection itself
is relatively innocuous and that the disease results from the
humoral antibody response to the virus with immune complex formation
and deposition.

GENETIC INFLUENCES ON VIRAL PERSISTENCE, DISEASE, AND DEATH

Aleutian disease workers have generally agreed that viruses of
high virulence, such as the Utah-1 and Guelph strains, are those
which can cause progressive disease and death in non-Aleutian mink,
while low virulence virus, such as the Pullman strain, produces
disease and death only in Aleutian mink. A number of additional viral
strains of each type are available, but they have not been widely
used. No antigenic or physicochemical differences have been noted
between high and low virulence virus strains, but only major
differences could be found with the techniques employed.

Mink of the Aleutian genotype apparently will all die of Aleutian
disease once infected with a virus of either high or low virulence.
A typical life table for Aleutian mink inoculated with the low
virulence Pullman strain is shown in the work of Eklund et al.[54],
and 50% of the mink were dead by 120 days. Non-Aleutian mink given
the same virus inoculum rarely develop disease. Similar observations
have been made by others[8,55], and the pattern of death losses due
to natural infection of mink on commercial ranches apparently follows
a similar pattern[56]. The Pullman strain of virus does replicate in
non-Aleutian mink and during the first 8 weeks post-inoculation,
specific anti-viral antibody titers are similar, but the titers fall
in non-Aleutian mink at later times[57]. Virus has been recovered in
low titer from the mesenteric lymph nodes of such non-Aleutian mink
22 months after viral inoculation[54].

High virulence virus kills Aleutian mink in approximately the
same time as low virulence virus, however, the high virulence virus
will cause lesions in about 75% of non-Aleutian mink[58]. Deaths in

non-Aleutian mink reach the 50% level by one year, and a few of
the mink with lesions will survive for their normal life-span of
7 to 8 years. Analysis of the non-Aleutian mink which did not
become hypergammaglobulinemic indicated that they did not have
viremia, and offspring of these mink showed the same frequency
of progressive and non-progressive infections as their parents,
indicating that a single gene was not responsible for host resistance[58]. It is clear that in areas in which high virulence virus
is circulating in the mink population all genetic types of mink
may develop severe disease[27]. In such areas inapparent but persistent Aleutian disease is found in non-Aleutian mink, and animals of
this type may represent the natural reservior of the virus[59].

The rate of progression of the tissue lesions in Aleutian mink
is appreciably faster than in non-Aleutian mink, and this has been
thoroughly studied for the glomerular lesion[33], and noted for the
other lesions[55].

The mechanism by which Aleutian disease virus persists is not
clear. The possibilities include the temperature-sensitive nature
of viral replication[18,19], failure of mink antibody to neutralize
infectivity[18,23] and reactivation of antibody-complexed virus by
macrophages[23]. Both viral and host genetics play major roles in
determining the outcome of infections by Aleutian disease virus.
Large numbers of mink with defined coat-color markers and breeding
records are available and formal genetic studies of the disease
could be readily carried out.

## ACKNOWLEDGEMENTS

The author's work was supported by grant AI-09476 from the
National Institute of Allergy and Infectious Diseases, National
Institutes of Health, and by the Mink Farmer's Research Foundation.

## REFERENCES

1. Hartsough, G.R. and Gorham, J.R. (1956) Nat. Fur News 28, 10-11.
2. Karstad, L. and Pridham, J.G. (1962) Canad. J. Comp. Med. Vet.
   Sci., 26, 97-102.
3. Henson, J.B., Gorham, J.R., Leader, R.W. and Wagner, B.M. (1962)
   J. Exp. Med., 116,357-364.
4. Trautwein, G.W. and Helmboldt, C.F. (1962) Am. J. Vet. Res., 23,
   1280-1288.

5. Ingram, D.G. and Cho, H.J. (1974) J. Rheumatol., 1, 74-92.

6. Porter, D.D. and Larsen, A.E. (1974) Prog. Med. Virol., 18, 32-47.

7. Gorham, J.R., Henson, J.B., Crawford, T.B., and Padgett, G.A. (1976) in Slow Virus Diseases of Animals and Man, Kimberlin, R.H., ed., North-Holland, Amsterdam, pp. 135-158.

8. Henson, J.B., Gorham, J.R., McGuire, T.C., and Crawford, T.B. (1976) In Slow Virus Diseases of Animals and Man, Kimberlin, R.H., ed., North-Holland, Amsterdam, pp. 175-205.

9. Cho, H.J. and Porter, D.D. (1978) V. 11 ICN-UCLA Symposia on Molecular and Cellular Biology, Stevens, J.G. and Todaro, G.J., eds., Academic Press, New York. In press.

10. Ewer, R.F. (1973) The Carnivores, Cornell University Press, Ithaca.

11. Hansson, A. (1947) Acta Zoologica, 28, 1-136.

12. Shackelford, R.M. (1952) Am. Nat., 86, 311-319.

13. Shackelford, R.M. (1950) Genetics of the Ranch Mink, Pilsbury Press, New York.

14. Bowness, E.R. (1968) Canad. Vet. J., 9, 103-106.

15. Johansson, I. and Venge, O. (1951) Acta Zoologica, 32, 255-258.

16. Davis, W.C., Spicer, S.S., Greene, W.B., and Padgett, G.A. (1971) Lab. Invest., 24, 303-317.

17. Burger, D., Gorham, J.R. and Leader, R.W. (1965) NINDB Monograph No. 2., Slow, Latent and Temperate Virus Infections, U.S. Govt. Printing Office, Washington, D.C., pp. 307-313.

18. Porter, D.D., Larsen, A.E., Cox, N.A., Porter, H.G., and Suffin, S.C. (1977) Intervirology, 8, 129-144.

19. Preble, O.T. and Youngner, J.S. (1975) J. Infect. Dis., 131, 467-473.

20. Cho, H.J. (1976) In Slow Virus Diseases of Animals and Man, Kimberlin, R.H., ed., North-Holland, Amsterdam, pp. 159-174.

21. Shahrabadi, M.S., Cho, H.J. and Marusyk, R.G. (1977) J. Virol., 23, 353-362.

22. Rose, J.A. (1974) In Comprehensive Virology, V. 3, Fraenkel-Conrat, H., Wagner, R.R., eds., Plenum Press, New York, pp. 1-61.

23. Porter, D.D., Larsen, A.E. and Porter, H.G. (1969) J. Exp. Med., 130, 575-593.

24. Cho, H.J. and Ingram, D.G. (1974) J. Immunol. Methods, 4, 217-228.

25. Shahrabadi, M.S. and Cho, H.J. (1977) Canad. J. Comp. Med., 41, 435-445.

26. Crawford, T.B., McGuire, T.C., Porter, D.D., and Cho, H.J. (1977) J. Immunol., 118, 1249-1251.

27. Porter, D.D. and Larsen, A.E. (1964) Am. J. Vet. Res., 25, 1226-1229.

28. Porter, D.D., Dixon, F.J. and Larsen, A.E. (1965) J. Exp. Med., 121, 889-900.

29. Tabel, H. and Ingram, D.G. (1970) Canad. J. Comp. Med., 34, 329-332.

30. Porter, D.D., Dixon, F.J. and Larsen, A.E. (1965) Blood, 25, 736-742.

31. Porter, D.D., Porter, H.G. and Larsen, A.E. (1977) Fed. Proc., 36, 1268.

32. Portis, J.L. and Coe, J.E. (1978) Fed. Proc., 37, 1558.

33. Johnson, M.I., Henson, J.B. and Gorham, J.R. (1975) Am. J. Pathol., 81, 321-337.

34. Porter, D.D. and Larsen, A.E. (1967) Proc. Soc. Exp. Biol. Med., 126, 680-682.

35. Tabel, H., Ingram, D.G., and Fletch, S.M. (1970) Canad. J. Comp. Med., 34, 320-324.

36. Trautwein, G., Schneider, P. and Ernst, E. (1974) Zbl. Vet. Med. B., 21, 467-479.

37. Lodmell, D.L., Hadlow, W.J., Munoz, J.J. and Whitford, H.W. (1970) J. Immunol., 104, 878-887.

38. Kenyon, A.J. (1966) Am. J. Vet. Res., 27, 1780-1782.

39. Porter, D.D. and Larsen, A.E. (1968) Perspectives in Virol., 6, 173-186.

40. Perryman, L.E., Banks, K.L., McGuire, T.C. (1975) J. Immunol., 115, 22-27.

41. Helmboldt, C.F. and Jungherr, E.L. (1958) Amer. J. Vet. Res., 19, 212-222.

42. Obel, A.-L. (1959) Am. J. Vet. Res., 20, 384-393.

43. Henson, J.B., Leader, R.W., Gorham, J.R. and Padgett, G.A. (1966) Pathol. Vet., 3, 289-314.

44. Henson, J.B., Gorham, J.R. and Tanaka, Y. (1967) Lab. Invest., 17, 123-139.

45. Kindig, D., Spargo, B. and Kirsten, W.H. (1967) Lab. Invest., 16, 436-443.

46. Henson, J.B., Gorham, J.R. and Padgett, G.A. (1969) Arch. Pathol., 87, 21-28.

47. Pan, I.C., Tsai, K.S. and Karstad, L. (1970) J. Pathol., 101, 119-127.

48. Cho, H.J. and Ingram, D.G. (1973) Infect. Immun., 8, 264-271.

49. Porter, D.D., Larsen, A.E. and Porter, H.G. (1973) Am. J. Pathol., 71, 331-344.

50. McKay, D.G., Phillips, L.L., Kaplan, H. and Henson, J.B. (1967) Am. J. Pathol., 50, 899-916.

51. Porter, D.D., Larsen, A.E. and Porter, H.G. (1972) J. Immunol., 109, 1-7.

52. Cheema, A., Henson, J.B., and Gorham, J.R. (1972) Am. J. Pathol., 66, 543-556.

53. Porter, D.D., Larsen, A.E. and Porter, H.G. (1977) J. Immunol., 119, 872-876.

54. Eklund, C.M., Hadlow, W.J., Kennedy, R.C., Boyle, C.C. and Jackson, T.A. (1968) J. Infect. Dis., 118, 510-526.

55. Padgett, G.A., Reiquam, C.W., Henson, J.B., Gorham, J.R. (1968) J. Pathol. Bacteriol., 95, 509-522.

56. Gorham, J.R., Leader, R.W., Padgett, G.A., Burger, D. and Henson, J.B. (1965) In NINDB Monograph No. 2, Slow, Latent and Temperate Virus Infections, U.S. Govt. Printing Office, Washington, D.C., pp. 279-285.

57. Bloom, M.E., Race, R.E., Hadlow, W.J., and Chesebro, B. (1975) J. Immunol., 115, 1034-1037.

58. Larsen, A.E. and Porter, D.D. (1975) Infect. Immun., 11, 92-94.

59. An, S.H. and Ingram, D.G. (1977) Am. J. Vet. Res., 38, 1619-1624.

# PART III
# INDUCED DISEASE IN ANIMALS
Noel R. Rose *and* Ian Mackay, *Co-Chairmen*

# THE BIOLOGICAL SIGNIFICANCE OF ALLOREACTIVITY

BARUJ BENACERRAF AND STEVEN J. BURAKOFF
Department of Pathology, Harvard Medical School, Boston, Massachusetts 02115

Among the most perplexing observations that have confronted cellular immunologists and students of transplantation immunity concern the immune response to major histocompatibility complex (MHC) antigens. These include the following: (1) the very high frequency of alloreactive T cells present in peripheral lymphoid organs of adult mammals, prior to immunization with allogeneic cells. The frequency of prekiller T cells capable of developing into cytotoxic T lymphocytes (CTL) after stimulation across a single MHC haplotype difference has been estimated at approximately 1-3%[1-3]. A similarly large proportion of alloreactive T cells are able to mount a proliferative response after stimulation by allogeneic cells[4]. (2) The finding that T cells are able to respond to allogeneic MHC antigenic stimulation in mixed leukocyte cultures without prior immunization[5]. (3) The observation that MHC antigens expressed on allogeneic cells generally stimulate stronger T cell responses than xenogeneic MHC antigens (i.e., antigens expressed on cells of other species)[6]. This latter point appears paradoxical, for it contrasts with the immunological dogma that antibody responses are strongest against antigens furthest phylogenetically from the responding individual. This dogma apparently does not hold for T cell responses to MHC antigens.

The overwhelming commitment of T cells to specificities expressed by the MHC antigens of their own species has justifiably concerned immunologists in view of the apparent biologic irrelevence and the absence of evolutionary advantage of these alloreactive T cells.

Experiments from various laboratories in recent years demonstrated that T cells, endowed with: (a) cytolytic activity against virally infected cells[7], (b) delayed type hypersensitivity reactivity[8] or (c) helper cell activity for thymus dependent antigens[9], only recognized foreign antigens in association with autologous MHC gene produts on the surface of cells. These observations provided a reasonable explanation for the biological significance of MHC antigens in defense mechanisms. They nevertheless left many questions unanswered concerning: (1) the commitment of the majority of T cells to MHC antigens and the process by which this commitment is achieved, (2) the manner in which the same T cells are concurrently specific for foreign antigens and MHC antigens and (3) the very high frequency of alloreactive T cells. In this communication, we propose to address these issues in turn and to illustrate how alloreactivity developed as an unnecessary byproduct of the emergence of a highly effective surveillance mechanism to detect modifications of self MHC antigens; these modifications occuring from the association of autologous MHC antigens with foreign antigens, such as viruses.

The commitment of T cells to MHC antigens

Jerne[10] originally proposed a theory which was further elaborated by ourselves[6,11] to explain the generation in the thymus of T cells specific for MHC antigens. According to the theory, in the first stage T cells initially specific for self MHC gene products are selected in the thymus to differentiate and proliferate. Then, in a second stage, only those T cells which bear low affinity receptors for self MHC antigens are allowed to mature and leave the thymus as functional T cells. Such T cells, having low reactivity for self MHC antigens, have concommitently high affinity for variants of self MHC antigens. These variants appear to be the same or similar to the allogeneic MHC antigens expressed in the same species. Weaker affinity for xenogeneic MHC antigens would thus be expected. Simultaneously and independently these T cells develop receptors for determinants expressed on conventional thymus dependent antigens.

The high degree of reactivity to MHC antigens which constitute the polymorphic population encountered in the same species (i.e. alloantigens) and the lower reactivity to xenogeneic MHC antigens maybe accounted for by the fact that low affinity receptors for self MHC antigens might be expected to react optimally with allogeneic MHC antigens, but much less so with xenogeneic antigens. This would account for the paradox that the strongest T cell responses are not elicited by antigens further removed phylogenetically from the responder. Two predictions from this theory would be: (a) that clones of T cells induced by xenogeneic MHC antigens should be highly cross-reactive with allogeneic MHC antigens, even to the extent that they may demonstrate a heteroclitic response. This has indeed been demonstrated when mouse anti-rat CTL were shown in our laboratory to be comprised of clones cross-reactive with allogeneic target cells[6]. (b) Alloreactive T cells should be expected to be highly cross-reactive with modified syngeneic cells. This was also shown to be the case when we observed considerable cross-reactivity by alloreactive cells for TNP conjugated target cells syngeneic to the responder[11].

Since the T cell repertoire for MHC specificities is normally determined by the self MHC antigens of the thymus, we should expect the T cell repertoire to vary according to the MHC of the thymus in which T cells differentiate. Recent experiments utilizing radiation chimeras by Zinkernagel and associates[12]

Bevan[13] and Billings and ourselves[14] have demonstrated this to be the case.
The dual specificity of the T cells for MHC antigens and foreign antigens.

When Zinkernagel and Doherty discovered that CTL specific for LCM virus infected cells only killed cells bearing these viral antigens which were also syngeneic with the immunizing cells, they proposed two distinct mechanisms to explain the specificity of the same T cells for both MHC antigens and viral antigens[7]. These were termed the altered self theory and dual interaction theory. According to the altered self theory, T cells bear only one receptor for all the possible variants of MHC self antigens. The interaction of the MHC molecules with the appropriate antigens cause the formation of new determinants to which the T cells bear receptors of high affinity. The dual interaction theory postulates that a T cell bears two receptors, one for self MHC antigens and the other independently expressed for conventional thymus dependent antigens.

In spite of considerable experimental work, it has not been possible to establish definitely which of these two theories is correct. Certain important findings nevertheless have been made which must be taken into account when a final explanation prevails. In several systems, interaction of viral antigens with MHC gene products on the cell membranes has been documented[15,16]. Furthermore, it would appear that the MHC antigens and the viral antigens, which stimulate CTL responses cannot be presented to the T cell on two different cell surfaces, but must be on the same cell to stimulate a response. Recently, Finberg, Mescher and Burakoff[17] in our laboratory, in studying the CTL response to Sendai virus in a syngeneic system, deomonstrated that H-2 proteins and Sendai virus proteins could be solubilized in detergent and then incorporated into liposomes by dialyzing out the detergent. These proteins retained their ability to stimulate a CTL response and it was observed that only when Sendai proteins and H-2 proteins were incorporated into the same liposome was the optimal CTL response stimulated. Equivalent amounts of H-2 proteins and Sendai virus proteins added to cultures in separate liposomes, were far less stimulatory. These results suggest that the H-2 antigens and Sendai proteins may form complexes which affect the tertiary structure of these molecules. This would be expected to be the case if the MHC proteins have evolved for the purpose of specific interactions with viral antigens and/or thymus dependent antigens or their fragments. These interactions may thus create the variants of self MHC antigens recognized by T cells (whose repertoire has evolved in the manner proposed in the previous section).
Alloreactive T cells are stimulated by immune responses to modified syngeneic cells.

The postulate that alloreactivity results from T cells differentiating in

the thymus that are strongly reactive for variants of self MHC antigens leads
to the expectation that immunization with virally infected syngeneic cells
should result in the stimulation of T cell clones that are reactive with the
virally infected syngeneic cells used to immunize and also reactive with unin-
fected allogeneic target cells.

We have recently shown that immunization of BALB/c ($H-2^d$) mice with Sendai
coated syngeneic cells stimulate CTL which lyse Sendai coated BALB/c target
cells but also lyse uncoated $H-2^b$, $H-2^q$, $H-2^k$, $H-2^s$ and $H-2^r$ allogeneic target
cells to an appreciable degree[18]. We further demonstrated by the cold target
inhibition technique that the same clones that lysed BALB/c coated Sendai
targets also crossreactively lysed the allogeneic targets. Furthermore, it
was observed that separate CTL clones lysed each of the different allogeneic
targets. In addition, there was significantly less lysis of target cells
bearing the $H-2^q$ haplotype than of target cells bearing the $H-2^k$ or $H-2^r$
haplotypes. This later finding suggests that the association of Sendai virus
antigens with the $H-2^d$ gene products of BALB/c mice creates determinants which
are more crossreactive with $H-2^k$ and $H-2^r$ than with $H-2^q$ gene products. Our
hypothesis would therefore predict that the T cell responses by BALB/c mice
to other viruses or to Sendai virus by mice of different haplotypes should
result in different patterns of crossreactivity for allogeneic targets.

In another system, Bevan has reported that stimulation of mouse spleen cells
by H-2 identical spleen cells bearing foreign minor H antigens results in the
generation of T cells that lyse H-2 identical targets bearing the appropriate
foreign minor antigens[19]. Recently, he has found that repeated stimulation
of such CTL with allogeneic cells results in selection of an effector population
which lyses both allogeneic cells as well as H-2 identical target cells bearing
the appropriate minor foreign H antigens with a similar efficiency[20]. These
observations are consistant with our observation; specifically autologous MHC
antigens in association with non MHC antigens are crossreactive with allogeneic
MHC antigens.

Recent experiments on the Ly phenotype of the precursors of alloreactive
CTL are also in agreement with our hypothesis. Precursors of alloreactive
CTL have been found to belong to the Ly 23 set, while precursors that response
to TNP coupled syngeneic cells were found to reside in the Ly 123 set[21]. We have
found that alloreactive prekiller CTL in young mice reside initially in the
Ly 123 pool but shift to the Ly 23 pool by 5 weeks of age[22]. This suggests
that the Ly 23 alloreactive prekiller T cells found in adult mice are actually
"memory" cells; these alloreactive precursor T cells shift from the Ly 123
pool to the Ly 23 pool when they encounter autologous MHC antigens in association
with foreign antigens, such as virus.

In conclusion, we feel that the proposed hypothesis can account for several of the observations that have puzzled immunologists previously. It should be pointed out, however, that these data do not permit firm conclusions on the critical issue of whether cross-reactive recognition of MHC products and conventional antigens is mediated by T cells carrying a single receptor or two receptors. However, if, as proposed by many investigators, T cells bear two distinct receptors, one specific for MHC gene products and the other for conventional antigens, an analysis of the specificity of the receptors must account appropriately for the large number of alloreactive T cell clones. In this context, the data we have discussed strongly suggest that in a two receptor model, the receptor specific for MHC antigen is specific for variants of self rather than self, as proposed by Zinkernagel et al[12] and that the two receptors function in close relationship.

REFERENCES

1. Skinner, M.A. and Marbrook, J. (1976) J. Exp. Med., 143, 1562-1567.

2. Lindahl, K.F. and Wilson, D.B. (1977) J. Exp. Med., 145, 508-521.

3. Miller, R.G., Teh, H.S., Harley, E. and Phillips, R.A. (1977) Immunol. Rev., 35, 38-58.

4. Wilson, D.B., Blyth, J. and Nowell, P.C. (1968) J. Exp. Med., 128, 1157-1182.

5. Cerottini, J.C. and Brunner, R.T. (1974) Adv. Immunol., 18, 67-132.

6. Burakoff, S.J., Ratnofsky, S.E., and Benacerraf, B. (1977) Proc. Natl. Acad. Sci. USA, 74, 4572-4576.

7. Doherty, P.C., Blanden, R.V. and Zinkernagel, R.M. (1976) Transplant. Rev., 29, 89-124.

8. Miller, J.F.A.P., Vadas, M.A., Whitelaw, A. and Gamble, J. (1975) Proc. Natl. Acad. Sci. USA, 72, 5095-5098.

9. Paul, W. and Benacerraf, B. (1977) Science, 195, 1293-1300.

10. Jerne, N.K. (1971) Eur. J. Immunol., 1, 1-9.

11. Lemmonier, F., Burakoff, S.J., Germain, R.N., and Benacerraf, B. (1977) Proc. Natl. Acad. Sic. USA, 74, 1229-1233.

12. Zinkernagel, R.M., Callahan, G.N., Althage, A., Cooper, S., Klein, P.A. and Klein, J. (1978) J. Exp. Med., 147, 882-896.

13. Bevan, M.J. (1978) Nature, 269, 417-418.

14. Billings, P., Burakoff, S.J., Dorf, M.E., and Benacerraf, B. (1978) J. Exp. Med. In press.

15. Schrader, J.W., Cunningham, B.A., and Edelman, G.M. (1975) Proc. Natl. Acad. Sci., 72, 5066-5070.

16. Bubbers, J.E., Steves, R.A. and Lilly, F. (1976) Proc. Am. Ass. Cancer Res., 17, 93.

17. Finberg, R., Mescher, M., and Burakoff, S.J. (1978) Submitted.

18. Finberg, R., Burakoff, S.J., Cantor, H. and Benacerraf, B. (1978) Proc. Natl. Acad. Sci. USA, In press.

19. Bevan, M.J. (1975) Nature, 256, 419-421

20. Bevan, M.J. (1977) Proc. Natl. Acad. Sci. USA, 74, 2094-2098.

21. Cantor, H. and Boyse, E.A. (1977) Cold Spring Harbor Sym. Qant. Biol., 41, 23-32.

22. Burakoff, S.J., Finberg, R., Glimcher, L., Lemmonier, F., Benacerraf, B. and Cantor, H. (1978) Submitted.

327

POSSIBLE ROLES OF AUTO-ANTI-IDIOTYPIC IMMUNITY IN AUTO-IMMUNE DISEASE

HANS WIGZELL[X], HANS BINZ[XX], HANNES FRISCHKNECHT[XX], PER PETERSON[XXX], and KARIN SEGE[XXX]
[X]Department of Immunology, Uppsala University, Uppsala, Sweden
[XX]Department of Medical Microbiology, Zürich University, Zürich, Switzerland
[XXX]Department of Cell Biology, Uppsala University, Uppsala, Sweden

ABSTRACT

Auto-anti-idiotypic immunity is a normal part of a conventional immune process and may have as a consequence potentiation or elimination of a select immune function if this is dependent on the presence of a given clone of idiotype-positive cells. The conditions under which auto-anti-idiotypic immunity has been found to function to obliterate immune reactivity will be described and discussed. Likewise, the opposite results, that is using auto-anti-idiotypic immunity to induce the activation and proliferation of idiotype-positive cells resulting in immunity against a defined antigen will be discussed. Finally, molecular mimicry as to steric configuration between antigen and anti-idiotypic antibodies produced against antibodies specific for this antigen has been found to endow the anti-idiotypic antibodies with some functions normally restricted to the antigenic molecules. The conclusion from the above data is that auto-anti-idiotypic immune reactions carry the potential to selective change the immune course in already immune individuals. Whether the change of course will be significant enough and of positive value in auto-immune diseases only experiments will tell.

INTRODUCTION

Auto-anti-idiotypic immune reactions could be visualized to function in two different ways of interest in realtion to auto-immune disease. The anti-idiotypic reactions may in themselves cause disease or affect the development of disease in a negative manner. Alternatively, auto-anti-idiotypic immune reactions may be used to selectively obliterate exactly those clones of immunocompetent lymphocytes which are causing the disease. Both aspects will be discussed as to underlying theory and exemplified as to experimental systems when available. However, in order to understand how auto-anti-idiotypic immunity may afflict an individual certain basic understandings as to how the immune function is being regulated must be established.

Most bodily constitutents being able to function as antigens belong to the group of antigens known as T-dependent immunogens, meaning that in order for B cells to be triggered helper T cells would normally be required for antibody synthesis. It is of course to be realized that T-independent, cross reacting microbial antigens may circumvent the need for helper T cells but there is little doubt that T lymphocytes play an important role in most auto-immune diseases at some stage of the

disease. It is thus in the present concept quite important to know about the idiotypic variations encountered on immunocompetent T and B lymphocytes reactive against
the "same" antigen. A few facts seem to exist in this regard: T lymphocytes like B
cells express idiotypic receptors with similar idiotypes as the B cells and their
products, the antibody molecules[1,2]. However, idiotypic variability of B cells may
well be in part distinct from those of T lymphocyte receptors as B cells have been
found in some experiments to express idiotypic determinants not to be found on the
corresponding T cells[3,4]. When analyzed as to underlying molecular basis it would seem
clear that the T cell receptor molecules do normally only express the idiotypes found
on the heavy chain of the B cell derived immunoglobulin molecules with similar antigen-
binding specificity. Also, purified heavy chains from idiotypic immunoglobulin molecules but not light chains can completely obliterate anti-T cell idiotypic radioimmuno
assays[5]. Further proof for the involvement of products of heavy chain immunoglobulin-
linked genes in the creation of antigen-binding T cell receptors stem from the results of genetic experiments. Here it could be shown that the inheritance pattern of
T cell idiotypes went in complete positive concordance with that of heavy chain Ig
allotypes in both mouse[6] and rat[7]. No evidence was found for the participation of
light chains in the T cell receptor as to biochemical build-up or idiotype composition[8,9] whereas controversy does exist as to the possible involvement of major histocompatibility gene products in the direct creation of idiotypic T cell receptors[10,11].
Finally, the presence of idiotypic receptors or molecules from a variety of subset
of T lymphocytes have been reported including MLC and killer T cells[12], suppressor
and helper T lymphocytes[13,14] and suppressor T cell recruiting factors[15] as well as
helper T cell factors[16]. It would thus seem fair to conclude that auto-anti-idiotypic
immunity reactions of positive and negative kinds are apt to express themselves with
vigour at the level of both T and B cell populations.

A second fact of importance is the growing understanding that auto-anti-idiotypic
reactions may well constitute a normal and maybe even essential part in most normal
immune reactions involving T and B lymphocytes[17]. Thus, immunization with several
kinds of antigen such as transplantation antigens[18] and polysackarides[19,20] have
been shown to result in the product of both idiotypic and anti-idiotypic antibodies.
It is also becoming evident that interactions between T cells as well as between T
and B cells in the actual generation of more (or less) immune reactivity under normal immunization procedures to a sizeable extent may be dependent upon auto-anti-
idiotypic reactions[21,22]. In a most convincing series of experiments it was thus
shown that administration of an antigen would lead to the increase of B lymphocytes in the body of the recipient carrying idiotypes typical of antigen-binding
reactivity towards the inoculated antigen[22]. However, several of such idiotype-
positive B lymphocytes could be shown to produce upon polyclonal activation

idiotype-positive but non-antigen binding molecules thus excluding that the antigen administered had had any direct binding power or ability to cause the increase in the number of these cells. The most logical explanation for such results would be that the administration of antigen caused the induction of anti-idiotypic reactions of a helper T cell type with specific ability to now stimulate idiotype-positive B cells into proliferation regardless of whether their idiotype-positive surface molecules had antigen-binding ability or not. Thus, anti-idiotypic drive into proliferation of idiotype positive cells could thus at least in this experimental system provide a stimulus to proliferation superceeding that of direct antigen-specific, selecting forces. It is thus clear that antigenic determinants of external origin and idiotypic and anti-idiotypic structures of internal origin interact in a quite delicate and complicated manner during the generation and regulation of immune reactions. It is to be assumed that antigenic determinants of auto-antigenic nature besides idiotypes would do likewise.

Finally, it should be realized that the immune system allows the individual to produce antibodies against a vast number of different s.c. antigenic determinants including the individual's own idiotypic determinants.

As idiotypes at least in part are to be found in or close to the actual antigen-combining sites, it is to be assumed that certain anti-idiotypic antibodies may in fact sterically resemble the antigen binding to the first idiotypic molecules. It may thus be wise to in a conceptual form visualize the auto-antigenic determinants in an individual as consisting of two major groups: One group being homogenous within each subgroup (for instance organ specific antigens) being parallelled by a second major, degenerate, heterogenous group of molecules with similar steric features and maybe functions, namely the variable or idiotypic regions of the antigen-binding receptors of the immune system. Examples as to how an anti-idiotypic antibody molecule in fact may function like the initial immunogen both as to immunogenic properties or endocrine functions will be offered in this article.

In conclusion, idiotypes and anti-idiotypes are most likely involved in intricate interactions during normal immunization procedures as well as in auto-immune disorders. The positive and negative potentials of such anti-idiotypic reactions will now be discussed.


RESULTS

The use of selective anti-idiotypic immunity to induce specific obliteration of the immune response against defined antigens

No actual experiments have yet been reported as to the possible use of the above approach in the treatment of auto-immune disorders but several experimental systems exist    demonstrating the validity of the concept. Some details as to findings and

detailed approaches in one particular antigenic system, that of differences with
regard to major histocompatibility antigens in mouse or rat will be presented.
In this system we have tried by the use of auto-anti-idiotypic immunity to deprive
the individual of exactly those clones that could recognize the major histocompatibility antigens of one inbred strains of animals within that species whilst main-
tanining normal reactivity against the histocompatibility antigens of other strains.
In our initial experiments we were using available anti-idiotypic antibodies in an
immunosorbant form to purify idiotypic antibodies produced in the same strain but
with specificity for major histocompatibility antigens of a second strain[23]. Such
antibodies or receptors could be recovered in high amounts from immune sera but as
in the MHC system the numbers of idiotype-positive lymphocytes in a normal individual is very high[24]. This enabled  the use or large volumes of normal serum from
adult individuals as an alternative source of idiotypic molecules. Such idiotype-
positive molecules were thus purified via filtration of the anti-idiotypic immunosorbants and were usbsequently polymerized with glutaraldehyde (to introduce new
antigenic determinants for helper T cells), emulsified with Freund´s complete adjuvant and inoculated into autologous, normal animals. In a certain number of animals this procedure could be shown to lead to gradual induction of selective unresponsiveness to the relevant alloantigens as measured in MLC, CML, alloantibody
production or allograft survival. Table I shows a summary of these results as obtained in the rat.

TABLE 1

AUTOIMMUNIZATION WITH ANTIGEN-SPECIFIC, IDIOTYPIC LEWIS-ANTI-DA RECEPTORS INDUCE
UNRESPONSIVENESS IN ADULT LEWIS RATS AGAINST DA ALLOANTIGENS

| Testsystem | | Test carried out against: | |
| --- | --- | --- | --- |
| | | DA | BN |
| MLC | I | 15.172 | 38.558 |
| | N | 152.324 | 40.826 |
| GvH | I | 14.4 | 26.4 |
| | N | 30.7 | 25.8 |
| Skin graft rejection | I | 24.2 | 10.8 |
| | N | 9.6 | 11.0 |
| Antibody titers | I | $2^{6.8}$ | $2^{11.2}$ |
| | N | $2^{13}$ | $2^{11.5}$ |

(See Footnote on opposite page)

Footnote:
I = Lewis adult rats immunized with Lewis-anti-DA reactive molecules in Freund´s
complete adjuvant (for details see      ).
N = Lewis normal adult rats.
MLC: One-way mixed lymphocyte culture, values given as mean of tritiated thymidine
uptake at peak day in cpm.
GvH: Local popliteal lymph node graft-versus-host assay induced in LewisxDA or
LewisxBN $F_1$. Mean values as expressed in mg.
Skin graft: Mean rejection time of DA or BN skin grafts expressed in days.
Antibody titers: Hemagglutination titers of sera obtained after immunization
with one skin graft followed by $5 \times 10^7$ spleen cells. Mean values.
For details see[23].

Certain features needs emphasizing: Once induced the unresponsiveness seemed to
persist for a very long time, probably life. No significant side effects were
noted. Screening of the kidneys for immune complex deposits were negative. The
degree of suppression went by and large parallel with the appearance of auto-
anti-idiotypic antibodies and a simultaneous disappearance of idiotype-positive
lymphocytes, thus supporting the view that the induced selective suppression was
indeed induced via an auto-anti-idiotypic immune reaction.

In the histocompatibility system it is difficult to achieve selective purifi-
cation of alloantibodies using conventional antigenic immunosorbants (fixed cells
are comparatively inefficient in this regard) and anti-idiotypic immunosorbants
were for some time the only available means of obtaining pure idiotypic alloantibodies for auto-immunization in this antigenic system. However, it was comparatively soon realized that selective proliferation induced by antigen (alloantigenic
cells) in vitro would allow the generation of alloantigen-specific, idiotype-
positive blasts that could then be obtained in a virtually pure form using 1-g
velocity sedimentation procedures[25]. Although this system involved a potential
complication (MLC reactive T blasts can be shown to frequently display stimulator
cell alloantigens on their surface) it  could be shown that autologous MLC blasts
in adjuvant could be used for the induction of auto-anti-idiotypic immunity as
well as specific elimination of immune reactivity against the relevant alloantigens. The results of such auto-blast immunization procedures in a mouse system
are depicted in table II (see next page).

Again, certain features need emphasizing: When blasts generated in MLC were
used for auto-immunization with care taken to remove the possibly contaminating
alloantigens (treatment with proteolytic enzymes plus recovery at $37^{\circ}C$ overnight)
we have never observed any complicating alloimmunity but only suppression. However, the degree of suppression does vary greatly from experiment to experiments
and many auto-blast immunized animals fail to develop any significant degree of
tolerance whilst others are completely and selectively suppressed. The reasons
for this variability are still being explored but as yet poorly understood.

Furthermore, the only adjuvant yet functioning with any degree of regulatory in this system is Freund´s complete adjuvant, an adjuvant with known severe side effects. The search for other, less harmful adjuvants is being pursued in a very active manner with so far no equal or better adjuvant being found hitherto. That the suppression when achieved is indeed due to auto-anti-idiotypic immunity has been shown in a conclusive manner using proper idiotypic immunosorbants, transfer of suppression with anti-idiotype reactive T cells and demonstration of killer T cells in autoblast immunized animals with specificity for the relevant, autologous T blasts. Thus, the auto-blast immunization principle is a valid one although the degree of penetrance as to successful, demonstrable suppression is still to be improved would one like to try this approach in a clinical transplantation system.

Auto-blast immunization attempts have also been made in a system using soluble antigens, namely the tuberculin system. Here it could be shown that T blasts generated in vitro from BCG immune animals via stimulation of purified T cells with PPD could be used in the auto-immunization procedure to achieve immune unresponsiveness in vitro against PPD as exemplified in table III (see next page).

TABLE II

AUTOBLAST IMMUNIZATION IN CBA/H MICE WITH CBA/H-ANTI-C57BL/6 BLASTS MAY CAUSE SELECTIVE UNRESPONSIVENESS AGAINST C57BL/6 AS MEASURED BY CML AND MLC AND ANTI-IDIOTYPIC ANTIBODIES AS MEASURED IN A PROTEIN A ASSAY

| | Donors of lymphocytes or serum for testing in MLC, CML or protein A tests | |
| | Normal CBA/H mice | CBA/H mice im. with anti-C57BL/6 blasts |
| --- | --- | --- |
| T cell activity | | |
| MLC-anti-C57BL/6 | 145.030±1.596 cpm | 6.116±1.185 cpm |
| MLC-anti-DBA/2 | 146.296±2.895 cpm | 135.688±3.663 cpm |
| CML-anti-C57BL/6 | 32.2±1.6 % | 4.6±1.4 % |
| CML-anti-DBA/2 | 18.9±3.4 % | 12.1±0.5 % |
| Serum acticity | | |
| CBA/H-anti-C57BL/6 blasts | 1.846±286 | 5.251±430 |
| CBA/H-anti-DBA/2 blasts | 1.831±174 | 2.456±212 |
| C57BL/6-anti-CBA/H blasts | 1.468±171 | 1.235± 24 |

For details see[26].

TABLE III

AUTO-BLAST IMMUNIZATION IN THE ANTI-PPD SYSTEM WILL ABROGATE T CELL POTENTIAL
REACTIVITY AGAINST TUBERCULIN AS MEASURED IN A PROLIFERATIVE ASSAY IN VITRO

Schedule: Immunize C57BL/6 mice with BCG. At intervals thereafter remove spleens,
purify T cells and stimulate with PPD in vitro. Purify responding T blasts and use
$10^7$ blasts per mouse for back-immunization in adjuvant into normal C57BL/6 mice.
Control mice received only adjuvant. This was repeated thrice. All animals were
then immunized with BCG and their spleen T cells tested for anti-tuberculin
reactivity 3 weeks later in vitro using PPD.

Results: Average peak values of DNA incorporation of thymidine around 5.000 cpm
using spleen T cells from the two groups in absence of PPD in vitro.
T cells from BCG-immunized animals had a peak value using PPD in vitro around
45.000 cpm.
T cells from animals immunized with BCG as well as autoblasts of anti-PPD type
gave a maximum response towards  PPD of around 11.000 cpm.

was

Finally, the auto-blast approach attempted in a system more like the situation encountered would one try to treat already on-going auto-immune diseases
with this approach using alloimmune animals as recipients of auto-blasts in an
attempt to break the already established immunity. It was indeed found possible
to at least in part break an immune stage by auto-blast immunization procedures
as exemplified in table IV.

TABLE IV

ABROGATION OF EXISTING IMMUNITY BY THE ADMINISTRATION OF ANTIGEN-SPECIFIC,
AUTOLOGOUS T BLASTS IN ADJUVANT

CBA/H mice were immunized with $10^7$ C57BL/6 spleen cells i.p. They were subsequently left untreated or immunized in 3 weeks intervals thrice with $10^7$
CBA/H-anti-C57BL/6 blasts. Their splenic T cells were screened for MLC reactivity 103 days after immunization.

| CBA/H spleen cell donors | MLC with stimulatory cells derived from:[x] | |
|---|---|---|
| | C57BL/6 mice | CBA/2 mice |
| Immunized 103 days before with C57BL/6 spleen cells | 105.406±9.036 cpm | 49.999±2.317 cpm |
| Immunized as above but also autoblast treated | 19.274±2.266 cpm | 50.881±2.620 cpm |

[x]MLC:s were measured at day 3. Results as to suppression of response against
C57BL/6 stimulatory cells equally marked at day 2. For details see[25].

To conclude this chapter we would summarize the following: It is possible at least in certain antigenic systems via auto-anti-idiotypic immunization procedures to achieve a longlasting, selective reduction or sometimes complete obliteration of the relevant idiotypic clones and accordingly their effectuated immune response. No significant side effect of negative nature have been encountered so far. Already existing immunity could at least in part be broken (some 70-80 % reduction) by this approach allowing the cautious hope that similar success may be encountered in other situations involving auto-immune disease. It is still well worth noticing, however, that the remarkable success in some of the presented experiments may be due to the particular antigens used (= allo-MHC antigens) as such structures are known to play comparatively unique roles in the self-regulation of immune reactivity.

### The use of selective anti-idiotypic immunity to induce specific immunity against defined antigens

In the previous chapter some examples were offered demonstrating how anti-idiotypic immunity could be used to selectively obliterate an immune reactivity via the elimination of antigen-specific, idiotype-positive clones. It is, however, also very clear from several experimental systems that anti-idiotypic antibodies may themselves function as stimulators for the idiotype-positive cells (see introduction) in a similar manner to the immunogen fitting the receptors on the latter cells. In the special system previously described with allo-MHC reactions and idiotypes involved it could be shown that anti-idiotypic antibodies might also function as specific <u>inducers</u> of proliferation of relevant idiotype-positive, antigen-specific cells in vitro both in primary as well as secondary MLC:s[26]. Examples of such experiments are given in table V (see next page).

It should be noted that the present induction of proliferation of relevant T cells in vitro by the use of the anti-idiotypic antibodies was occurring in the absence of complement. Presence of complement in such situation would have resulted in the exactly opposite, namely obliteration of the specific MLC potential via cytolysis of the idiotypic cells by antibodies and complement. It is thus obvious that the same molecule, an anti-idiotypic (or idiotypic) antibody molecule may exert diametrically opposite effects in the presence or absence of an additional functional system, here the complement system. Would thus for any reason insufficient complement be around in a certain local tissue in vivo when anti-idiotypic antibodies react with idiotype-positive cells stimulation rather than obliteration of the idiotype-positive cells may follow. Other examples of the complexity in these anti-idiotypic regulatory networks can be given by the findings in another system that anti-idiotypic antibodies of the same specificity but of different immunoglobulin classes could be shown to have directly opposite results on the idiotypic cells = suppression or immunity resulted depending on immuno-globulin class[2,3]. Thus, the present results should be considered as yet another reminder of the complexity of the "natural" regulation of the immune system and the accordingly similar complexity encountered when one tries to interfer with such a system in any precise manner.

TABLE V

STIMULATION AND RESTIMULATION OF C57BL/6 T CELLS WITH SPECIFICITY FOR CBA/H ALLOANTIGENS USING AUTO-ANTI-IDIOTYPIC ANTISERA

| Primary MLC | Peak DNA synthesis |
|---|---|
| Stimulating agent | cpm |
| O | 7.000 |
| C57BL/6, normal serum | 7.500 |
| Auto-anti-C57BL/6 anti-CBA serum | 155.000[x] |
| CBA/H spleen cells | 211.000 |
| DBA/2 spleen cells | 245.000 |
| Secondary MLC[xx] | |
| Stimulating agent | |
| O | 3.000 |
| C57BL, normal serum | 7.000 |
| Auto-anti-C57BL/6 anti-CBA serum | 225.000 |
| CBA/H spleen cells | 325.000 |
| DBA/2 spleen cells | 125.000 |

[x] Cells generated specifically lytic for CBA/h target cells.
[xx] Secondary MLC against CBA/H, for details see[26].

## Anti-idiotypic antibodies may share non-immunological functions with the immunogen that induced the original idiotypic antibodies

The previous chapter did emphasize the possibility of anti-idiotypic antibodies to substitute for the original immunogen in the induction of antigen-specific, idiotype-positive lymphocytes. Such molecular mimicry in a superficial sense may also be anticipated to be encountered to extend to other functions if the original immunogen had the possibility to exert such. Three different reports suggesting this possibility have now been published in systems involving receptors for neuro-transmitters, vitamin-binding proteins or hormones as the original immunogen[27,28,29]. In two of these systems only results as to molecular mimicry in the form of binding studies have been reported whereas in the hormonal systems anti-idiotypic antibodies could be shown to express both the binding as well as the functional properties of the hormone. A summary of the latter results will be presented here. The hormone

used was insulin and anti-insulin antibodies were produced in one species of
animals whereafter the purified anti-insulin antibodies were used to induce anti-
idiotypic antibodies in a second species. Some of the anti-idiotypic antibodies
could be shown to behave like insulin in several species and functions tested as
exemplified in table VI.

TABLE VI
ANTI-ANTI-INSULIN ANTIBODIES MAY ACT LIKE INSULIN

---

Blocking of insulin binding to insulin receptors on fat cells: In a typical experiment 2 microgrammesper ml of anti-anti-insulin antibodies would cause 50 % inhibition of insulin binding whereas 20 microgrammes per ml caused around 80 % inhibition. Control antibodies failed to cause any detectable blockage at highest dose
tested = 40 microgrammes per ml.

Induction of amino-butyric acid in thymocytes in vitro: Compared to control levels
= 100 % both insulin and anti-anti-insulin antibodies caused an uptake of acid at
a level of between 150-160 %. Two control antisera (one anti-beta$_2$ microglobulin,
one anti-H-2) failed to have any inducing impact.

Reduction in blood sugar levels in diabetic mice: The administration of insulin
caused a reduction in blood sugar reaching a peak at 1/2 hour after administration
with 60 % reduction reaching "normal" levels after two hours. Anti-anti-insulin
antibodies caused a reduction in blood sugar levels reaching its peak around 3
hours after administration (40 % reduction) and returned to normal levels after
6 hours. Irrelevant antisera failed to have any impact on blood sugar levels.

---

For details see[29].

This group of results indicate two points of interest: a) anti-antibodies
may in fact take over some of the non-immunological functions of the original
immunogenic molecules and b) this may mean that for instance the presence of an
anti-receptor antibody in the serum of an individual may have had as its initiation point either the receptor in itself being immunogenic or an antibody specifically reactive with the normal ligand for that receptor as the antigen. It is
obvious that this latter possibility is of interest in such auto-immune diseases,
where antibodies are known to interfer with endocrine functions.

Concluding remarks

Auto-anti-idiotypic immunity has a strong power to interact with and regulate
the immune system in a specific way. The potential of this selective auto-immunity
in the beneficial regulation of auto-immune disorders remains to be explored but

the concept has been already shown valid in "normal" immune regulation. Possibilities exist, however, that auto-anti-idiotypic immune reactions may also inflict negative reactions in potentiating a reaction negative for the individual or actually in themselves be responsible for the symtoms of the disorder. Only further experiments will tell whether auto-anti-idiotypic immunity can be used in a regular, positive manner for the individual afflicted with auto-immune disease or if such immunity will remain to be of interest only in laboratories studying basic immunology.

ACKNOWLEDGEMENTS

The present work was supported by the Swedish Cancer Society, the Swedish Medical Research Countil, the Swiss National Science Foundation grant 3.688-0.76, NIH grant AI.CA.13485-01, and the Nordiska Insulinfonden.

REFERENCES

1. Binz, H. and Wigzell, H. (1975) J.Exp.Med., 142, 197-211.

2. Eichmann, K. and Rajewsky, K. (1975) Eur.J.Immunol., 5, 661-666.

3. Krawinkel, U., Cramer, M., Berek, C., Hämmerling, G., Black, S.J., Rajewsky, K. and Eichmann, K. (1976) Cold Spring Harbor Symp. Quant. Biol., 16, 285-294.

4. Binz, H., Lindenmann, J. and Wigzell, H. (1974) J.Exp.Med., 140, 731-742.

5. Binz, H., Frischknecht, H. and Wigzell, H. Exp.Cell Biol., in press.

6. Hämmerling, G.J., Black, S.J., Berek, C., Eichmann, K. and Rajewsky, K. (1976) J.Exp.Med., 148, 861-872.

7. Binz, H., Wigzell, H. and Bazin, H. (1976) Nature, 264, 639-670.

8. Binz, H. and Wigzell, H. (1976) Scand.J.Immunol., 5, 559-571.

9. Krawinkel, U. and Rajewsky, K. (1976) Eur.J.Immunol., 6, 529-537.

10. Kramer, P. and Eichmann, K. (1977) Nature, 269, 733-735.

11. Binz, H. and Wigzell, H. Scand.J.Immunol.

12. Binz, H. and Wigzell, H. (1978) J.Exp.Med. 147, 63-76.

13. Black, S.J., Eichmann, K., Hämmerling, G. and Rajewsky, K. (1975) in Membrane receptors of lymphocytes, Seligmann, M. ed., North Holland, pp. 117-124.

14. Eichmann, K. (1975) Eur.J.Immunol., 5, 511-519.

15. Benecerraf, B., personal communication.

16. Mozes, E. and Kontiainen, S., personal communication.

17. Jerne, N.K. (1974) Ann.Immunol., 125, 373-381.

18. McKearn, T.J., Stuart, F.P. and Fitch, F.W. (1974) J.Immunol., 113, 1876-1882.

19. Kluskens, L. and Köhler, J. (1974) Proc.Nat.Acad.Sci., US, 71, 5083-5087.

20. Rodkey, L.L. (1974) J.Exp.Med., 139, 713-724.

21. Ward, K. and Cantor, H. (1977) in "Regulation of the immune system: Genes and the cells in which they function", Herzenberg, L.A., Sercarz, E.E. and Fox, C.F. eds., Acad.Press, in press.

338

22. Eichmann, K., Coutinho, A. and Melchers, F. (1977) J.Exp.Med., 146, 1436-1449.

23. Binz, H. and Wigzell, H. (1976) J.Exp.Med., 144, 1438-1450.

24. Binz, H. and Wigzell, H. (1975) J.Exp.Med., 142, 1218-1230.

25. Andersson, L., Auguet, M., Wight, E., Andersson, R., Binz, H. and Wigzell, H. (1977) J.Exp.Med., 146, 1124-1137.

26. Frischknecht, H., Binz, H., and Wigzell, H. (1978) J.Exp.Med., 147, 500-514.

27. Sege, K. and Peterson, P. (1978) Nature, 271, 167-168.

28. Sege, K. and Peterson, P. (1978) Proc.Nat.Acad.Sci., U.S., 75, 2443-2447.

29. Schwartz, M., Novick, D., Givol, D. and Fuchs, S. (1978) Nature, 273, 543-545.

DISCUSSION

WEKERLE:  Do anti-self idiotypic killer cells have to recognize the idiotype plus self MHC determinants?

WIGZELL:  I would guess not.  This is because anti-Ig killer T cells generated in mice against human IgM or IgG can kill IgM or IgG positive Burkitt lymphoma cells even if such target cells fail to express HLA antigens.  This work was done by Dr. Perini.  Thus, I would believe that immunoglobulin molecules on the surface of a target cell may substitute for MHC structures.

LENNON:  You have suggested that anti-idiotype antibodies are physiological regulators of immune responses, and can in some circumstances mimic autoantigens.  You also stated that anti-idiotype antibodies can stimulate rather than delete idiotype-bearing immunocompetent cells in the absence of complement.  Could this explain the increased incidence of autoimmune diseases, such as SLE, in patients with deficiencies of defined complement components?

WIGZELL:  I would consider this quite a reasonable possibility.

MACKAY:  The significance of your interesting experimental approach may still not be fully appreciated clinically.  Firstly, anti-idiotypic antibody might represent the "natural defense" against autoimmunity and the fluctuating course of autoimmune disease seems consistent with this.  Secondly, can we consider how the "defensive" anti-idiotypic system can be potentiated, and whether cytotoxic immunosuppressive drugs may at times suppress the anti-idiotypic response and in this respect be harmful?
Thirdly, a "cascade" of idiotype-anti-idiotype responses may explain the curious hypergammaglobulinemia of certain immunopathic diseases, including chronic active hepatitis.  My question concerns the use of Freund's complete adjuvant in raising anti-idiotypic antibody, in this respect making this approach rather "artificial".

KOHLER:  The need to use Freund's adjuvant to produce auto-anti-idiotypic antibodies is of course a drawback for the use in humans.  However, Klaus recently has demonstrated in mice that soluble complexes of antigen and antibodies given intravenously can induce auto-anti-idiotypic antibodies.

WIGZELL:  Attempts to produce high titered auto-anti-idiotypic immunity allowing the in vivo wipe-out of the relevant idiotypic cells do normally fail unless Freund's complete adjuvant is used.  We have tried several other adjuvants such as various polymerizing agents and Bordetello pertussis, but found them inefficient.  We are searching for other adjuvants as such would be required should one ever come to use this approach in the clinic.  I would also support the possibility that anti-idiotypic-idiotypic complexes may explain the fluctuation observed in certain autoimmune diseases.  Such complexes may in fact themselves cause sizeable damage.

ROITT:  It is more difficult to induce anti-idiotypic responses to polyclonal specificities than to restricted specificities.  Does your relative success with T-blasts suggest that T cells express fewer idiotypes than B cells for a given antigen response?  If the responses in human autoimmune diseases are polyclonal, it may prove quite difficult to try and suppress them by inducing anti-idiotypic responses.

WIGZELL: I would fully agree that complexity as to idiotype diversity is a major challenge when trying to wipe out such idiotype-positive clones by auto-anti-idiotypes. Our data as to diversity of idiotypes on T versus B receptors for antigen would support the view that T cells are generally more restricted when analyzed at the level of simple receptor molecules. Our data in the histocompatibility systems do not, however, justify the conclusion that idiotypic blasts are better than idiotypic immunoglobulin molecules in inducing auto-anti-idiotypic immunity. Roughly speaking, they would seem to be comparable in this regard.

ROITT: Our early studies with Lynn Nye have shown considerable complexity for the spectrotypes of anti-thyroglobulin in Hashimoto patients' sera. We still have to identify the contribution of isotype to this complexity. Even if these should prove to be polyclonal, it is possible that each clonotype may share a restricted number of idiotypes.

DATTA: Do memory cells arise because they are relatively resistant to this idiotype-anti-autoidiotype regulation, during a conventional immune response? In other words, if idiotype-carrying T blasts that are specific for an antigen are eliminated by the anti-idiotype regulatory mechanism, how do you explain the heightened secondary responses on repeated immunization?

WIGZELL: When giving antigen you change the balance of the immune system as to idiotypes via the creation of an immunogenic complex of antigen plus idiotype-positive receptor(s). This will normally lead to an increase of idiotype-positive, antigen-specific cells as well as an increase in anti-idiotypic cells. Why this normally results in an immunity of conventional type and not in a return to normal reactivity (or cancelling out of idiotypes versus idiotypes) is, however, not due any particular resistance in memory cells to anti-idiotype mediated suppression. They seem to be as susceptible as are virginal immunocompetent cells.

KINCADE: To what extent does the isotype (major class) of anti-idiotype correlate with preferential behavior either as a tolerogen or as activator of B or T cells?

WIGZELL: Your question is one of great interest. The works of Eichmann et al. are the only ones so far indicating such a dependance in a defined manner. I would deem it quite possible that isotypes play an important role as to what the idiotypic determinants may cause as to induction of pharmacologically distinct lymphocyte subgroups.

GERSHWIN: Are there any differences between strains or any known relationship with the MHC in production of anti-idiotypic antibodies?

WIGZELL: In the mouse strains we have investigated with regard to the auto-anti-idiotypic reactions towards H-2 associated antigen we have failed to note any significant difference. In the rat systems, Lewis rats seem especially prone to produce "successful" auto-anti-idiotypic antibodies in our particular test system.

WARNER: Would you consider it reasonable that in autoimmunity a generalized escalation in the production of auto-anti-idiotypic antibody directed against autoantibodies may occur, thus escalating immune complex deposition and the actual pathogenesis of autoimmune disease. Does this not complicate the potential elimination of auto-immunity through anti-idiotypic immunization?

WIGZELL: The possibility of immune complex disease in the idio-type-anti-idiotype immunization approach is obvious. We have screened some successful mice and rats (+auto-anti-idiotype producing, MLC suppressed) for possible immune complex deposition up to a year with maintained selective MLC suppression. Here, we failed to find signs of complex deposits in the kidneys of these animals. However, it may well be that when using immune instead of normal animals when trying to induce auto-anti-idiotype mediated immune suppression, concentrations of idiotypic-anti-idiotypic complexes may reach harmful levels. Only experiments will give the answer.

CHUSED: How stable is the tolerance induced by anti-idiotype immunization to challenge by skin grafts or the like?

WIGZELL: Once induced, the auto-anti-idiotype mediated suppression in the histocompatibility system is quite stable. It may last for life, maybe via the internal boosting device constituted for by the continuous new production of idiotypic cells from stem cells.

ROITT: Following Dr. Warner's point, aside from microbial infections, attempts to detect autoantigens in circulating complexes in human autoimmune diseases have not been conspiciously successful. DNA or other nuclear components have not so far been demonstrable in the circulating complexes in SLE and David Male in our laboratory has found them to be of apparently simple composition; essentially IgG and complement components which would be consistent with idiotype-anti-idiotype complexes with an optional contribution by antiglobulins (i.e., anti-Fcγ). (Note: deposited complexes in SLE such as those in the kidney, do have antigen but this is interpreted as evidence in favor of locally formed as distinct from deposited complexes.) The presence of complexes in autoimmune insulin-dependent diabetes long after the destruction of the islet cells might be the expression of a similar phenomenon as we discussed after Dr. Irvine's paper.

What sort of complexes would idiotype-anti-idiotype form? Would they be linear or circular?

WIGZELL: That would mean that there are only single idiotypes around. There is nothing to say that; in fact, there is quite great likelihood that there is more than one idiotypic determinant present on several of these molecules. So you may get a sort of linear/circular precipitate, but you may also get some very classical precipitates. You can get precipitates in Ouchterlony; with some of them, very nicely.

ROITT: Are the idiotypic determinants then in the variable region far enough apart to enable two antibody molecules to combine with each variable region or is your precipitation reaction due to rheumatoid factor? This could be checked by using $F(ab')_2$ reagents.

WIGZELL: That I don't know.

SCHWARTZ:  Is there differential sensitivity of T or B cells to anti-idiotypic antibodies?

Your results imply the possibility that antibodies against a receptor (or receptors) common to many clonotypes of helper T cells could arise. Such anti-receptor antibodies, by stimulating helper cells, could lead to the production of autoantibodies in a manner analogous to certain lymphocyte mitogens.  Is there evidence that autoantibodies are produced following, for example, immunization with alloantigens?

WIGZELL:  We have not made parallel assays comparing T and B cells in vitro in this regard.  Idiotype-positive T cells were, however, quite susceptible to anti-idiotypic antibodies and complement in vitro.  In view of other workers' difficulties in killing B cells with anti-Ig antibodies plus complement, I would not be surprised if T are more susceptible than B in this cytolytic assay.  As to your second question:  we have frequently seen autoantibodies after administration of allogeneic cells, be they of lymphoid or non-lymphoid nature.  We have no reasons, thus, to believe these latter auto-antibodies are of the nature you indicated.  However, the hetero-geneity of T cells carrying a given idiotypic determinant may be considerable and cause surprising "additional" immune perturbations so far not discovered.

SCHWARTZ:  Have you tested animals hyperimmunized with allogeneic cells for antinuclear or anti-DNA antibodies?

WIGZELL:  No.

# AUTO-IMMUNE-LIKE REACTIONS DURING ONTOGENY AND OPERATION OF THE IMMUNE RESPONSE

HEINZ KÖHLER

La Rabida Institute, University of Chicago and Dept. of Pathology

## ABSTRACT

Auto-immune reactions are based on the principle of self-recognition. Self-recognition occurs "normally" during immune responses and is defined as an idiotype-anti-idiotype interaction. Examples of this phenomenon are the development and the interaction of immune clones responding to phosphorylcholine (PC). The ontogeny of the major PC-responsive clone, T15, is under idiotype-specific regulation. The precision in complementary idiotype interactions becomes evident from two findings: (i) the isologous anti-idiotypic antibody response is of clonal restriction and (ii) an anti-anti-idiotypic antibody, induced by using purified anti-idiotype as antigen, crossreacts idiotypically with the primary idiotype. From these data the concept of a functional response unit consisting of a doublet of complementary idiotype is developed. Auto-immune disorders can originate from a failure of regulatory complementary circuits in the immune response.

## INTRODUCTION

The term auto-immunity has received its present meaning from the observation of pathological conditions. Thus, auto-immunity carries the attribute of something which is harmful and undesirable. However, recent changes in the appreciation of immune mechanisms led to a more objective view of what has been collectively termed auto-immunity. The key to this new understanding came from perceiving the common underlying principle in immunity which we have called "self-recognition"[1]. Self-recognition is based on complementarity of interacting structures. By this definition we can see self-recognizing molecules functioning as the glue in biological systems including the specialized surveillance system, the immune system. It is evident that under this viewpoint the meaning of auto-immunity loses its negative aspects.

Our interest in the phenomenon of self-recognition or auto-immunity developed during studies of regulatory mechanisms in the immune response[2]. Working with a clonally restricted response against the small epitope phosphorylcholine (PC)[3] we observed auto-immune-like reactions which are based on the complementarity of idiotypes[4]. Similar findings were made in other systems[5,6,7,8]. The possible benefits from these studies for the specific problems of auto-immune pathology cannot be defined at this time but the breakdown of the self-recognizing immune regulation appears as the principle causative event.

The development of immune clones is controlled by idiotypic interactions

The response of Balb/c mice to phosphorylcholine is unique through its natural restriction to one major clonotype, the T15/H8 idiotype[3]. Balb/c mice respond to immunization with T-dependent and T-independent forms of PC-antigens with antibodies which are 95% of the T15 idiotype despite that the available repertoire of PC-specific clones is extensive[9]. To account for the restriction of the anti-PC response to the T15 idiotype one has to postulate that some sort of selection process must occur which produces the T15 clonal dominance. We have argued that the selection of clonal dominance happens during the early neonatal period of the Balb/c mouse[10]. The data for this hypothesis are shown in Table 1.

TABLE 1

CLONAL DOMINANCE AFTER ADOPTIVE TRANSFER

| Recipient[a] | Donor[b] | Plaque forming response[c] | | |
|---|---|---|---|---|
| | | Anti-TNP | Anti-PC | H8 idiotype |
| Normal Balb/c | Adult Balb/c | 4.18 ± 0.09 | 3.94 ± 0.19 | 96% |
| Normal Balb/c | Neonatal Balb/c | 4.83 ± 0.16 | 3.99 ± 0.17 | 26% |
| Neon. Suppr. Balb/c | Neonatal Balb/c | 4.38 ± 0.11 | 3.89 ± 0.16 | 95% |

[a]Normal or neonatally suppressed Balb/c mice of at least 3 months of age were lethally irradiated (800r) and reconstituted with syngeneic adult spleen or neonatal liver cells.

[b]$1 \times 10^7$ adult or neonatal cells were used to reconstitute the irradiated recipients.

[c]Three weeks after reconstitution the animals were immunized with TNP-Ficoll and R36A and the response was measured 5 days later. The log of the geometric mean and standard error are given. Anti-idiotypic plaque inhibition of the anti-PC plaque forming response was used to determine the % of the H8 positive PFC's.

Adult spleen cells or neonatal liver cells are transferred into syngeneic lethally radiated recipients. During the third week after reconstitution the animals are immunized with TNP and PC and the plaque forming response is measured five days later. Using normal Balb/c mice as recipients for neonatal cells, the response to PC is "abnormal" in its loss of the T15 clonal dominance. With adult donor cells the reconstituted anti-PC response is, as expected, T15 dominant. However, if neonatally suppressed[11] mice are used as recipients, the T15 dominance can develop from the transferred neonatal liver cells. It appears that the environment is important for the maturation of young immunogenerative cells. Inasmuch as the neonatal suppression represents an idiotype-specific manipulation of the recipient idiotype-specific recognizing events between developing cells and the host must occur and determine the outcome of clonal dominance.

The sequential appearance of anti-idiotype and idiotype during the neonatal period (see Figure 1) can be taken as additional evidence for the occurrence of self-recognizing idiotype interactions. When normal neonatal Balb/c mice are analyzed with a sensitive RIA for anti-idiotype, an early anti-idiotype activity in the serum can be detected which disappears by day 5; then the T15 idiotype appears reaching normal levels by 6 weeks. The role of the early anti-idiotypic antibody or its origin are not known but it is tempting to speculate that it is involved in the developmental process of the T15 dominance.

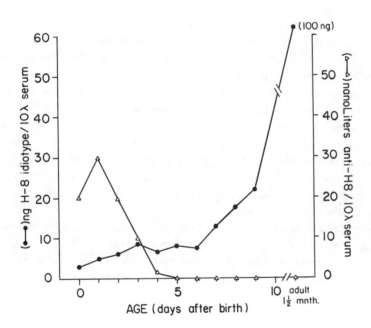

Fig. 1. Neonatal Balb/c mice of different ages were assayed for the presence of the H8 idiotype and anti-H8 antibody using solid phase radioimmunoassays.

## Self-recognizing immunity operates with high fidelity

Idiotype-anti-idiotype reactions have been described not only in the PC-system[3] but also in several other immune responses[5,8,12]. In all these situations the implication is strongly supported that these auto-anti-idiotypic reactions are manifestations of regulation of the immune response. The experimental

evidence from idiotype suppression in adult[3,13,14] and neonatal animals[11] demonstrates clearly the suppressive mechanism of such idiotype interactions, but there
is also experimental support for idiotype stimulation[15,16]. An important rule
in complementary idiotype-reactions appears to be the "priority of the first induced response"[17]. We shall not go further into the discussion of these data
nor describe any of these systems in more detail but rather focus on the precision and fidelity in complementary idiotype interactions.

Considering the biological importance of self-recognition in the immune system,
manifested in idiotype interactions, the question of how specific this process is
becomes crucial. If idiotype interactions are of exquisite specificity the
interaction of cells and antibodies is based upon a perfect fit of complementary
structures of binding sites. Assume that a cell of receptor idiotype with
specificity "A" will find only one idiotypic counterpart in the immune repertoire
which carries the complementary mirror-image structure, which shall be called
"B". Thus, the fit of the "A-B" type is the best possible one; but other kinds
of fits of less perfect complementarity must be considered, and it becomes a
matter of setting the cut-off point in this gradient of differing complementarities. As we later attempt to define the cut-off by functional criteria, let
us adopt for the moment an idealized attitude and describe the idiotype interactions as being of extreme specificity. Inevitably, the cells or antibodies
involved in a given idiotype interaction become a pair of two perfectly complementary entities. We shall call this a DOUBLET OF COMPLEMENTARY IDIOTYPES. In
contrast, if idiotypic interactions are possible among cells and antibodies
having only remotely complementing structures, the number of involved idiotypes
increases exponentially. Evidently, the question of the degree of specificity
which occurs in idiotype interaction will dictate the choice between the two
models, the complementary idiotype doublet[18] or the extensive idiotype network[19].

In pursuit of this important question we shall consider now some pertinent
experimental observations. This implies that we restrict ourself to arguments
of a functional nature and shall neglect theoretical reasoning. Furthermore,
the inherent limitation of the sensitivity of assays will provide us with an
operational cut-off point. However, anything beyond detection by rather sensitive assays approaches a level which is probably remote from biological significance.

Observing the idiotype-anti-idiotype interaction in the test tube and in
the animal demonstrates a high degree of specificity. Homologous or isologous
anti-idiotypic antibodies bind to the idiotype that was used to induce them in
an exquisitely specific manner. In the PC-system[20] the A/He or Balb/c anti-H8
can distinguish between the T15/H8 idiotypes and other idiotypes, such as the
603, 167 or 511 which are PC-binding myeloma proteins of different idiotypic
specificity[21]. Or, homologous anti-H8 selectively suppresses the anti-PC response

of the H8 and leaves other anti-PC-clones of non-H8 idiotype responsive[22].

Another experimental approach to the question of fidelity in complementary idiotypy can be taken by asking this question: How many different kinds of complementary clones can a given idiotype stimulate? If the specificity of idiotypes is high the number of different responding idiotypes will not be much greater than the number of different inducing idiotypes. Again, in the PC-system we can find some answers to this question. The number of clones which produce an anti-167 response in Balb/c mice upon immunization with the 167 idiotype is restricted[23]. Similarly, the number of different clones in A/He and Balb/c mice responding to the dominant T15 is also limited[24]. We have purified H8 (primary idiotype) and anti-H8 (secondary idiotype) and labeled and subjected to isoelectric focusing. Both antibodies focus within a narrow range of pH indicating clonal restriction (Figure 2).

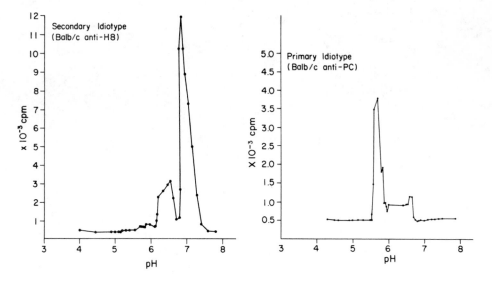

Fig. 2. Purified H8 (primary idiotype) and isologous anti-H8 (secondary idiotype) were labeled and subjected to isoelectric focusing in gel tubes. The radioactivity is plotted against the pH.

The experimental investigation into the question of fidelity has recently been pursued further by Urbain et al.[25] and Cazenave[26] who studied the functional and idiotypic properties of anti-anti-idiotypic antibodies in outbred rabbits. These authors prepared such antibodies according to the scheme $Ab_1 \rightarrow Ab_2 \rightarrow Ab_3$. Surprisingly, $Ab_3$ contained idiotypic species which were very similar to the $Ab_1$ idiotype. We have used a similar protocol in Balb/c mice in the PC-system. Purified anti-H8 was used as antigen in A/He and Balb/c mice. Since the T15/H8 idiotype in A-strain mice is not dominant, attempts were made to increase the T15 clone in the response to PC by immunization with the complementary idiotype. As seen in Table 2 the immunization regimen with purified A/He anti-H8 raises the portion of the T15 clone in the subsequently induced anti-Pc response to about 50%. This effect is of short duration and the original response pattern returns within 2 months after the immunization on the anti-H8.

TABLE 2

EFFECTS OF IMMUNIZATION WITH COMPLEMENTARY IDIOTYPE IN A/HeJ MICE

| 1st Immun.[a] | 2nd Immun.[b] | H8 Id.[c] μg/ml | α-PC antibody[c] μg/ml | Ratio H8-α-PC | α-PC PFC/spleen |
|---|---|---|---|---|---|
| — | | .064 | 5.16 | .012 | |
| — | R36A | 1.61 | 92.1 | .017 | 4.16 ± .3 |
| A/HeJ α-H8 6 times | R36A, 6 wks after 1st immun. | 18.3 | 32.2 | .56 | 3.60 ± .3 |
| " | R36A, 9 wks after 1st immun. | 1.3 | 17.2 | .07 | 2.56 ± .1 |
| " | R36A, 12 wks after 1st immun. | 1.2 | 47.9 | .024 | 4.48 ± .3 |

[a] A/HeJ were immunized with 125 λ of A/HeJ α-H8 in complete Freund's adjuvant, followed by injections 10 days apart in incomplete Freund's.

[b] Animals received $5 \times 10^8$ R36A organisms in .2 μl saline i.v. Assays were run 5 days post-immunization.

[c] As determined by solid phase radioimmunoassay.

In Balb/c mice the T15 clone is dominant and therefore we used neonatally suppressed Balb/c mice in experiments to promote the suppressed T15 clone by immunization with the complementary idiotype. The data in Table 3 show that immunization with anti-H8 stimulates the T15 clone raising it to about half of the level found in normal Balb/c. In both experiments the amount of anti-PC antibody is evidently not increased by the immunization with anti-H8. Thus, induced $Ab_3$ is identical or cross-reactive with $Ab_1$ but seems to lack the Ag-binding site of $Ab_1$. We conclude from these experiments that the secondary idiotype, $Ab_2$, can induce its mirror-image idiotype in a remarkably faithful fashion.

TABLE 3

EFFECTS OF IMMUNIZATION WITH COMPLEMENTARY IDIOTYPE IN BALB/C

| Animals[a] | Immunization with:[b] | H8 idiotype[c] | Anti-PC[d] |
|---|---|---|---|
| Normal Balb/c | —— | 25.3 | 26.93 |
| Neon. Suppr. Balb/c | —— | 0.28 | 6.88 |
| Normal Balb/c | Purified Balb/c anti-H8 | 29.92 | 26.04 |
| Neon. Suppr. Balb/c | Purified Balb/c anti-H8 | 13.4 | 5.28 |

[a]Normal Balb/c or Balb/c mice which had been suppressed as neonates with anti-H8 were used.

[b]Animals were immunized with 10 µg of purified Balb/c anti-H8 three times at weekly intervals in complete and incomplete Freund's adjuvant.

[c]The H8 idiotype was measured by a solid phase RIA using purified H8 as standard inhibitor. The amount is expressed in µg/ml serum.

[d]Similarly, the anti-PC antibody activity was measured by RIA using purified H8 as standard inhibitor. The amount is expressed in µg/ml serum.

### Discussion:  Auto-immune disease - Lack of fidelity

Niels Jerne has described the immune system as a network of idiotypes[19]. The scope of this model is broad enough to accommodate every observable or imaginary immune reaction. Without stating some basic rules by which the immune network operates, the network idea becomes of limited usefulness as a model for an understanding and an experimental approach of immunity. We have presented experimental data which demonstrate a high degree of precision and fidelity in idiotype-anti-idiotype reactions. Under a strictly functional viewpoint idiotype-specific reactions appear to be restricted to a doublet of complementary idiotypes. This doublet represents a response unit which is self-contained and self-regulating. The essential feature of our model is the exclusion of cascades of idiotypic-anti-idiotypic responses. However, even with this simplified model the immune system is highly complex by virtue of its different isotypic antibodies and classes and subclasses of interacting cells. At this level the immune system can be seen as a network with high connectivity.

As any other complex and highly organized biological system the immune system is expected to be prone to error and failure. We can easily point to at least three mechanisms which can lead to mistakes. The first is the specific loss of one partner in the idiotype doublet. This "point mutation" would uncouple the feed-back control for the remaining idiotype. The second error-prone mechanism can develop from the simultaneous occurrence of complementary responses which are without reciprocal regulation[27]. These dual responses may be enhanced and circulating complexes can be formed. Thirdly, cross-connections of the immune system to the endocrine and neuronal system are feasible and have been already experimentally documented[28,29]. Antibodies specific for epitopic structures of hormones or neurotransmitter can interfere with the function of hormone-sensitive cells by neutralizing or mimicking hormone activity. Equally possible are interference by antibodies with enzyme activities. From this kind of consideration, we come to a unifying view of biological systems. The basis of this concept is the sharing of complementarity structures among different specialized systems, such as the immune, the endocrine system, or the enzymes. Within the universe of complementarity (see Figure 3) multiple cross-connections are feasible and may give rise to bizarre abnormalities which might be the underlying cause for many auto-immune disorders.

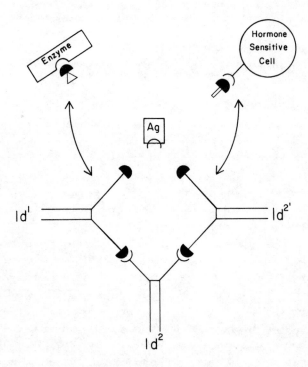

Fig. 3. Schematic representation of shared binding site complementarities in the immune system, the hormone systems and the enzymes.

ACKNOWLEDGMENTS

The original data, reported in this article, were obtained with the support of USPHS Grant AI-11080. The helpful advice of Dr. D.A. Rowley for the preparation of this manuscript was greatly appreciated.

REFERENCES

1. Köhler, H. and Rowley, D.A. (1977) in "Autoimmunity", ed. N. Talal, Academic Press, 267.

2. Rowley, D.A., Fitch, F.W., Stuart, F.P., Köhler, H. and Cosenza, H. (1973) Science 181, 1133.

3. Köhler, H. (1975) Transplant Rev. 27, 26.

4. Kluskens, L. and Köhler, H. (1974) Proc. Natl. Acad. Sci. USA 71, 5083.

5. McKearn, T.J. (1974) Science 183, 94.

6. Bankert, R.B. and Pressman, D. (1976) J. Immunol. 117,457.

7. Geczy, A.F. (1977) J. Exp. Med. 147, 1093.

8. Bona, C., Lieberman, R., Chien, C.C., Mond, J., House, S., Green, I. and Paul, W.E. (1978) J. Immunol. 120,1436.

9. Sigal, N.H. Pickard, A.R., Metcalf, E.S., Gerhart, P.J. and Klinman, N. (1977) J. Exp. Med. 146,933.

10. Kaplan, D.R., Quintáns, J. and Köhler, H. (1978) Proc. Natl. Acad. Sci. USA 75, 1967.

11. Strayer, D.S., Lee, W., Rowley, D.A. and Köhler, H. (1975) J. Immunol. 114, 722.

12. Binz, H. and Wigzell, H. (1977) Cont. Top. Immunobiol. 7, 111.

13. Hart, D.A., Wang, A.L., Pawlak, L.L. and Nisonoff, A. (1972) J. Exp. Med. 135, 1293.

14. Eichmann, K. (1975) Eur. J. Immunol. 5, 511.

15. Eichmann, K. and Rajewski, K. (1975) Eur. J. Immunol. 5, 661.

16. Janeway, C.A., Sakato, N. and Eisen, H. (1975) Proc. Nat. Acad. Sci. USA 72, 2357.

17. Rowley, D.A., Köhler, H., Schreiber, H. and Lorbach, I. (1976) J. Exp. Med. 144, 946.

18. Köhler, H., Rowley, D.A., DuClos, T. and Richardson, B. (1977) Fed. Proc. 36,221.

19. Jerne, N.K. (1973) Sci. American 229, 52.

20. Köhler, H., Richardson, B.C. and Smyk, S. (1978) J. Immunol. 120,233.

21. Potter, M. and Lieberman, R. (1970) J. Exp. Med. 132, 737.

22. Augustin, A. and Cosenza, H. (1976) Eur. J. Immunol. 6, 497.

23. Sakato, N., Hall, S.H. and Eisen, H. (1977) Immunochem. 14, 621.

24. Köhler, H. and Smyk, S. (1978) Cell. Immunol., in press.

25. Urbain, J., Wikler, M., Franssen, J.D. and Collignnon, T. (1977) Proc. Natl. Acad. Sci. USA 74, 5126.

26. Cazenave, P.-A. (1977) Proc. Natl. Acad. Sci. USA 74, 5122.

27. Rowley, D.A., Miller, G. and Lorbach, I. (1978) J. Exp. Med., in press.
28. Jacobs, S., Chang, K.J. and Cuatrecasas, P. (1978) Science 200, 1283.
29. Sege, K. and Peterson, P.A. (1978) Proc. Natl. Acad. Sci. USA 75, 2443.

DISCUSSION

KONG: Are the subclasses of anti-idiotypic antibody to phosphoryl-choline always the same as indicated by restricted heterogeneity from studies using isoelectric focusing? Does conjugating the phosphoryl-choline to a T-dependent antigen influence the subclasses which result?

KÖHLER: Both T-dependent and T-independent forms of phosphoryl-choline antigens induce the same idiotype Tl5 dominance and the same predominance of the IgM class. The anti-idiotypic antibodies were always induced with the IgA Tl5 myeloma and the induced subclass of anti-idiotypic antibody is about 80% IgG.

WARNER: The concept of complementary idiotypes is certainly com-patible with your diagram. However, I can envision situations where the anti-idiotypic antibody ($id_2$) binds to a determinant that is not itself in the antigen binding site of the antibody molecule ($id_1$). Hence the reverse binding in this case would not be a true anti-idiotype, since the actual binding site of $id_1$ was not involved. Since idiotypes can be subdivided in terms of hapten inhibition, I wonder if you have studied your particular cases for hapten inhibition to determine whether the reciprocal $id_1$-$id_2$ types are hapten inhibit-able for $id_1$?

KÖHLER: In our system the binding of the Tl5 idiotype to the isolo-gous anti-H8 anti-idiotype is not inhibited by the free hapten phos-phorylcholine. Thus we conclude that the idiotypic determinant recognized by the anti-idiotypic antibody is outside the antigen-binding site.

KINCADE: Is it reasonable to assume that the buffering capacity of idiotype networks is directly related to the heterogeneity of immune responses? If so, simplification of networks might occur in partial immune deficiencies and result in diminished regulation. This would explain the occurrence of autoimmunity in immunodeficient patients. Are there examples of autoantibodies which are restricted in heterogeneity?

KÖHLER: The study of idiotype-anti-idiotype interactions depends very much on the availability of well-defined antigens and identified idiotypes. For these technical reasons, this means that only re-stricted responses have been studied in sufficient detail to demon-strate complementary idiotype interactions. There is one example of autoantibodies in humans. Certain cold agglutinins, studied by Kunkel and Capra, are idiotypically cross-reacting and of re-stricted heterogeneity.

WARNER: I would like to comment concerning the question raised by Kincade on clonality of autoantibodies. It is indeed generally concluded that autoantibodies are polyclonal and this could render some difficulties for the application of anti-idiotype control. However, in an earlier study with NZB $F_1$ hybrids, we showed that the usual rule was that in individual mice, the anti-erythrocyte autoantibodies were of only one parental allotype, but could be of either allotype. These data are compatible with a restricted origin of autoantibody producing clones, and perhaps more comprehensive studies; for example using IEF, might show less polyclonality of autoantibodies than is currently assumed.

MACKAY: The polyclonality of autoantibodies is a feature of autoimmunity. Could this be explained by loss of fidelity in idiotype-anti-idiotype complementarity?

Regarding potential autoantibodies to enzymes (for which the anti-idiotype would be equivalent to substrate), do we know of examples of pathology due to anti-enzyme autoantibodies?

KÖHLER: Certainly, in the proposed model a "point mutation" of the regulating complementary idiotype clone would be a loss of fidelity in the proposed feedback loop. The consequence of such an event could lead to a breakdown in the control of auto-reactive clones and to a polyclonal auto-reactive response.

To my knowlege there are no examples of interference of enzyme activities of immune reactions. But this possibility should be considered at least on theoretical ground.

SCHWARTZ: There have been several well-documented examples of monoclonal cold agglutinins and rheumatoid factors in humans. Individual specimens of these autoantibodies have been found to possess idiotypes that cross react with the corresponding autoantibodies from other persons. Thus, there is evidence for cross-reactive idiotypes in certain autoantibodies.

INFLUENCE OF MHC-LINKED AND NON-MHC-LINKED GENES ON SUSCEPTIBILITY OF
RATS TO EXPERIMENTAL ALLERGIC ENCEPHALOMYELITIS

DAVID L. GASSER, STAVROS J. BALOYANNIS AND NICHOLAS K. GONATAS
Department of Human Genetics and Division of Neuropathology (Department
of Pathology), University of Pennsylvania, School of Medicine,
Philadelphia, Pennsylvania 19104

ABSTRACT

Although AgB-linked control of EAE susceptibility can be clearly
demonstrated in crosses between Lewis and BN strains of rats, there are
additional genes which influence this response in other strains.
Because of these other genes, strains with the $AgB^4$ haplotype may be
very susceptible to EAE (DA) or completely resistant (AVN). Therefore
the susceptibility of rats to EAE induced by spinal cord or myelin
basic protein of guinea pigs must be considered a polygenic trait, with
non-MHC genes being capable of strongly modifying the effects of MHC-
linked responder or nonresponder genes.

INTRODUCTION

Experimental allergic encephalomyelitis (EAE) induced by guinea pig
myelin has proven to be a very useful model for the study of the
genetic control of autoimmune disease. When Lewis ($AgB^1$) and BN ($AgB^3$)
strains are crossed, susceptibility is largely determined by a single
dominant gene closely linked to the AgB locus,[1,2] which is the major
histocompatibility complex (MHC) of rats. Strains with the $AgB^4$
haplotype show varying degrees of susceptibility[3,4], and it has been
reported that the nonsusceptibility of the AVN strain is transmitted
as a dominant trait[4]. We have re-examined the genetics of susceptibility of $AgB^4$ rats, and have begun experiments to localize the part
of the MHC that plays a role in determining susceptibility. To do this,
we have studied the $AgB^1$ Black Hooded strain and the $AgB^7$ KGH strain,
which is compatible with $AgB^1$ in the mixed lymphocyte culture (MLC)
test, but differs serologically from this haplotype[5,6].

MATERIALS AND METHODS

Rats. Lewis, BN, DA and Black Hooded rats were obtained from the
animal colonies of the Departments of Human Genetics and Pathology at
the University of Pennsylvania. The AVN and LEW.AVN strains were
generously provided by Dr. Jane Schultz of the Veterans Administration

Hospital, Ann Arbor, Michigan. The CAR, CAS and NBR strains were
obtained from the NIH, and Dr. Donald Cramer of the Department of
Pathology, University of Pittsburgh, kindly provided the KGH strain.
The F.BN strain is a homozygous congenic line derived by backcrossing
the $AgB^3$ haplotype of BN for 10 generations onto the F.344 (Fischer)
background. BN.B4 was developed at the Wistar Institute by backcrossing
the $AgB^4$ haplotype of DA onto the BN background.

Immunization. Animals were injected in each hind footpad with
guinea pig spinal cord or guinea pig myelin basic protein emulsified in
Freund's complete adjuvant. Spinal cords were obtained from Pel-Freeze
Biologicals (Rogers, Arkansas), and myelin basic protein was extracted
by the method of Diebler et. al.[7] Rats receiving homogenized spinal
cord were given a dose of 110 mg of fresh tissue homogenate whereas
those injected with basic protein each received 50 micrograms.

Histology. The animals were killed 17-18 days after injection, and
longitudinal sections of the entire spinal cord including spinal roots,
and coronal sections of cerebral hemispheres, brain stem and cerebellum
were processed for histologic study. Paraffin-embedded sections were
stained with haematoxylin and eosin. All sections were serially studied
at 25x and 100x by two of us who did not know the identity of the
slides. Scoring for EAE was based on the identification of perivascular (perivenular) infiltrates of mononuclear cells and lymphocytes.

RESULTS AND DISCUSSION
    In Table 1 are shown the results of testing a variety of inbred
strains and backcross generations using guinea pig spinal cord for
sensitization. Since disease in some cases could be caused by neural
antigens other than myelin basic protein, additional tests were done
using purified basic protein for sensitization. These results are
shown in Table 2, in which the distribution of lesions throughout the
central nervous system is also summarized. These results show that the
greatest amount of infiltration was always in the spinal cord, and
lesions tended to appear in the brain only in strains with substantial
spinal cord involvement. Although the cerebellum was affected slightly
more than the spinal cord in a few individual rats, this was not a
consistent finding in any strain examined. When the responses to
basic protein and spinal cord homogenate were compared, we observed
only rather minor differences.
    Several conclusions emerge from these observations. When the Lewis
and BN strains were crossed and the BN.Lewis backcross generation
obtained, susceptibility was determined almost entirely by an AgB-

linked gene[1,2]. The F.BN congenic strain was only weakly susceptible
when injected with guinea pig spinal cord and not at all susceptible
when injected with myelin basic protein. Thus an MHC-linked gene is
of primary importance when BN is compared with either Lewis or F.344.

The $AgB^4$ haplotype however, is somewhat more complicated. Although
DA is fairly susceptible, this "intermediate responder" trait is not
strongly associated with the AgB haplotype. The BN.B4 congenic strain
was almost completely resistant to the disease, and the $AgB^4/AgB^3$
heterozygotes in the first backcross generation were only slightly
more susceptible than $AgB^3/AgB^3$ homozygotes.

It has been reported that the AVN strain, which is also $AgB^4$, is not
susceptible to EAE and that (Lewis x AVN)$F_1$ hybrids are also non-
susceptible[4]. This is potentially very important, because a dominant
nonresponder gene in this system would represent an exception to the
general rule and would imply that cross-tolerance was involved. We
have therefore studied AVN, LEW.AVN and $F_1$ hybrids between these
strains and Lewis. As shown in the tables, we have confirmed that AVN
is indeed unresponsive to guinea pig spinal cord, but we have not
confirmed that this is transmitted as a dominant trait. Only 3 of 11
(Lewis x AVN)$F_1$ hybrids were unresponsive to guinea pig spinal cord.
To a fairly substantial degree, the unresponsiveness of AVN can be
attributed to an AgB-linked gene, since 8 out of 10 LEW.AVN congenic
rats were unresponsive to guinea pig spinal cord. All 4 of the (Lewis
x LEW.AVN)$F_1$ hybrids injected with myelin basic protein developed
histological EAE, so the low responsiveness associated with the $AgB^4$
haplotype is transmitted as a recessive trait. The high level of
reactivity of the DA strain must therefore be attributed to non-MHC
responder genes. DA is a responder in spite of being $AgB^4$ rather than
because of it.

It is interesting to note that all of the $F_1$ hybrid combinations
that we tested (Lewis x BN, DA x BN, Lewis x AVN and Lewis x LEW.AVN)
reacted at a level that was intermediate between the levels observed in
the two parents. The average EAE score was not necessarily half-way
between those obtained in the two parental strains, but it was nonetheless intermediate in every case.

In mice, the susceptibility to EAE induced by mouse spinal cord
homogenates is controlled by an H-2-linked gene which has not been
localized to the I (immune response) region. The B10.BR strain ($H-2^k$)
is resistant to the development of EAE, while SJL and A.SW (both $H-2^s$)
are highly sensitive. But the recombinant strain A.TL is also highly
sensitive. This strain has an s allele in H-2K but a k allele in all

TABLE 1

SEVERITY OF EAE IN RATS OF VARIOUS TYPES AFTER SENSITIZATION WITH GUINEA PIG SPINAL CORD*

| Strain or Generation | AgB Type | No. of Rats | Histological EAE Score** | | | | | | Average EAE Score |
|---|---|---|---|---|---|---|---|---|---|
| | | | 0 | ± | 1 | 2 | 3 | 4 | |
| BN | 3/3 | 10 | 10 | 0 | 0 | 0 | 0 | 0 | 0 |
| Lewis | 1/1 | 27 | 0 | 0 | 4 | 8 | 14 | 1 | 3.59 |
| (Lewis x BN)F$_1$ | 1/3 | 14 | 0 | 0 | 2 | 9 | 2 | 1 | 2.14 |
| (Lewis x BN)F$_1$ x BN | 3/3 | 11 | 11 | 0 | 0 | 0 | 0 | 0 | 0 |
| (Lewis x BN)F$_1$ x BN | 1/3 | 15 | 3 | 0 | 1 | 4 | 5 | 2 | 2.13 |
| DA | 4/4 | 14 | 2 | 1 | 2 | 2 | 6 | 1 | 2.04 |
| (DA x BN)F$_1$ | 4/3 | 11 | 6 | 0 | 4 | 1 | 0 | 0 | 0.55 |
| (DA x BN)F$_1$ x BN | 3/3 | 18 | 18 | 0 | 0 | 0 | 0 | 0 | 0 |
| (DA x BN)F$_1$ x BN | 4/3 | 14 | 10 | 3 | 1 | 0 | 0 | 0 | 0.18 |
| Black Hooded | 1/1 | 6 | 2 | 0 | 2 | 2 | 0 | 0 | 1.00 |
| NBR | 1/1 | 7 | 2 | 0 | 3 | 2 | 0 | 0 | 1.00 |
| CAS | 1/1 | 11 | 0 | 0 | 3 | 6 | 2 | 0 | 2.55 |
| CAR | 1/1 | 9 | 1 | 0 | 1 | 1 | 5 | 1 | 2.44 |
| F.344 (Fischer) | 1/1 | 5 | 0 | 0 | 0 | 1 | 3 | 1 | 3.00 |
| F.BN | 3/3 | 10 | 7 | 0 | 0 | 0 | 3 | 0 | 0.90 |
| KGH | 7/7 | 10 | 7 | 0 | 0 | 3 | 0 | 0 | 0.60 |

TABLE 1, Cont.

SEVERITY OF EAE IN RATS OF VARIOUS TYPES AFTER SENSITIZATION WITH GUINEA PIG SPINAL CORD*

| Strain or Generation | AgB Type | No. of Rats | Histological EAE Score** | | | | | | Average EAE Score |
|---|---|---|---|---|---|---|---|---|---|
| | | | 0 | ± | 1 | 2 | 3 | 4 | |
| AVN | 4/4 | 6 | 6 | 0 | 0 | 0 | 0 | 0 | 0 |
| (Lewis x AVN)$F_1$ | 1/4 | 11 | 3 | 0 | 0 | 6 | 2 | 0 | 2.27 |
| LEW.AVN | 4/4 | 10 | 8 | 0 | 0 | 2 | 0 | 0 | 0.40 |
| BN.B4 | 4/4 | 14 | 13 | 1 | 0 | 0 | 0 | 0 | 0.04 |

*Some of these results were published previously in references 1 and 3.

**The degree of severity of EAE was scored as follows: 0, no perivascular, meningeal, or intra-parenchymal infiltrates of mononuclear cells; ±, only one cellular infiltrate in a spinal root or in leptomeninges; 1, several infiltrates in spinal roots and occasionally in spinal cord; 2, many infiltrates in the roots and at least one infiltrate in the cord at x25 field; 3, confluent infiltrates in the roots, many infiltrates in the cord, occasional infiltrate in the cerebrum; 4, confluent infiltrates in roots and spinal cord, several infiltrates in cerebrum.

TABLE 2

HISTOLOGICAL DISTRIBUTION AND SEVERITY OF LESIONS IN RATS OF VARIOUS TYPES AFTER SENSITIZATION WITH GUINEA PIG MYELIN BASIC PROTEIN

| Strain | AgB Type | No. of Rats | Histological EAE Score | | | | | | Average Intensity of Infiltration* | | | | Overall EAE Score |
|---|---|---|---|---|---|---|---|---|---|---|---|---|---|
| | | | 0 | ± | 1 | 2 | 3 | 4 | Cerebrum | Cerebellum | Brain Stem | Spinal Cord | |
| KGH | 7/7 | 10 | 9 | 0 | 1 | 0 | 0 | 0 | 0.15 | 0 | 0 | 0.30 | 0.30 |
| Lewis | 1/1 | 6 | 0 | 0 | 0 | 0 | 1 | 5 | 1.50 | 1.33 | 2.83 | 4.00 | 3.83 |
| DA | 4/4 | 7 | 4 | 0 | 0 | 1 | 1 | 1 | 0 | 0 | 0 | 1.29 | 1.29 |
| Black Hooded | 1/1 | 5 | 5 | 0 | 0 | 0 | 0 | 0 | 0 | 0 | 0 | 0 | 0 |
| (Lewis x LEW.AVN)F₁ | 1/4 | 4 | 0 | 0 | 0 | 1 | 2 | 1 | 0.50 | 1.25 | 0.50 | 3.00 | 3.00 |
| BN | 3/3 | 3 | 3 | 0 | 0 | 0 | 0 | 0 | 0 | 0 | 0 | 0 | 0 |
| F.BN | 3/3 | 10 | 10 | 0 | 0 | 0 | 0 | 0 | 0 | 0 | 0 | 0 | 0 |

*Intensity of infiltration was scored on an arbitrary scale of 1 to 4. Values shown represent the means for all the animals in that group. 1: a rare perivascular infiltrate seen after screening 10 or more fields at x25; 2: one infiltrate per x25 field; 3: a few infiltrates per x25 field; 4: infiltrates become confluent at x25.

other known regions of H-2. Thus the gene for susceptibility seems to map either within H-2K or very close to it, a short distance to the left or to the right[8]. We have begun studying the KGH strain in the hope of determining where the AgB-linked gene for EAE susceptibility of rats maps in relation to serologically defined and MLC-defined loci. KGH is believed to possess a recombinant chromosome, since it is compatible with AgB[1] strains in the mixed lymphocyte culture test but is very different from AgB[1] serologically[5,6]. As shown in Tables 1 and 2, KGH responds very poorly to both guinea pig spinal cord and myelin basic protein. Since this is so different from the behavior of Lewis, it appears that the gene for response to guinea pig basic protein may not be within the MLR-determining part of the rat MHC.

The unresponsiveness of the Black Hooded strain to myelin basic protein is even more problematic, since this strain has an AgB[1] haplotype that is indistinguishable from that of Lewis. There are two possible explanations for these results: (a) both KGH and Black Hooded have the gene for high responsiveness, but the genetic background in each case overcomes the influence of the MHC-linked Ir gene; or (b) the Ir gene in question maps outside of the MHC as it is currently defined. Experiments are in progress to determine which of these alternatives is correct.

ACKNOWLEDGEMENTS

We thank Mrs. Jacqueline Gonatas for preparing the basic protein and Ms. Anna Stieber for technical assistance. We are especially grateful to Ms. Dorothea Brusstar and Ms. Eileen Heatherby for their heroic performace in completing a large volume of histological work. This study was supported by Grant # RG 1161-A-3 of the National Multiple Sclerosis Society of American and Grant # CA 15146 of the National Institutes of Health.

REFERENCES

1. Gasser, D. L., Newlin, C. M., Palm, J. and Gonatas, N. K. (1973) Science, 181, 872-873.
2. Williams, R. M. and Moore, M. J. (1973) J. Exp. Med., 138, 775-783.
3. Gasser, D. L., Palm, J. and Gonatas, N. K. (1975) J. Immunol., 115, 431-433.
4. Perlik, F. and Zidek, Z. (1974) Z. Immun. Forsch. Bd., 147, 191-194.
5. Cramer, D. V., Shonnard, J. W. and Gill, T. J. (1974) J. Immunogenet. 1, 421-428.
6. Stark, O., Gunther, E., Kunz, H. and Gill, T. J. (1977) J. Immunogenet. 4, 405-410.

362

7. Deibler, G. E., Martenson, R. E. and Kies, M. W. (1972)
   Preparative Biochemistry 2, 139-165.
8. Bernard, C. C. A. (1976) J. Immunogenet. 3, 263-274.

DISCUSSION

MACKAY: The SJL mouse (H-2$^S$) has been found to be susceptible to EAE, and F$_1$ hybrids are as susceptible as the parent strain. My colleague, Claude Bernard, has shown that responsiveness is MHC-linked and the susceptibility locus (Ir gene site?) is in the K-IA subregion.

My question relates to possible genetic influences over the adjuvant influence of Freund's complete adjuvant.

GASSER: It is quite possible that there are important genetic differences in responses to Freund's adjuvant, but we don't have any data on this point.

In your work on EAE susceptibility of mice, did you rule out the possibility that the Ir gene involved might map to the left of K?

MACKAY: No.

WARNER: Regarding the possibility of other background genes that may negate the susceptibility of certain strains to EAE, could you tell me whether the F$_1$ hybrid of LEW and Black Hooded is susceptible, and secondly, whether there are any MHC congenic rat strains of l/l type derived from Black Hooded?

GASSER: We do not have any data yet on Lewis X BH hybrids, and there are no congenic strains involving Black Hooded. The way that we have to do the experiment is to cross Black Hooded with LEW.AVN, since there are no AgB-linked markers to distinguish the Lewis and Black Hooded MHC chromosomes. This will allow us to place the Black Hooded MHC haplotype on what is essentially the Lewis background.

GERSHON: Factors other than the immune response to brain antigens can control the development of severity of EAE. One important factor could be the ability of cells to cross the blood brain barrier. Thus, it is possible that non-MHC genes might control factors which regulate lymphocyte traffic into the CNS (the number or quality of local mast cells, for example). Have you tested whether EAE non-responder rats make an immune response to brain antigens by parameters other than development of EAE, such as DTH?

GASSER: No, we have not. As far as the BN strain is concerned, we don't think these factors could explain the unresponsiveness. It was reported by Levine and confirmed by Dr. Lennon that BN rats can get EAE if pertussis is included in the adjuvant. It was also reported by Ortiz-Ortiz and Weigle that T cells which bind basic protein can be identified in Lewis but not BN rats.

GERSHON: Pertussis, besides acting as an adjuvant, increases the response of sensitive cells to vasoactive amines and therefore could theoretically increase disease by means other than increasing immune responses.

LENNON: Dale McFarlin and co-workers (J. Exp. Med., 1974) have clearly demonstrated that the insusceptibility of BN rats to EAE is due to nonresponsiveness at the T cell level to the myelin basic protein's major encephalitogenic determinant for rats.

In our own experiments (Lennon et al., Eur. J. Immunol., 1976), we found that EAE can be induced in BN rats using basic protein as immunogen (with CFA + B. pertussis), but in contrast to Lewis rats, large amounts of basic protein were required (200 µg) and rat and guinea pig basic proteins were equivalently encephalitogenic. This suggests that another (minor) encephalitogenic determinant of the basic protein was acting as immunogen and is entirely consistent with the BN rat's lack of T cell responsiveness to the major encephalitogenic determinant.

SHULTZ: Have you looked for differences in complement levels between susceptible and nonsusceptible rats?

GASSER: No.

GERSHWIN: There are several papers in the literature demonstrating anti-lymphocyte antibodies in human patients with multiple sclerosis. Do comparable antibodies exist in EAE?

ARNON: MS patients have elevated levels of lymphocytes but they don't have antibodies against lymphocytes.

KONG: I realize that EAE is considered a cell-mediated disease but did you measure antibody levels to determine if the levels also corresponded with the MHC-linked susceptibility? The reason I asked this is that in the mouse thyroiditis model, by changing the adjuvant from CFA to LPS, the differences in antibody levels are much more noticeable, although the level still corresponds with H-2-linked susceptibility which is T cell based.

GASSER: The BN rats are able to make antibodies to basic protein, but these antibodies are not directed against the specific fragment that is encephalitogenic. McFarlin and colleagues have shown that Lewis rats can make a good antibody to the encephalitogenic fragment, but BN rats cannot.

BIGAZZI: Dr. Mackay's comment on the importance of adjuvants is very pertinent. Our data on sperm autoantibody production are obtained without the use of any adjuvant and we do not find a correlation with MHC genotypes.

ROSE: We have found that experimental thyroiditis can be induced in the mouse with bacterial lipopolysaccharide (LPS) as adjuvant as well as with complete Freund's adjuvant (CFA). Good responder (e.g., $H-2^k$) strains using CFA are also good responders when LPS is used, showing that H-2-linked genetic control applies to the thyroid antigen rather than to the adjuvant used. On the other hand $C_3$ H/HeJ ($H-2^k$) mice, which have a mutation making them nonsusceptible to LPS, fail to develop thyroiditis when it is used as adjuvant with thyroglobulin.

LENNON: There is no evidence that complement plays any role in the pathogenesis of EAE. However, complement could be involved in the induction of autoimmunity to basic protein since it is a thymus-dependent response.

GERSHON: In relation to control of lymphocyte traffic to the brain, I am reminded of an experiment by Bogdanove and Clark reported in the '50's. They used a technique to physically break the blood brain barrier in rodents immunized against brain determinants. This increased disease symptomatology. Perhaps you could use this technique to distinguish between immune and other factors which may determine disease severity.

GENETIC CONTROL OF SUSCEPTIBILITY TO EXPERIMENTAL ALLERGIC
ENCEPHALOMYELITIS - IMMUNOLOGICAL STUDIES

DVORA TEITELBAUM, ZEEV LANDO AND RUTH ARNON
Department of Chemical Immunology, The Weizmann Institute of Science,
Rehovot (Israel)

ABSTRACT

The genetic control of susceptibility to experimental allergic encephalomyelitis was studies in guinea pigs and mice.  In guinea pigs
susceptibility is linked to the major histocompatibility (MHC) locus,
and is probably controlled by two genes that segregate independently.
In mice the susceptibility to disease is also genetically controlled,
but is governed by genes located outside of the MHC, and not by H-2
linked Ir genes.  Correlation was observed between the sensitivity to
disease and the cellular immune response toward the major encephalitogenic determinants in the basic protein.

INTRODUCTION

It is well established that immune responses to many antigens are
regulated by distinct immune response (Ir) genes.  Many such Ir genes
are associated with the major histocompatibility complex (MHC) of the
particular species and mapped within the I region.  In addition, Ir
genes not linked to the MHC were demonstrated to influence the immune
response to a variety of antigens[1,2].  Thus, the genetics of the immune
response may be rather complicated and involve several genes and cooperation between genes[3,4], and may also vary between species[5].  Moreover,
the immune response to different determinants on the same molecule may
be under different genetic control[6-8].

Genetic differences in immune response may play an important role in
susceptibility to a variety of diseases, including autoimmune diseases,
in both animals and man.  This was demonstrated for experimental allergic encephalomyelitis and experimental allergic thyroiditis in several
species, e.g. rats, guinea pigs and mice[1,9,10].

Experimental allergic encephalomyelitis (EAE) is an acute neurologi-
cal autoimmune disease which is well characterized on the molecular and
cellular level.  It is induced by a single injection of a basic encephalitogenic protein (BE) which is a constituent of the myelin sheath.
Particular regions in the BE molecule were identified as responsible
for the induction of the disease with pronounced species variability.

The primary immunological mechanism which is involved in this disease is a cell mediated immune response[11].

Genetic studies of EAE in rats demonstrated that the genetic control is not identical in all strain combinations. In one case (BN x Lewis) it is controlled by a single autosomal gene which, although closely linked to the MHC, is rather distinct from it[12,13]. In other strain combinations such as (BN x DA) it is completely independent of the MHC[14]. Studies in mice yielded controversial results concerning the role of the H-2 complex in susceptibility to EAE[10,15], while in guinea pigs the relation of the MHC to EAE induction has not been investigated hitherto.

We have investigated the mode of the genetic control of susceptibility to EAE in both guinea pigs and mice, as well as the immune response to BE in correlation with the variance in disease susceptibility (our results in guinea pigs were partially summarized previously[16,17]).

STUDIES IN GUINEA PIGS

Strain differences in susceptibility to EAE

Already in 1962 Stone[9] has demonstrated that inbred strain guinea pigs 2 and 13 differ in their susceptibility to EAE: Strain 2 guinea pigs are resistant to EAE, while strain 13 animals are susceptible. In those studies whole spinal cord tissue was used to induce the disease. We have further investigated this phenomenon using the basic encephalitogenic protein of either bovine or guinea pig origin, purified by chromatography on sulphoethyl sephadex[18]. Strain 2, strain 13 and their $F_1$ hybrids were sensitized to BE. The results summarized in Table 1 demonstrate that strain 13 animals were susceptible to both clinical and histological symptoms of the disease, whereas strain 2 animals were completely resistant to clinical manifestations of the disease. Histological changes of a mild nature were found in only 12% of strain 2 animals. The $F_1$ hybrids showed the same degree of sensitivity as the parent strain 13. Thus, susceptibility to EAE is inherited as a dominant trait. We have also investigated the susceptibility to EAE using the synthetic encephalitogenic nonapeptide, Phe-Ser-Trp-Gly-Ala-Glu-Gly-Gln,Lys. This sequence of the BE molecule has been demonstrated as the major encephalitogenic determinant in guinea pigs[19]. The results obtained by sensitization with the encephalitogenic peptide (EP) (Table 1) were essentially identical to those obtained with the intact BE. Hence, the same differences in susceptibility to EAE pertain with different sensitizing agents - whole spinal cord tissue, heterologous or homologous BE, and the EP.

TABLE 1

STRAIN DIFFERENCES IN SUSCEPTIBILITY TO EAE IN GUINEA PIGS

| | Induction with BE | | Induction with EP | |
| --- | --- | --- | --- | --- |
| | Clinical incidence | Histological changes | Clinical incidence | Histological changes |
| Strain 13 | 52/72 | 72/72 +++[*] | 7/10 | 8/10 +++ |
| Strain 2 | 0/40 | 5/40 + | 0/20 | 0/20 - |
| F1 | 14/18 | 16/17 +++ | 9/10 | 9/10 +++ |

* Degree of histological changes[20].

### The humoral response to BE in the different strains

The ability of the different strains to produce humoral antibodies to BE was tested using the sensitive passive cutaneous anaphylaxis (PCA) assay that measures the 7S-$\gamma$1 antibody[21]. It was found that strain 2 guinea pigs, which are resistant to the induction of the disease, produce this antibody in a titer 40-50 fold higher than guinea pigs of strain 13 which are sensitive to EAE[16].

It appears, therefore, that an inverse relationship exists between the capacity to develop EAE and the capacity to produce 7S-$\gamma$1 antibody to BE, a finding which is in accordance with the evidence that 7S-$\gamma$1 antibody has a protective role in EAE[22]. It seems, however, that this alone does not account for the differences in susceptibility, as indicated by the following results.

### The cellular immune response to BE and EP

The cellular immune response of strain 2 and 13 guinea pigs to BE was evaluated both *in vivo* by means of a delayed hypersensitivity skin test and *in vitro* by means of lymphocyte transformation, as measured by stimulation of incorporation of labeled thymidine. As shown in Table 2, both strains responded equally well to the homologous protein in the skin test. Furthermore, there was virtually complete cross-reaction between the guinea pig basic protein (GPBE) and the bovine basic protein (BBE). The results were confirmed by the *in vitro* lymphocyte transformation test. Maximum response was obtained at a concentration of 5-10µg of BE/culture in both strains. There was no difference between the two strains, neither in the incidence of nor in the degree of response. Cross-stimulation with bovine BE demonstrated a high degree of cross-reactivity with the homologous antigen.

TABLE 2

CELLULAR IMMUNE RESPONSE TO BE IN INBRED STRAINS 2 AND 13

| Sensitization with 10 μg of GPBE | Skin test (mean diameter in mm) | | Lymphocyte transformation stimulation index (mean ± SE) [b] | |
|---|---|---|---|---|
| | GPBE (50 μg) | BBE (50 μg) | GPBE | BBE |
| Strain 2 | 8.5 | 9 (10/10) [a] | 3.8 ± 1.9 | 3.6 ± 1.2 (6/10) |
| Strain 13 | 8.9 | 8.5 (10/10) | 3.7 ± 1.8 | 3.6 ± 1.6 (7/10) |

[a] The numbers in brackets represent the incidence of reacting animals.

[b] The mean values were calculated from the maximal stimulation indices obtained in the responding animals only. Maximum stimulation was obtained at a concentration of 5-10 μg/culture.

Though the cellular response to BE does not differ significantly in the two strains, the specificity of the response could be directed towards different determinants on the molecule. Strain differences in response to different antigenic determinants were observed using both synthetic polypeptides[6] and natural proteins e.g. insulin[7] and lysozyme[8]. In order to test this hypothesis, we have analyzed the cellular immune response to the EP in the various guinea pig strains.

As shown by the results in Table 3, in the delayed hypersensitivity skin test strain 13 animals responded without exception, though the reaction was weak due to the small size of the test antigen. None of the strain 2 animals responded with the slightest reaction. Similar results were obtained *in vitro* using lymph node cells from animals sensitized with the peptide. Strain 13 animals yielded lymph node cells that responded well to the peptide, whereas strain 2 cells showed no specific stimulation of incorporation of radioactive thymidine over the background level. $F_1$ hybrids responded similarly to the strain 13 animals, both in the delayed hypersensitivity skin test and in the lymphocyte-transformation technique.

TABLE 3

CELLULAR IMMUNE RESPONSE TO THE ENCEPHALITOGENIC
NONAPEPTIDE IN INBRED STRAINS 2 AND 13

| Sensitization with 20μg of peptide | Skin test with 100μg of peptide | | Lymphocyte transformation |
|---|---|---|---|
| | Erythema (mm) | Induration (mm) | Stimulation index $\pm$ SE[a] |
| Strain 2 | 0 | 0 (0/15) | 1.6 $\pm$ 0.4 (0/5) |
| Strain 13 | 5.3 | 5.1 (15/15) | 4.2 $\pm$ 2.0 (5/5) |
| $F_1$ | 5.5 | 5.2 (10/10) | 5.3 $\pm$ 2.2 (4/5) |

[a]The mean values were calculated from the maximal stimulation indices obtainied in the responding animals only. Maximum stimulation was obtained at a concentration of 5-10μg/culture.

We have thus demonstrated that strain differences in susceptibility to EAE in guinea pigs correspond to the response of this species to the major encephalitogenic determinant. These results may serve as a basis for explaining the variance in disease susceptibility observed in these strains. It is noteworthy that a similar pattern of response was also observed in rats. In BN rats which are resistant to EAE, cellular immunity to the intact BE can be demonstrated, although there is no detectable response to the encephalitogenic fragment for that species. On the other hand, the EAE susceptible Lewis rats, develop cellular immunity to both the intact molecule and the encephalitogenic fragment 23,24.

Genetic analysis of susceptibility to EAE

In order to investigate the genetic basis of susceptibility to. EAE in guinea pigs, namely the mode of inheritance, the number of loci involved and the possible linkage to the major histocompatibility complex (MHC), we studied the $F_1$ hybrids of strain 2 and 13 and the backcrosses of the $F_1$ to the two parental strains.

The MHC genotype of the various offsprings was typed using specific alloantisera. The alloantisera (strain 13 anti-strain 2, and strain 2 anti-13) were prepared as described by Shevach et al.[25] by cross immunization of strains 2 and 13 with homogenate of lymph node and spleen cells. The alloantisera were tested for their activity and specificity on spleen lymphocytes purified on Ficoll gradient, of strain 2, strain 13 and F1 using the [51]Cr release cytotoxic assay[25]. The results demonstrated that indeed anti-strain 2 antisera reacted with only strain 2 lymphocytes, and anti-strain 13 reacted only with strain 13 lymphocytes, while both antisera reacted with $F_1$ lymphocytes.

EAE was induced in the various guinea pigs by sensitization with 20 µg of bovine BE in complete Freund's adjuvant in the hind foot pads. All animals were typed for their MHC specificity, and their brains were subjected to histological examinations that were performed under code, either when paralysis was observed, or after 4 weeks in animals that were not paralyzed. The results are summarized in Table 4.

TABLE 4

GENETIC ANALYSIS OF SUSCEPTIBILITY TO EAS IN GUINEA PIGS

| Strain | Genotype* | Clinical Incidence | Histological changes Incidence | Degree |
|--------|-----------|--------------------|-------------------------------|--------|
| Strain 13 | 13/13 | 52/72 | 72/72 | ++,+++ |
| Strain 2 | 2/2 | 0/40 | 5/40 | + |
| Fl (2 x 13) | 2/13 | 14/18 | 16/17 | ++,+++ |
| Fl x Strain 13 | | 36/43 | 42/43 | ++,+++ |
| | 2/13 | 17/20 | 20/20 | ++,+++ |
| | 13/13 | 19/23 | 22/23 | ++,+++ |
| Fl Strain 2 | | 8/40 | 13/40 | |
| | 2/13 | 8/21 | 12/21 | ++,++ |
| | 2/2 | 0/19 | 1/19 | + |
| F2 (Fl x Fl) | | 6/34 | 19/34 | |
| | 2/13 | 3/20 | 12/20 | ++,+++ |
| | 13/13 | 3/7 | 5/7 | ++,+++ |
| | 2/2 | 0/7 | 1/7 | + |

* Determined by [51]Cr release assay with alloantisera[25].

As described before, strain 13 is susceptible to EAE (72% clinical disease and 100% histological disease) while strain 2 is resistant (no clinical disease and only mild histological lesions in few animals). $F_1$ hybrids developed EAE to the same extent as strain 13. Thus susceptibility to disease is inherited as a dominant trait. The progeny of the backcross of $F_1$ to strain 13 also developed EAE similarly to the susceptible parent strain 13, as expected. On the other hand, when the backcrosses of the $F_1$ to the resistant strain 2 were studied only 20% developed clinical symptoms and 32% had histological changes. All the animals that developed EAE were of the genotype 2/13 while among those of the genotype 2/2 no animal developed clinical disease, and

only one out of 19 had mild histological lesions. In the $F_2$ generation
the incidence of clinical EAE was 18%, and 56% had histological
changes. All the animals that developed clinical EAE were either 2/13
or 13/13, with no disease in 2/2 offsprings. The ratio of the various
genotypes in the two backcrosses and in $F_2$ generation are very close to
that expected according to Mendelian genetic segregation.

Thus, the results demonstrate a clear linkage of susceptibility to
EAE with the MHC of strain 13. On the other hand, the low incidence of
disease in the backcrosses of $F_1$ and strain 2 and in the $F_2$ generation
indicate that more than one gene is involved. According to Mendelian
inheritance laws we could expect in the backcross to the resistant
strain 50% in case of involvement of one gene, and 25% in case of 2
genes. In the $F_2$ generation we would expect 25% if one gene is involved and 56% if there are two genes. Our results both with the backcross $F_1$ x strain 2 and with $F_2$ generation, mainly as manifested by the
histological findings, are compatible with two genes inheritance.

Our genetic analysis indicate therefore, the existance of two Ir
genes to EAE in strain 13 guinea pigs. One is linked to the MHC of
this species and located probably in the I region, as strains 2 and 13
differ only in the I region of their MHC[26]. The other gene is located
outside the MHC and its linkage group is unknown. These two genes segregate independently and both of them must be expressed to render the
animal susceptible to EAE. T cell responses were demonstrated to be
MHC linked[2]. It is thus possible that the MHC-Ir gene controls immune
response at the T cell level to the encephalitogenic determinant as
demonstrated before. The results of Ben-Nun and Cohen[27] that the
response to this determinant in $F_1$ can be initiated only when BE is
present on strain 13 macrophages are in accordance with this assumption. The other background gene, might control the immune response at
a different level.

GENETIC STUDIES OF EAE IN MICE

Susceptibility to EAE in various inbred strains and $F_1$ hybrids

Studies on susceptibility to EAE in mice were performed both by
Levine and Sowinsky[15] and by Bernard[10]. While the first authors concluded that the MHC does not control susceptibility to EAE in mice,
Bernard claimed that susceptibility is inherited as a dominant trait
and is controlled in part by genes linked to the $H-2^S$ and $H-2^b$ haplotypes.

In order to study whether susceptibility to EAE is always dominant
we studied EAE in various $F_1$ combinations of the reportedly susceptible

strain SJL/J with various resistant strains (Table 5).

TABLE 5

EAE IN PARENTAL AND Fl HYBRIDS MICE

| Strain | H-2 Type | Clinical disease | | Histological changes | |
|---|---|---|---|---|---|
| | | Incidence | Mean score* | Incidence | Degree |
| SJL/J | s | 13/27 | 1.9 | 9/10 | +,++ |
| BALB/C | d | 1/200 | 2.0 | 1/30 | + |
| (SJL/J X BALB/C)Fl | s/d | 48/50 | 2.2 | 10/10 | +,++ |
| BALB.B10 | b | 1/23 | 1 | 1/5 | + |
| (SJL/J X BALB.B10)Fl | s/b | 21/23 | 2.1 | 6/7 | +,++ |
| BALB/C3H | k | 1/15 | 2.0 | 1/11 | +,++ |
| (SJL/J X BALBC3H)Fl | s/k | 7/7 | 1.9 | 7/7 | +,++ |
| NZB | d | 0/7 | 0 | 0/5 | |
| (SJL/J X NZB)Fl | s/d | 16/16 | 2.94 | 10/10 | ++,+++ |
| DBA/2 | d | 0/6 | 0 | 0/6 | |
| (SJL/J X DBA/2)Fl | s/d | 2/19 | 1.0 | 9/19 | +,++ |
| C57BL/6J | b | 1/19 | 1.0 | 2/19 | + |
| (SJL/J X C57BL/6J)Fl | s/b | 3/10 | 1.7 | 3/10 | + |
| ASW | s | 0/22 | | 3/20 | (15%) |
| (ASW X BALB/C)Fl | s/d | 0/4 | | 0/4 | |
| B10S | s | 0/6 | | 0/6 | |
| (B10S X BALB/C)Fl | s/d | 0/21 | | 0/21 | |

*Score: 1. Loss of weight and loss of tail tonicity; 2. Mild Disease - paresis or paralysis of hind legs; 3. Severe paralysis leading often to death.

EAE was induced with mouse spinal cord homogenate by a single injection to the footpads with complete Freund's adjuvant, followed by pertussis vaccine injections as described by Bernard[10]. In the SJL/J strain we obtained 48% clinical incidence of disease and all the mice showed histological changes in the brain. The $F_1$ hybrid of SJL/J with the resistant strain Balb/c was found to be much more sensitive to EAE than the parent strain SJL/J. Crossing SJL/J with the congenic strains of Balb/c (Balb/C$_3$H and Balb/B10) which are of the same genetic background but differ in their H-2 complex resulted in $F_1$ with high sensitivity to EAE. Similarly $F_1$ of SJL/J x NZB is very sensitive as manifested both in severity and incidence of clinical and histological disease. On the other hand, crossing SJL/J with DBA/2 which is H-$2^d$ as Balb/c and NZB, or with C57BL/6J which is H-$2^b$ as Balb/B10, resulted in $F_1$ which were less sensitive to EAE than the parental strain SJL/J.

Our results demonstrate therefore that there is no pure dominant Mendelian inheritance of susceptibility to EAE in mice. In the crosses with the various resistant Balb strains and NZB strain there is probably complementation with another gene. This gene is located outside the H-2 complex, as several congenic strains of Balb/c give the same complementation, while DBA/2 and C57BL/6J with the same H-2 but with another background do not give such complementation. The fact that the $F_1$ in the latter crosses is less sensitive may indicate a suppressive effect of the resistant parent, the nature of which is unclear.

To investigate the possible linkage of EAE susceptibility to the H-2$^S$ haplotype we tried to induce the disease in two additional strains that possess the H-2$^S$ haplotype, namely ASW and B10.S, as well as in their $F_1$ hybrids with Balb/c. In all these strain combinations we did not succeed to induce the disease (Table 5). The results with ASW differ from those of Bernard but are compatible with those obtained by Yasuda et al.[28] and Levine and Sowinsky[15], as are the results with B.10S. It is thus apparent that susceptibility to EAE is not linked only to the H-2$^S$. Bernard[10] has demonstrated in the $F_2$ generation of SJL/J x Balb/c sensitivity to disease in mice that were of the genotype d/d. Hence, there is no linkage at all to the H-2$^S$, and the gene determining susceptibility to EAE is probably located outside the H-2 complex. The differences between our results in ASW to those of Bernard can therefore be explained by differences in the genetic background of the strains that originated from different sources.

Our results indicate that mice in which the sensitivity to EAE is not controlled by H-2 linked gene differ from guinea pigs in which the involvement of an Ir gene linked to the MHC was demonstrated. Such differences in the genetic linkage of Ir gene in various species was demonstrated in other cases. The immune response to the synthetic antigen T,G-(Pro--L) in mice is not linked to the H-2 complex[29], while in rats linkage to the MHC was demonstrated[5]. Non-MHC linked Ir genes were demonstrated to regulate the immune response to a variety of antigens. Such Ir genes were found to be linked to several linkage groups such as minor histocompatibility genes, genes of immunoglobulin allotypes or the X-chromosomes[1]. However, to our knowledge no antigen specific T cell immune responses were found to be regulated by non-MHC linked Ir genes. In this respect it would be interesting to find out whether the gene involved in susceptibility to EAE regulate T cell function or operates on another level.

## The effect of low dose of cyclophosphamide on EAE

It has been recognized recently that suppressor cells play a critical role in immune regulation, e.g., maintenance of tolerance, antigenic competition, and genetic control of immune responses. It has been demonstrated that loss of effective suppressive T cell function can lead to autoimmune diseases[30]. In order to examine the possibility that resistance to EAE is due to the presence of suppressor cells in the resistant strains, we treated mice with low dose of cyclophosphamide (20 mg/Kg) 2 days before induction of disease. This dose of cyclophosphamide was reported to eliminate specifically the population of T suppressor cells in other systems[31].

When the effect of cyclophosphamide (CY) was tested in the susceptible mice SJL/J and (SJL/J x Balb/c)Fl, no differences were observed in the incidence or severity of EAE (Table 6).

TABLE 6

EFFECT OF LOW DOSE CYCLOPHOSPHAMIDE ON EAE INDUCTION

| Strain | H-2 Type | Mice with clinical EAE | | | |
| | | Without Cyclophosphamide | | With Cyclophosphamide | |
| | | Incidence | Mean score | Incidence | Mean score |
|---|---|---|---|---|---|
| SJL/J | s | 10/16 | (2.0) | 7/16 | (1.8) |
| (SJL/J X BALB/C)Fl | s/d | 18/20 | (1.8) | 20/22 | (2.2) |
| BALB/C | d | 0/40 | | 19/32 | (1.7) |
| BALB/B10 | b | 1/23 | (1.0) | 2/5 | (1) |
| NZB | d | 0/7 | | 2/8 | (2.0) |
| DBA/2 | d | 0/6 | | 0/5 | |
| A/J | a | 0/6 | | 0/6 | |
| B10A | a | 0/7 | | 0/9 | |
| AKR | k | 0/19 | | 0/17 | |
| C57BL/6J | b | 1/19 | (1.0) | 0/18 | |

On the other hand, it has a dramatic effect in the resistant strain Balb/c. Balb/c mice treated with CY developed EAE of the same severity and in the incidence as the susceptible SJL/J mice. We tested the effect of CY in additional resistant strains. The results, although preliminary, suggest that in Balb/B10 and NZB strains there is a similar effect of CY on susceptibility to EAE, whereas in the other strains tested that were of various H-2 types, CY had no effect on the development of EAE.

It seems therefore, that there are two types of inbred strains that are resistant to EAE. CY can convert one type from resistant into susceptible, while it has no effect on the other type. In strains of the first type a gene that is responsible for susceptibility to EAE probably exists, but these strains maintain a natural high level of suppressor cells which prevent the manifestations of disease. Following treatment with CY these suppressor cells are eliminated and disease is overt. In the $F_1$ hybrid, the balance between effector cells and suppressor cells, is shifted towards the effector cell function, and therefore EAE can be induced. In addition there is complementation of genes from the two parents and the resultant $F_1$ may be even more sensitive. In rare cases, some animals of such strains may possess a lower level of suppressor cells leading to the very low incidence of EAE observed in three resistant strains, even without CY treatment. In the second type of resistant strains such as DBA/2 etc., there is no Ir gene for EAE and therefore CY has no effect. The low incidence of disease observed in C57Bl/6J (1/6), on which CY has no effect may be explained by response to other minor encephalitogenic determinants.

Immunological response to the small mouse BE

In guinea pigs and rats susceptibility to EAE was correlated with cellular immune response to the encephalitogenic determinant of these species. It was therefore of interest to find out whether this phenomenon exists also in mice. For this purpose we evaluated the delayed hypersensitivity (DHS) response to the small mouse basic encephalitogen (SMBE), which is encephalitogenic in mice[32]. The DHS response was measured using the *in vivo* radiometric ear skin test[33]. We followed the kinetics of the DHS response to SMBE in (SJL/J x Balb/c)$F_1$ mice sensitized either with spinal cord homogenate or purified SMBE. The results revealed that maximum response was obtained 9 days after sensitization, with a decrease on days 10-12, which correlated with appearance of clinical symptoms. Unprimed mice or mice sensitized to lysozyme did not demonstrate any non-specific reactions to SMBE. We then tested the DHS response in two resistant strains, Balb/c and A/J. The Balb/c mice injected with spinal cord homogenate developed DHS response to the same extent as the $F_1$ mice. While in the A/J strain only 1 out of 6 animals responded and only at a marginal level. It thus appears that Balb/c mice that according to the experiment with cyclophosphamide possess Ir gene to EAE, are capable of developing cell mediated immune response to the SMBE, while A/J mice which do not possess Ir gene to EAE at all, do not develop cellular response to the encephalitogenic protein. Additional strains of mice have to be tested to ascertain whether this correlation is indeed valid.

ACKNOWLEDGEMENTS

    We wish to thank Dr. Edna Mozes for many helpful discussions and to Dr. A. Meshorer for evaluating the histological examinations.

REFERENCES

1.  Gasser, D.L. and Silvers, W.K. (1974) Adv. in Immunol., 18, 1.

2.  Benacerraf, B. and Katz, D.H. (1975) Adv. in Cancer Res., 21, 121.

3.  Dorf, M.E., Dunham, E.K., Johnson, J.P. and Benacerraf, B. (1974) J. Immunol., 112, 1329.

4.  Kolsch, E. and Falkenberg, F.W. (1978) J. Immunol., 120, 6.

5.  Gunther, E., Mozes, E., Rude, E. and Sela, M. (1976) J. Immunol, 117, 2047.

6.  Mozes, E., McDevitt, H.O., Jaton, J.C. and Sela, M. (1969) J. Exp. Med., 130, 1263.

7.  Arguilla, E.R., Miles, P., Knapp, S., Hamlin, J. and Bromer, J. (1967) Vox. Sang, 13, 321.

8.  Maron, E., Scher, H.J., Mozes, E., Arnon, R. and Sela, M. (1973) J. Immunol., 111, 101.

9.  Stone, S.M. (1962) Int. Arch. Allergy, 20, 193.

10. Bernard, C.C.A. (1976) Immunogenetics, 3, 263.

11. Paterson, P.Y. (1976) in Textbook of Immunopathology, Miescher, P.A. and Muller-Eberhard, H.J. eds., Grune & Stratton, Inc., New York, vol. I, p. 701.

12. Williams, R.M. and Moore, N.J. (1973) J. Exp. Med., 138, 775.

13. Gasser, D.L., Newlin, C.M., Palm, J. and Gonatas, N.K. (1973) Science, 181, 872.

14. Gasser, D.L., Palm, J. and Gonatas, N.K. (1975) J. Immunol., 115, 431.

15. Levine, S. and Sowinski, R. (1974) Immunogenetics, 1, 352.

16. Webb, C., Teitelbaum, D., Arnon, R. and Sela, M. (1973) Immunol. Commun., 2, 185.

17. Teitelbaum, D., Webb, C., Arnon, R. and Sela, M. (1977) Cell. Immunol., 29, 265.

18. Hirshfeld, T., Teitelbaum, D., Arnon, R. and Sela, M. (1970) Febs. Lett., 7, 317.

19. Westall, F.C., Robinson, A.B., Caccam, J. and Eylar, E.M. (1971) Nature, 229, 22.

20. Teitelbaum, D., Meshorer, A., Hirshfeld, T., Arnon, R. and Sela, M. (1971) Eur. J. Immunol., 1, 242.

21. Ovary, Z. (1958) Progr. Allergy, 5, 459.

22. Lebar, R. and Voisin, R.A. (1974) Int. Arch. Allergy Appl. Immunol. 46, 82.

23. McFarlin, D.E., Hsu, S.C-L., Clemendo, S.B., Chou, F.C.-M. and Kibler, R.F. (1975) J. Exp. Med., 141, 72.

24. McFarlin, D.E., Hsu, S.C-L., Clemendo, S.B., Chou, F.C-M. and Kibler, R.F. (1975) J. Immunol., 115, 1456.

25. Shevach, E.M., Rosenreich, D.L. and Green, I. (1973) Transplantation, 16, 126.

26. Geczy, A.F. and de Weck, A.L. (1977) Prog. in Allergy, 22, 148.

27. Ben-Nun, A. and Cohen, I., Nature, in press.

28. Yasuda, T., Tsumita, T., Nagai, Y., Mitsuzawa E. and Ohtani, S. (1975) Jap. J. Exp. Med., 45, 423.

29. Mozes, E., McDevitt, H.O., Jaton, J.-C. and Sela, M. (1969) J. Exp. Med., 130, 1263.

30. Waldmann, T.A. and Broder, S. (1977) Prog. in Clinical Immunology, 3, 155.

31. Chiorazzi, N., Fox, D.A. and Katz, D.H. (1977) J. Immunol., 118, 48.

32. Bernard, C.C.A. and Carnegie, P.R. (1975) J. Immunol., 114, 1537.

33. Vadas, M.A., Miller, J.F.A.P., Gamble, J. and Whitelaw, A. (1975) Int. Arch. Allergy, 49, 610.

DISCUSSION

MACKAY: I note the different results obtained in your laboratory compared with ours in regard to susceptibility to EAE, especially in ASW mice. Could I comment again on possible genetic influences over responses to adjuvants, including FCA and also pertussis which we are currently studying. I draw attention to an interesting modification of the Miller-Vadas "ear assay" based on [125]IUdR incorporation into replicating monocytes. Instead of removing the ear for gamma counting, one can remove the entire organ, i.e., brain in the case of EAE, and count it in a gamma counter. My colleague, Dr. Scott Lenthicum, has found that when brain and spinal cord are counted in groups of mice sequentially after indication of EAE, counts sharply increase in the brain at ten days, and in the spinal cord at 12 days, after which signs of EAE appear. This objective quantified radio-isotope system eliminates subjective assessment of clinical disease or histology.

DUPONT: Have you measured complement C4 levels and C4 phenotypes in the resistant and susceptible guinea pigs? And the second question related to this: Is there a sex difference in susceptibility to EAE in guinea pigs and mice?

ARNON: No, we have not investigated the C4 levels and their effect on susceptibility and resistance of guinea pigs to EAE.
And to your second question, we have not observed any difference between males and females in sensitivity to EAE, either in guinea pigs or in mice.

WARNER: In view of the contrast of your data with that of Bernard, have you tested ASW mice in comparison to A/J mice, following cyclophosphamide treatment, for susceptibility to EAE?
Secondly, may I comment that your data indicating a susceptibility of BALB/c, NZB and SJL mice are paralleled by studies on the refractoriness to immunological tolerance as measured by the technique of Talal et al. with deaggregated gamma globulin. These are the three strains that are relatively resistant to such tolerance induction, in comparison to many other strains, and this scheme of tolerance induction is associated with suppressor activation.

ARNON: We have not tested yet the effect of cyclophosphamide on the susceptibility of ASW mice, but intend to do so in the near future.

ROITT: Was the delayed sensitivity produced in BALB/c mice untreated with cyclophosphamide?

ARNON: Yes.

ROITT: Then if cyclophosphamide is inhibiting a suppressor cell, it is not one which is affecting the expression of delayed hypersensitivity to the whole encephalitogen. It might affect delayed hypersensitivity to a crucial epitope (compare the nonapeptide) or cells giving rise to other features of the immune response, IgE for example, or even non-immunological factors.

ARNON: Yes, this is possible. As far as the specificity is concerned, we have tested the small mouse basic protein which, although smaller than the intact basic protein, is still a large molecule. We do not have yet a smaller fragment which is encephalitogenic in mice.

ENGLEMAN: Do you have, or do you know of any direct demonstration of the existence of suppressor cells in immunized BALB/c mice? For example, can disease protection be transferred via lymphocytes from sensitized BALB/c mice to susceptible strains?

ARNON: This experiment has not been done yet.

GERSHON: Two comments. First, cyclophosphamide (Cy) does not kill only suppressor cells. It is toxic for some cells in the suppressor circuit (phenotype Ly123) but is not specific for this cell set. Second, the Vadas-Miller technique is not a true test of DTH. It measures cell trapping which is necessary but not sufficient for a DTH reaction to occur. After effector cells arrive at DTH sites, other events must occur at the site for tissue damage that leads to the subsequent swelling and induration, two important parameters of DTH, which are not measured by the Vadas-Miller technique.

Determining whether animals mount a Vadas-Miller response may have no more bearing on whether pathogenic mechanisms are present than determining whether antibody is made.

ARNON: This is true, but the method can give us information about the immune response to a particular antigen in different strains of mice. It could thus distinguish between BALB/c and A/J, in a manner corroborating the susceptibility of these two strains to EAE after pretreatment with cyclophosphamide.

MACKAY: Could I suggest that the "whole organ" radiometric assay, as opposed to the ear assay, might give an indication of the effector component of a delayed type hypersensitivity response.

GERSHON: Our experience in trying to correlate the localization of bone marrow-derived effector cells and disease symptoms in EAE has not given us any clear-cut answers. However even if correlations can be shown to be present in one situation, this will not guarantee that the correlations will hold in other situations.

INTERACTION BETWEEN LYMPHOCYTES AND MACROPHAGES IN GENETIC CONTROL OF
EXPERIMENTAL AUTOIMMUNE ENCEPHALOMYELITIS

AVRAHAM BEN-NUN and IRUN R. COHEN
Department of Cell Biology, The Weizmann Institute of Science, Rehovot, Israel

ABSTRACT

We studied mechanisms of the genetic control of experimental autoimmune
encephalomyelitis (EAE) in inbred strains 2, 13 and $F_1$ hybrid guinea pigs. By
using bone marrow chimeras of resistant cells in susceptible animals or vice
versa, we found that genetic control of EAE was primarily a property of lympho-
hematopoietic cells. The cellular mechanism of genetic control appeared to be
at the level of interaction between macrophages presenting myelin basic protein
(BP) and responding lymphocytes. $F_1$ hybrid animals responded to the encephalitogenic nonapeptide (NP) determinant when whole BP was presented by macrophages
from the susceptible parental strain. In contrast, macrophages from the resistant parental strain could present other determinants of the BP molecule but not
the critical NP determinant.

INTRODUCTION

Susceptibility to induction of EAE in guinea pigs has been found to be under
genetic control. Inbred strain 13, but not strain 2 guinea pigs can be induced
to develop EAE by injection of myelin basic protein (BP) in a suitable adjuvant[1].
Two findings suggest that genetic control of EAE is expressed primarily through
the immune system. Firstly, resistance or susceptibility to EAE in guinea pigs
has been related to the response to the nonapeptide (NP) determinant of the BP
molecule. Susceptible strain 13 animals respond to NP, while resistant strain 2
respond to other determinants but not to NP[2]. Secondly, susceptibility to EAE
was found to be linked to genes in the major histocompatibility complex (MHC)
which conceivably may be immune response (Ir) genes[3].

These findings demonstrate that the proper genotype of the cells in the immune
system is a necessity for induction of EAE. Although Ir genes may be necessary
for expression of EAE, they may not be sufficient by themselves to determine
induction or resistance to EAE. Other factors, such as the blood-brain barrier
or the expression of BP in the nervous system, might also play a role in genetic
control of EAE. Furthermore, although EAE is probably mediated by T lymphocytes[4],
we do not know the cells through which genetic control is exerted within the
immune system. At what cellular level in the response to BP could Ir genes
control susceptibility to EAE?

The experiments reported in this paper were designed to study these questions. By using bone marrow chimeras between susceptible and resistant guinea pigs, we found that the genetic origin of the lympho-hematopoietic cells was sufficient by itself to regulate susceptibility or resistance to EAE. In addition, we observed that the immune response to the encephalitogenic NP determinant of BP was controlled at the level of presentation of BP by peritoneal exudate macrophages.

MATERIALS AND METHODS

Guinea pigs. Inbred strains 2, 13 and $F_1$ hybrid animals were supplied by the Animal Breeding Center of this Institute. Genetic constitution of the strains was confirmed by differential response to reference antigens[5].

Bone marrow chimeras. Animals 2 months old were lethally irradiated by 650 R from a cobalt source and reconstituted intravenously with $3 \times 10^7$ bone marrow cells. Two months later they were injected in each hind foot pad with 0.05 ml of an emulsion of Freund's complete adjuvant containing 10 µg of bovine BP to induce EAE. Diagnosis of EAE was made by development of clinical paralysis and histologic evidence of lesions.

Response to macrophages fed with BP. Peritoneal exudates were stimulated in strains 2, 13 or (2x13)$F_1$ guinea pigs, by injecting them intraperitoneally with 20 ml sterile light mineral oil (Drakeol 6-VR, Penneco, Butler, Pennsylvania). Macrophage-enriched exudates were collected 3-4 days later by lavaging the peritoneum with phosphate buffered saline (PBS) + heparin (5 units/ml$^{-1}$). The cells (85-90% macrophages) were washed by centrifugation x3, and $10^7$ ml$^{-1}$ in RPMI-1640 were incubated at 37°C with BP, 100 µg ml$^{-1}$ for 45 min. The cells were washed again x3 in large volumes of PBS, to remove free BP. Groups of three (2x13)$F_1$ guinea pigs were injected subcutaneously into two hind foot pads with $5 \times 10^6$ BP-fed exudate cells per foot pad in 0.1 ml of PBS.

The draining popliteal lymph nodes were removed 9 days after injection and suspensions of lymph node lymphocytes were pooled from each group and were seeded in flat-bottom microtiter plates (Falcon), in quadruplicate wells containing $5 \times 10^5$ lymphocytes in 200 µl RPMI-1640 + 1.25% heat-inactivated fetal calf serum, L-glutamine (300 mg/ml$^{-1}$) and 25 mM HEPES buffer. The cultures were incubated for 72 h with various concentrations of PPD, BP, or NP (prepared and donated by Dr. D. Teitelbaum or obtained from Peninsula Labs, San Carlos, California). Each well was then pulsed with 1 µCi $^3$H-thymidine (Amersham) for 4 h, the cells were collected on glass fiber filters and incorporation was measured in a liquid scintillation counter. The proliferative response was measured as the stimulation index: cpm test/cpm control. Standard deviations of cpm were 10-15% of the mean in almost all cases.

RESULTS

Chimeras

Table 1 shows that normal strain 2 guinea pigs generally were resistant to induction of EAE, while strain 13 or $F_1$ animals were susceptible. To test whether induction of EAE was determined by lymphoid cells alone, chimeras were made by injecting lethally irradiated resistant or susceptible animals with bone marrow cells of resistant or susceptible origin. We found that resistance or susceptibility to EAE was a function of the genetic origin of the bone marrow cells and not of the recipient animals. Resistant strain 2 animals developed disease when outfitted with $F_1$ lymphoid cells, and susceptible $F_1$ animals resisted EAE when outfitted with resistant strain 2 cells. Thus, genetic control of EAE was determined by genes expressed in the lympho-hematopoietic system and not by those in the nervous or circulatory systems.

TABLE 1

LYMPHOID CELLS DETERMINE GENETIC CONTROL OF SUSCEPTIBILITY TO EAE IN CHIMERAS

| Guinea pig strain | Origin of bone marrow cells | Incidence of EAE after injection of BP + adjuvant |
|---|---|---|
| Normal | | |
| 2 | — | 1/20 |
| 13 | — | 18/20 |
| (2x13)$F_1$ | — | 19/20 |
| Chimeras | | |
| $F_1$ | $F_1$ | 5/5 |
| $F_1$ | 2 | 0/7 |
| 13 | $F_1$ | 4/4 |
| 2 | 2 | 1/5 |
| 2 | $F_1$ | 8/10 |

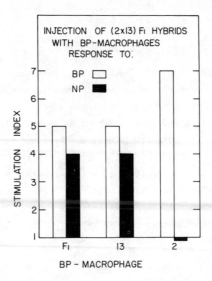

INJECTION OF (2x13) F₁ HYBRIDS
WITH BP-MACROPHAGES
RESPONSE TO:

Fig. 1. Genetic origin of injected macrophages determines response of $F_1$ hybrids to NP determinant.

Response to NP

Experiments were done to investigate the cellular level of expression of genetic control of the response to the encephalitogenic NP determinant of BP. To study genetic factors in antigen presentation, we used $F_1$ hybrid guinea pigs that responded to NP and developed EAE. Such animals were injected with peritoneal exudate macrophages of strains 2, 13 or $F_1$ hybrid origin that had been fed with BP. Nine days later, lymphocytes from the draining lymph nodes were studied for T lymphocyte proliferative response to whole BP or to the NP encephalitogenic determinant. The results are shown in Fig. 1.

We found that the T lymphocytes of the $F_1$ animals responded to the NP determinant as well as to undefined determinants of BP, when sensitized by injection of BP-fed macrophages originating from strain 13 or from $F_1$ hybrids. In contrast, the $F_1$ animals did not respond to the NP determinant when injected with BP-fed macrophages originating from resistant strain 2 animals. Nevertheless, they did respond well to other determinants on the BP molecule presented by strain 2 macrophages. Therefore, the NP determinant was immunogenic for receptor-bearing T lymphocytes only in association with macrophages of the susceptible MHC genotype.

In the course of these experiments, we were able to observe a number of animals for signs of disease. Paralysis leading to death developed in 2 out of

5 animals injected with strain 13 macrophages + BP, while none of 8 animals showed any clinical or histologic signs of EAE after injection with strain 2 macrophages + BP. Therefore, the response to the NP determinant measured in vitro was correlated with appearance of EAE in vivo. In addition, this finding demonstrated that the presence of macrophages fed with BP can be sufficient to induce auto-immunity, without injecting adjuvants. This confirms in vivo the immunogenicity for self lymphocytes of the macrophage-BP complex that we previously observed in vitro[6].

DISCUSSION

These results suggest that the response to the encephalitogenic NP determinant and EAE is controlled by gene products involved in cooperation between macrophages and T lymphocytes. There are several possible mechanisms that might explain why $F_1$ hybrid T lymphocytes with receptors for the NP determinant did not react to NP presented on macrophages of the strain 2 parent.

(a) Specific catabolism: It is possible that strain 2 macrophages differ from strain 13 macrophages in selective destruction of the NP determinant of BP. However, this is unlikely, since it would require an unprecedented degree of specificity from the proteolytic mechanism of the strain 2 macrophage coded for by an Ir gene in the MHC.

(b) Specific determinant selection[7]: Ir gene products expressed on macro-phages might be required to render the NP determinant immunogenic and the strain 2 MHC alleles could be unsuitable for this purpose.

(c) Discordance in dual recognition: Activation of T lymphocytes requires recognition of the particular antigen in association with products of MHC genes[8]. It is possible that $F_1$ T lymphocytes that recognize strain 2 MHC products lack the receptor for the NP determinant.

(d) Suppression[9]: The NP determinant in association with strain 2 MHC products might trigger suppressor T lymphocytes to a greater degree than they do effector T lymphocytes.

These and other mechanisms have been considered to act in the Ir gene regulation of responses to non-self antigens. At present, there is no evidence which can unequivocally decide between these alternatives. It is likely that the various Ir genes will be found to control activation of T lymphocytes at a number of different points in the immune response: T-T cell interactions[10,11], as well as macrophage-T cell interactions. In addition, it remains to be seen whether or not there are substantive differences between genetic control involving self as compared to foreign antigens. In any case, it is evident that one mechanism of regulation of the response to a self component involves cooperation between cells of the immune system and that this cooperation is determined by genes in the MHC.

ACKNOWLEDGEMENT

We thank Dr. Dvora Teitelbaum for donating BP and NP. This work was supported by grants from the Stiftung Volkswagenwerk and the United States-Israel Binational Science Foundation, Jerusalem.

REFERENCES

1. Paterson, P.Y. (1977), in "Autoimmunity: Genetic, Immunologic, Virologic and Clinical Aspects," Talal, N., ed., Academic Press, New York, pp. 643-692.

2. Teitelbaum, D., Webb, C., Arnon, R., and Sela, M. (1977), Cell. Immunol. 29, 265-271.

3. Teitelbaum, D., and Arnon, R. (1978), in "Genetic Control of Autoimmune Dissease," Rose, N.R., Bigazzi, P.E. and Warner, N.L., eds., Elsevier, Amsterdam, in press.

4. Ortiz-Ortiz, I., and Weigle, W.O. (1976), J. Exp. Med. 144, 604-616.

5. Schwartz, B.D., Paul, W.E., and Shevach, E.M. (1976), Transplant. Rev. 30, 174-196.

6. Steinman, L., Cohen, I.R., Teitelbaum, D., and Arnon, R. (1977), Nature 265, 173-175.

7. Rosenthal, A.S., and Shevach, E.M. (1976), Contemp. Top. Immunobiol. 5, 47-90.

8. Paul, W.E., Shevach, E.M., Pickeral, S., Thomas, D.W., and Rosenthal, A.S. (1977), J. Exp. Med. 145, 618-630.

9. Pierce, C.W., Germain, R.N., Kapp, J.A., and Benacerraf, B. (1977), J. Exp. Med. 146, 1827-1832.

10. Cohen, I.R., and Livnat, S. (1976), Transplant. Rev. 29, 24-58.

11. Zinkernagel, R.M., Callahan, G.N., Althage, A., Cooper, S., Streilein, J.W., and Klein, J. (1978), J. Exp. Med. 146, 897-911.

DISCUSSION

MULLEN: The recruitment assay that you described using thyro-globulin was done using the EAT susceptible strain 2 guinea pigs. Have you done these experiments with the EAT resistant (nonresponder) strain 13 guinea pigs?

COHEN: We have triggered strain 2 guinea pig thymus lymphocytes to respond to guinea pig thyroglobulin in vitro by using syngeneic macrophages that have been fed with the thyroglobulin. This triggering by the macrophage-thyroglobulin complex was inhibited by the presence of soluble thyroglobulin (Steinman, Cohen, and Teitelbaum, submitted for publication). Such studies have not yet been done with strain 13 guinea pigs resistant to induction of thyroiditis.

SVEJGAARD: If the difference between strains 2 and 13 is due to differences in the degree of presentation by the macrophages, this could be due to differences in their ability to bind the antigen. Do you have the nonapeptide in labeled form, and do you know if strain 13 macrophages bind more nonapeptide than strain 2 macrophages?

COHEN: We have not yet carried out such studies.

WARNER: I would tend to disagree with some of your conclusions drawn on the basis of the chimera experiments. The failure of strain 2 bone marrow cells to render the F1 susceptible is uncontrolled because you have not shown whether strain 13 bone marrow cells could do so. Secondly, since bone marrow cells can also result in repopula-tion of such chimeras with monocytic/macrophage cells, you cannot conclude that Ir control is expressed at the lymphocyte level as you stated.
Concerning the experiments in vitro with cytotoxic T cell generation and sensitization followed by host cell recruitment, have you performed these with the nonapeptide rather than with soluble BP, since it is clear that this is the only relevant antigen in disease induction.

COHEN: You rightly point out a missing control in this experiment. However, the conclusions remain valid that F1 cells in a resistant strain 2 guinea pig produce disease. This indicates that genetic control is not expressed at the level of the brain or elsewhere beyond the lympho-hematopoietic system.
The lymph-hematopoietic system of course includes macrophages as well as lymphocytes and other cells.
For technical reasons, it has been difficult to measure cytotoxicity against the nonapeptide determinant associated with macrophages.

GERSHON: Why is it that low responder animals do not get disease when they have the same number of initiator and cytotoxic T cells against all brain antigens except the nonapeptide? Is the nonapeptide the only antigenic portion exposed in the brain?

COHEN: For as yet unkown reasons, the induction of EAE in guinea pigs is linked to the specific response to the nonapeptide (NP) determinant of whole myelin basic protein. Our experiments were designed to investigate the conditions in which NP is immunogenic, and not to answer the question of how the response to NP actually produces the pathology of EAE.
Our results suggest that the NP determinant will trigger specific T lymphocytes only when associated with or "carried" by the proper gene products of the self-MHC. Furthermore, these MHC genes seem to function during the interaction of T lymphocytes with macrophages

bearing the whole BP molecule. The basic question of how this response to NP produces EAE remains to be investigated. This question is more pressing in view of the fact that no disease results from T cell responses to other determinants on the same BP molecule.

ARNON: For induction of EAE by the intact basic protein the region of the nonapeptide has to be accessible. If the tryptophane residue in it is modified, the molecule is not encephalitogenic any more. Thus, the exposure of other regions in the basic protein will not induce encephalitis unless the nonapeptide is exposed too.

ROITT: In considering whether soluble proteins can act to inhibit autosensitization in vivo, could you give us some idea of the range of concentrations of thyroglobulin which inhibit sensitization in vitro? For comparison, I should say that we have shown in normal humans that the circulating concentration of thyroglobulin is of the order of 100 ng/ml.

COHEN: We have used concentrations of 50-100 μg/ml in vitro to inhibit primary autosensitization aganist macrophages pulsed with either basic protein or thyroglobulin. This is a highly artificial system and I would hesitate to conclude anything about the quantitative aspects of inhibition by soluble antigen. In vitro systems can only teach us about possibilities; they may not reflect the nature of the mechanisms functioning in vitro. The blocking effect of soluble self antigen in vitro only tells us that such blocking is possible. Whether it really plays a role in vivo is the critical question.

ACTIVE IMMUNOREGULATION OF AUTOIMMUNE ENCEPHALOMYELITIS

ROBERT H. SWANBORG and ANDREW M. WELCH
Department of Immunology and Microbiology, Wayne State University
School of Medicine, Detroit, Michigan   48201 (U.S.A.)

ABSTRACT

Suppressor cells which regulate the development of experimental
autoimmune encephalomyelitis (EAE) can be demonstrated in an adoptive
transfer system in Lewis rats rendered tolerant by pretreatment with
myelin basic protein antigen (BP) in nonencephalitogenic form, and in
rats that have recovered from disease.

INTRODUCTION

When susceptible Lewis rats are immunized with guinea pig BP in
Freund's complete adjuvant (CFA), they develop EAE.  The disease is
characterized by hind limb paralysis 10-12 days post-challenge; by
day 18 most rats recover.  If these animals are pretreated with BP in
a nonencephalitogenic form (i.e., in incomplete adjuvant), they are
rendered tolerant to EAE as evidenced by failure to develop disease
when subsequently challenged.  Thus, EAE provides a unique model system for studies of immunologic self-tolerance.

We have previously reported that BP-pretreated rats develop suppressor cells which can be found in lymph nodes and spleen[1-3], and
now show that these cell populations differ with respect to adherence
properties.  Moreover, we present preliminary data to suggest that
recovery from EAE is associated with a switch from effector cells to
suppressor cells.

MATERIALS AND METHODS

EAE was induced by injection of guinea pig BP + CFA into the hind
foot pads.  Rats were made tolerant by pretreatment with BP in incomplete adjuvant, as previously described[1].  The preparation and transfer of lymph node and spleen cells and removal of adherent cells on
glass wool columns have also been described[1-3].  Where indicated, cell
recipients were challenged with BP + CFA 24 hours after cell transfer.
Clinical EAE was graded on a scale of 0 (no disease) to 3 (complete
paralysis), and the disease index represents the mean group score
(6-8 rats/group).

390

In studies of tolerant (i.e., pretreated) rats, we observed a
marked difference between the lymph node and splenic suppressor cells
(Figure 1). The suppressor lymph node cells are true antigen-specific
T lymphocytes, which do not adhere to glass wool, are susceptible to
antithymocyte serum + complement, bear receptors for *Helix pomatia*
lectin (which is a rat T cell marker[4]), and lack surface immunoglobulin[2]. As few as $3 \times 10^7$ suppressor lymph node cells protect recipients[3]. In contrast, the suppressor spleen cells adhere to glass wool;
thus, suppression can be abrogated by passing the spleen cells through
glass wool columns (Figure 1). On the basis of *in vitro* testing, we
have determined that the splenic suppressor cell is probably a macrophage, and that it nonspecifically inhibits T cell responses to mitogens[5].

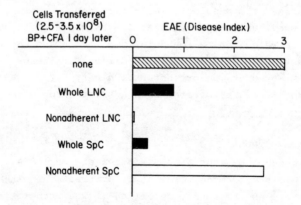

Fig. 1. Adoptive transfer of unresponsiveness to EAE with lymph node
cells (LNC) and spleen cells (SpC) from tolerant syngeneic donor rats.
Adherent cells were removed by passage through glass wool; 6-8 recipi-
ents/group.

Since Lewis rats spontaneously recover from EAE approximately 18 days after challenge, we were also interested in the possibility that suppressor cells might account for recovery. Accordingly, we challenged groups of rats with BP + complete adjuvant, and at various times thereafter we transferred viable lymph node cells to normal syngeneic recipients which were challenged one day later. As shown in Table 1, viable but not heat-killed lymph node cells transferred 21 or 34 days after donor challenge (when donors had recovered) protected the recipients. Cells transferred before, or at time of onset, were ineffective with respect to suppressor function, although they were able to transfer EAE (i.e., they have effector function), since recipients developed EAE without challenge. Thus, a balance apparently exists between effector and suppressor LNC during the course of disease.

TABLE 1

ALTERNATE EFFECTOR AND SUPPRESSOR ACTIVITY DURING THE COURSE OF EAE

| LNC transfer[a] | Challenge[b] | EAE disease index |
|---|---|---|
| Day 9 | no | 2.2 |
|  | yes | 2.5 |
| Day 12 | yes | 3.0 |
| Day 21 | no | 0.5 |
|  | yes | 1.4 |
| Day 34 | yes | 0.8 |
| Day 21 (heated killed) | yes | 3.0 |
| No LNC | yes | 3.0 |

[a]$5 \times 10^8$ LNC
[b]BP + CFA 1 day after LNC transfer.

DISCUSSION

Injection of Lewis rats with BP in incomplete adjuvant leads to the production of two types of suppressor cells. Adherent cells which nonspecifically inhibit T cell function are present in the spleen[5]. These splenic suppressor cells, which are probably macrophages, may provide an additional fail-safe mechanism to prevent undesirable autoimmune responses, or they may merely represent an epiphenomenon. Others have also demonstrated that rat adherent spleen cells suppress immune responses nonspecifically[6].

On the other hand, the lymph node suppressor cells are antigen-specific T lymphocytes[1,2]. Since these inhibit autoimmune encephalomyelitis, we conclude that immunologic self-tolerance may be regulated by suppressor T cells elicited in response to tolerogenic forms of autoantigen.

Of particular interest is the observation that recovery from EAE is associated with the appearance of suppressor cells (Table 1). Similar results have recently been reported by Adda et al.[7] These findings may have clinical relevance with respect to the remissions and exacerbations seen in patients with multiple sclerosis, which conceivably could reflect changes in the balance of regulatory vs. effector lymphocytes.

In conclusion, our findings support the hypothesis that suppressor T cells regulate self-tolerance. Accordingly, autoimmunity could arise as a consequence of a shift in suppressor-effector cell equilibrium. Investigations concerning factors which may cause such shifts might be fruitful with respect to elucidating how autoimmune diseases are initiated.

ACKNOWLEDGMENTS

This investigation was supported by NIH grant NS-06985-11 and National Multiple Sclerosis Society grant 1073-A-4.

REFERENCES

1.  Swierkosz, J.E. and Swanborg, R.H. (1975) J. Immunol., 115, 631-633.

2.  Welch, A.M. and Swanborg, R.H. (1976) Eur. J. Immunol., 6, 910-912.

3.  Swierkosz, J.E. and Swanborg, R.H. (1977) J. Immunol., 119, 1501-1506.

4.  Swanborg, R.H., Hellström, U., Perlmann, H., Hammarström, S. and Perlmann, P. (1977) Scand. J. Immunol, 6, 235-239.

5.  Welch, A.M., Swierkosz, J.E. and Swanborg, R.H. (1978) J. Immunol., in press.

6.  Weiss, A. and Fitch, F.W. (1977) J. Immunol. 119, 510-512.

7.  Adda, D.H., Beraud, E. and Depieds, R. (1977) Eur. J. Immunol., 7, 620-623.

AUTOIMMUNE RESPONSE TO F ANTIGEN

R.F. ANDERS, P.C. COOPER, M.F. COFFEY AND I.R. MACKAY
Clinical Research Unit, The Walter and Eliza Hall Institute of
Medical Research, Post Office, Royal Melbourne Hospital, Victoria
3050, Australia

GENETIC REGULATION OF AUTOANTIBODY RESPONSE TO F ANTIGEN

In 1968 Fravi and Lindenmann[1] showed that certain strains of mice
immunized with liver extracts from some other strains of mice produced antibodies to a 'liver-specific' protein which they designated
F antigen. Among mouse strains there are two immunogenic types of
F antigen, Types 1 and 2, and a liver extract induces antibody to F
antigen only in strains of the opposite F antigen type. However the
antibody raised to F antigen reacts with autologous liver extract
and thus behaves as an autoantibody. Notably not all strains of
mice can be successfully immunized, so that responder and non-
responder strains of each type can be recognized[1,2].

TABLE 1

RESPONSE TO F ANTIGEN IN RECOMBINANT AND CONGENIC MICE

| Strain | H-2 type | Composition of H-2 region | | | | | | Immunodiffusion positivity | no. positive no. injected* |
|--------|------|---|-----|-----|-----|---|---|-------------------|------------------|
|        |      | K | I-A | I-B | I-C | S | D |                   |                  |
| CBA       | k  | k | k | k | k | k | k | ++ | 4/4  |
| B10.BR    | k  | k | k | k | k | k | k | +  | 3/7  |
| A/J       | a  | k | k | k | d | d | d | ++ | 3/3  |
| B10.A     | a  | k | k | k | d | d | d | +  | 5/10 |
| B10.A(2R) | h2 | k | k | k | d | d | b | +  | 1/5  |
| B10.A(4R) | h4 | k | k | b | b | b | b | ++ | 4/5  |
| B10.A(5R) | i5 | b | b | b | d | d | d | 0  | 0/8  |
| C57Bl/6J  | b  | b | b | b | b | b | b | 0  | 0/7  |
| BALB/c    | d  | d | d | d | d | d | d | 0  | 0/6  |
| SJL/J     | s  | s | s | s | s | s | s | 0  | 0/5  |
| A.SW      | s  | s | s | s | s | s | s | 0  | 0/5  |
| A.TH      | t2 | s | s | s | s | s | d | 0  | 0/15 |
| A.TL      | tl | s | k | k | k | k | d | 0  | 0/5  |

* Source of F antigen was CBA mice except that CBA mice were injected
  with A/J liver (data of Long et al.[5]).

There appear to be at least two genes involved in the regulation
of the autoantibody response to F antigen and there is recessive
rather than dominant inheritance of responsiveness[3]. Experiments
with congenic or H-2 recombinant mice have established that one of

the genes regulating the response to F antigen is linked to the H-2 locus[4,5], as indicated in Table 1. Only mice carrying the H-2K[k] allele were capable of responding to F antigen but, with the recombinant mice available, it was not possible to distinguish between the K subregion or the K end of the I-A subregion for the location of the regulatory gene[4]. The involvement of a second gene, not linked to the H-2 complex, in the regulation of the response to F antigen is indicated by the range of antibody responses in backcross mice[3] or certain recombinant strains carrying the H-2K[k] allele[5].

ELECTROPHORETIC TYPING OF F ANTIGEN

During attempts to isolate and characterize murine F antigen, we observed that preparations had one of two markedly different isoelectric points, 7.3 and 6.8, and these corresponded to the two immunogenic types, Types 1 and 2 respectively. Thus, when purified F antigen or unfractionated liver extracts are subjected to immunoelectrophoresis, F antigen of Type 1 migrates more cathodally than does F antigen of Type 2. Liver extracts prepared from a large number of mouse strains available from The Walter and Eliza Hall Institute animal facility have been analysed for F antigen type by immunoelectrophoretic mobility; only two electrophoretic mobilities were seen, so that all strains examined could thereby be classified as having F antigen of either Type 1 or Type 2. Type 1 was present in most strains, namely C3H/He, CBA, DBA/2, DBA/1, NZO, NZC, NZW, ASW, AKR, AQR and 129, whilst Type 2 was present in A/J, BALB/c, C57BL/6, SJL and NZB.

The same procedure was used to examine the electrophoretic mobility of F antigen in extracts of human liver, 18 specimens being examined. There were no differences in electrophoretic mobility of human F antigen, and the mobility in all cases corresponded closely to Type 1 in the mouse.

DELAYED-TYPE HYPERSENSITIVITY RESPONSE TO F ANTIGEN

It has been postulated that induction of autoantibodies to F antigen involves collaboration between helper T cells recognizing the alloantigenic region of the molecule, acting as a "carrier" determinant, and B cells with specificity for "haptenic" determinants common to both types of F antigen[2]. Much of the interest in F antigen as a model antigen for studies on autoimmunity resides in this possibility that an autoimmune response depends upon T cell recognition of what could be a readily definable determinant in an otherwise complex antigen. To approach the problem of T cell recognition of F antigen more directly, we have investigated whether purified F antigen could elicit a T cell response manifested by an in vivo delayed-type hypersensitivity (DTH) reaction. The procedure used was the radiometric "ear-assay" described by Vadas and Miller[6]. For this, F antigen was injected intradermally into the ear of mice primed 5 days earlier with liver homogenate of appropriate F antigen type emulsified in Freund's complete adjuvant. The DTH response to F antigen was then assessed by measuring the uptake into the ear of [125]I-UdR subsequently injected intraperitoneally. Using this assay, strong DTH reactions have been obtained to purified Type 2 F antigen in responder mice carrying Type 1 F antigen, but no responses were obtained to Type 1 F antigen in any strain of mouse. When primed CBA (Type 1) mice were tested for DTH with A/J (Type 2) F antigen, a significant DTH response was obtained in those mice sensitized with liver homogenates from A/J, BALB/c or C57BL/6 (strains with Type 2 F antigen), but not in mice sensitized with liver homogenates from CBA, C3H/He or DBA/2 (strains with Type 1 F antigen). Thus the DTH response distinguishes the same two immunogenic forms of F antigen as were originally defined by the humoral immune response.

TABLE 2

DTH RESPONSE OF CBA MICE TO ALLOGENEIC LIVER EXTRACTS

| Liver extract used to sensitize* | Challenged With | L/R [125]I-dUdR Uptake[+] | | |
|---|---|---|---|---|
| A/J | A/J F Ag | 2.98 | ± | 0.25 |
| BALB/c | " | 3.12 | ± | 0.45 |
| C57BL/6 | " | 2.73 | ± | 0.07 |
| CBA | " | 1.13 | ± | 0.12 |
| C3H/He | " | 1.19 | ± | 0.07 |
| DBA/2 | " | 1.33 | ± | 0.09 |
| – | " | 1.12 | ± | 0.07 |

* Mice sensitized with dose corresponding to 25 µl crude liver extract[5].
+ Ratio of counts in the left ear into which the challenge dose (10 µg) of F antigen was injected to that in the right uninjected ear. Values are arithmetic means ± 1 standard error. 5 mice per group.

RADIOIMMUNOASSAY FOR F ANTIGEN

A solid phase radioimmunoassay for F antigen has been developed using polyvinylchloride microtiter plates (Cooke Laboratory Products, Alexandria, Virginia) coated with immunoglobulin isolated by affinity chromatography on a protein A-Sepharose column from CBA anti A/J F antigen antiserum. This assay detected less than 0.5 ng F antigen, an increase in sensitivity of over 100-fold compared to the radio-immunoassay described by Silver and Lane[7].

The inhibition curves in Figure 1 show that the binding of radio-iodinated F antigen to the alloantibodies was inhibited to the same degree by both allotypes of F, confirming that the antibody response induced by alloimmunization is totally directed to autologous F antigen[4]. Using this radioimmunoassay, the question of organ specificity of F antigen was re-examined. The findings were that F antigen is not in fact liver-specific: Whereas the highest concentration of F antigen was found in liver, high concentrations were also found in kidney which contained over 10% of the concentration in the liver, and lower concentrations were also found in spleen, heart and lung extracts, as well as in normal mouse serum. No differences were detected in the concentration of F antigen in the organs of a re-sponder (CBA) and non-responder strain (DBA/2) of mouse. Consistent with the high concentration of F antigen in the kidney found by radioimmunoassay was the detection of F antigen in kidney extracts by immunodiffusion (Figure 2), as reported by Fravi and Lindenmann when first describing F antigen[1].

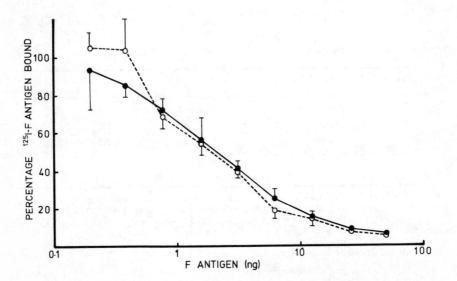

Fig. 1. Standard curves for the F antigen radioimmunoassay showing equivalent inhibition of binding of $^{125}$I-F antigen (A/J) by Type 1 (CBA, O----O) and Type 2 (A/J, ●——●) F antigen.

Mouse alloantisera to F antigen have previously been reported to detect a related antigen in liver extracts of other mammalian species[1]. As shown in Figure 2 there is only partial cross-reaction between mouse and human F antigen.

Fig. 2. Immunodiffusion showing F antigen detectable in kidney homogenates and partial cross-reactivity between mouse and human F antigen. 1 and 4, mouse liver homogenate; 2, mouse kidney homogenate; 3, human kidney homogenate; 5, human liver homogenate; 6, purified mouse (CBA) F antigen.

## COMMENT

Organ-specific antigens are of great interest as potential target antigens for autoimmune disease. Our finding that F antigen is quite widely distributed in the body, and the lack of any pathology suggesting an autoimmune disease in animals with high levels of autoantibodies, argue against a role for F antigen in causing autoimmune liver disease. Even so, F antigen remains of interest as a model for investigating tolerance to those self antigens for which tolerance is maintained solely at the level of the T cell. Our current work in the dissection of the response to F antigen, using the radioimmunoassay and the DTH radiometric ear assay to define the regions of the molecule recognized by B and T cells respectively, will allow greater insight into the mechanisms of tolerance and non-responsiveness to this prototype autoantigen.

## ACKNOWLEDGEMENT

This study was aided by a grant from the National Health and Medical Research Council of Australia. The assistance of Dr. J.F.A.P. Miller and Miss J. Gamble with the DTH experiments is gratefully acknowledged.

REFERENCES

1. Fravi, G and Lindenmann J. (1968) Nature, 218, 141-143.

2. Iverson, G.M. and Lindenmann, J. (1972) Eur. J. Immunol. 2, 195-197.

3. Silver, D.M. and Lane, D.P. (1975) J. Exp. Med. 142, 1455-1461.

4. Silver, D.M. and Lane, D.P. (1977) Immunogenetics, 4, 295-299.

5. Long, G.W., Bernard, C.C.A., Mackay, I.R., Whittingham, S. and Bhathal, P.S. (1978) Tissue Antigens, 11, 45-49.

6. Vadas, M.A., Miller, J.F.A.P., Gamble, J. and Whitelaw, A. (1975) Int. Arch. Allergy, 49, 670-692.

7. Lane, D.P. and Silver, D.M. (1976) Eur. J. Immunol. 6, 480-485.

DISCUSSION

ROITT:    The inability of complete Freund's adjuvant to help autologous F protein to induce autoantibodies (in contrast to the situation
with autologous thyroglobulin) may be attributable to the presence of
T cells recognizing thyroglobulin but not F protein.  Freund's would
appear to be able to act by denaturing the antigen and creating new
determinants, so sidestepping the need for autoreactive T cells.
One may argue that F protein resisted such change within the Freund
emulsion.

LENNON:  With regard to importance of the nature of the immunogen
in determining the induction of autoimmunity to a given autoantigen,
an excellent illustration of this point comes from immunochemical
studies of hyperacute EAE (Westall et al., Nature, 1977).  When Lewis
rats were injected, using identical adjuvants (CFA + B. pertussis),
with a 14 amino acid peptide containing the major encephalitogenic
determinant for rats, a striking difference in clinical and histologic outcome ensued depending on whether the peptide immunogen
(native or synthetic) corresponded to the rat (i.e., the autoantigen)
or the guinea pig basic protein's sequence.  The sole difference in
the guinea pig immunogen is a single serine for threonine substitution
in the peptide, i.e., the absence of a single methyl group in the
immunogen changes the autoimmune response in the rat from one of
ordinary EAE (onset day 12, transient disease) to hyperacute EAE
(onset day 8-10, rapidly fatal).  The "serine" (guinea pig) peptide
was potent at µg doses, whereas greater than 20-fold higher doses
of the "threonine" peptide induced only ordinary EAE.

KONG:  Is there any evidence that the second controlling locus is
inside or outside of the MHC?  Could the second locus be at the other
end of the MHC such as at the D-end or TL region?

MACKAY:  My interpretation of the available evidence is that the
second locus is non-MHC associated.

CELLULAR AND GENETIC CONTROL OF SUSCEPTIBILITY TO EXPERIMENTAL AUTO-
IMMUNE THYROIDITIS IN GUINEA PIGS

HELEN BRALEY-MULLEN, GORDON C. SHARP, MICHAEL KYRIAKOS AND PETER JEPSEN

Departments of Medicine and Microbiology, University of Missouri,
Columbia, Mo.  65212 and Department of Surgical Pathology, Washington
University School of Medicine, St. Louis, Missouri  63110

ABSTRACT
   These experiments were designed to determine the basis for the
difference in susceptibility to experimental autoimmune thyroiditis
(EAT) between strain 2 and strain 13 guinea pigs.  Strain 13 guinea
pigs immunized with a low dose (100 ug) of guinea pig thyroglobulin
(GPTG) do not develop EAT whereas similarly immunized strain 2 guinea
pigs do develop EAT.  Despite the genetic difference in susceptibility
to EAT, both strains were found to develop similar cell-mediated
responses to GPTG as determined by similar in vitro proliferative
responses to GPTG.  It was also found that suppressor cells could be
induced in both strains  which suppressed the development of EAT after
transfer to normal recipients.  There was however, no indication that
the 'low responder' strain 13 guinea pigs had a greater propensity
to develop suppressor cells that the 'high responder' strain 2 guinea
pigs.  In an attempt to alter the course of EAT by neonatal thymectomy
it was observed that a cell necessary for the induction of EAT was
eliminated by thymectomy in both strains.

INTRODUCTION
   Experimentally induced animal models of autoimmune diseases have
provided important tools for studying the mechanisms involved in the
induction, maintenance and loss of tolerance to "self".  Experimental
autoimmune thyroiditis (EAT) is typical of organ specific autoimmune
diseases.  EAT can be readily induced in a variety of experimental
animals by immunization with thyroglobulin  in complete Freund's
adjuvant (CFA).
   The ability of animals to respond to many antigens, including thyroglobulin, is known to be controlled by immune response (Ir) genes which
are closely associated with the major histocompatibility complex (MHC)
of the species.[1-4]  We have previously shown that when guinea pigs
were immunized with a low dose (100 ug) of guinea pig thyroglobulin

(GPTG) in CFA, strain 2 and (2x13)$F_1$ guinea pigs readily developed
EAT whereas similarly immunized strain 13 animals failed to develop
EAT.[3] The difference in susceptibility to EAT of strain 2 and strain
13 guinea pigs was shown to be associated with the MHC in guinea pigs.[4]

In more recent studies we have been investigating the cellular basis
for the difference in susceptibility to EAT of strain 2 and strain 13
guinea pigs. In particular, we have focused on the question of whether
the relative resistance of strain 13 guinea pigs to develop EAT
might be due to a preferential activation of suppressor T cells (Ts)
after immunization with GPTG in strain 13, but not in strain 2,
guinea pigs. It has been shown in other systems that genetic low
responders to some antigens preferentially develop Ts after immunization [5-7] and Ts have been shown to be involved in the regulation
of several autoimmune diseases.[8-12] These studies will be described
in this communication.

MATERIALS AND METHODS

Animals

Inbred strain 2 and strain 13 guinea pigs were obtained from the
University of Missouri breeding colonies. Both male and female
animals 300-700 g, were used for these experiments.

Neonatal Thymectomy

Guinea pigs were thymectomized when less than 24 hours old by
careful isolation and extirpation of the bilobed cervical thymus.
Thymectomized (Tx) and sham Tx animals were immunized with GPTG at
6-10 wk. of age.

Preparation of GPTG

GPTG was prepared from pooled strain 2 and strain 13 thyroids by
differential ultracentrifugation.[13]

Immunization

Guinea pigs were immunized with various amounts of GPTG in CFA
containing 2 mg/ml mycobacterium $H_{37}$ Rv to induce EAT. Each animal
received the GPTG-CFA in a total volume 1 ml given intradermally
in the four footpads. Twenty two days after immunization animals
were skin tested with 10 or 50 ug GPTG and on day 23 animals were
bled and the thyroids removed for histologic examination.

## Suppression of EAT

To suppress the induction of EAT, guinea pigs were pretreated with GPTG in incomplete Freund's adjuvant (IFA)[14]. Each animal received a total of 700 ug GPTG given in 7 injections (100 ug each time) over a two week period. Pretreated animals were then immunized with GPTG-CFA as described above. In some experiments $6 \times 10^8$ lymph node cells from pretreated animals were transferred i.p. to normal syngeneic recipients which were then immunized with GPTG-CFA[14].

## Antibody Determinations

Antibody to GPTG was measured by a radioactive antigen binding assay using [125]I-GPTG as described previously.[14] Results are expressed as the percent [125] I-GPTG bound by a 1:2 dilution of serum relative to a 100% control ([125]I-GPTG precipitated by 10% trichloracetic acid).

## Skin tests

Delayed hypersensitivity to GPTG was assessed 24 hours after an intradermal injection of 10 or 50 ug GPTG. Results are expressed as the diameter in mm. of induration and erythema.

## Mitogen Responses

Mitogen induced proliferative responses of peripheral lymph node cells were determined by culturing $2 \times 10^5$ lymph node cells in a flat bottom microtiter plate for 72 hours in the presence of 1 ug Concanavalin-A (Con A), 0.5 ug phytohemagglutinin (PHA) or 100 ug lipopolysaccharide (LPS). At 24 hours before harvest, cells were pulsed with 0.5 uCi [3]H thymidine ([3]HTdR) (6.7 Ci/mmol) and cells were harvested on a multiple automated sample harvester. Results are expressed as mean cpm of incorporated [3]HTdR of triplicate cultures.

## Antigen-Induced Proliferative Responses

Peritoneal exudate T lymphocytes were obtained by passing oil-induced peritoneal exudate cells from immunized guinea pigs over a nylon wool column.[15] The nylon nonadherent cells were then cultured with mitogens or antigen (20 or 100 ug GPTG) as described above.

## Histologic Criteria for Severity of EAT

Serial sections of thyroid tissue were stained with hematoxylin and eosin. A significant amount of chronic leucocyte infiltrate (lymphoctyes and histiocytes) served to establish the presence of

chronic autoimmune thyroiditis. The severity of thyroiditis was
graded on a scale of 1 to 4 in accordance with the following criteria:
1+ thyroiditis consisted of an infiltrate of at least 125 cells in
one or several foci or 5-10 smaller foci of the size of an average
follicle; Ten to 20 foci of infiltration each occupying an area the
size of several follicles constituted a 2+ reading; 2+ thyroiditis
includes infiltration of up to one-fourth of the gland, 3+ thyroiditis
up to one-half of the gland and 4+ thyroiditis over one-half of the
gland.

RESULTS

Differential Susceptibility to EAT of Strain 2 and Strain 13 Guinea Pigs

The differential susceptibility to EAT of strain 2 and strain 13
guinea pigs is demonstrated by the data shown in Table 1. The relative
resistance to EAT of strain 13 as compared to strain 2 guinea pigs
is evident even after immunization with a high dose (2 mg) of GPTG
but is most striking after immunization with a low dose (100 ug) of
GPTG where strain 13 guinea pigs fail to develop any evidence of EAT.
We have shown previously that strain 2 animals produce higher amounts
of antithyroglobulin antibody than the strain 13 animals while delayed
hypersensitivity reactions to GPTG are comparable in the two strains.[3]

TABLE 1
DIFFERENTIAL SUSCEPTIBILITY OF STRAIN 2 AND STRAIN 13 GUINEA PIGS TO
INDUCTION OF EAT

|  | Immunizing Dose | Severity of Lesions [a] | | | | |
|---|---|---|---|---|---|---|
|  |  | 0+ | 1+ | 2+ | 3+ | 4+ |
| Strain 2 | 2mg | 0 | 0 | 4 | 2 | 3 |
| Strain 13 | 2mg | 6 | 6 | 1 | 2 | 1 |
| Strain 2 | 100ug | 2 | 4 | 9 | 4 | 1 |
| Strain 13 | 100ug | 20 | 0 | 0 | 0 | 0 |

[a]Number of animals in each group with various degrees of severity of
EAT 23 days after immunization with 2 mg or 100 ug GPTG-CFA.

In Vitro Proliferative Responses to GPTG; Strain 2 and Strain 13
Guinea Pigs

In the case of most MHC-linked Ir gene-controlled responses it
has been found that strains which are genetic low responders to a
particular antigen are unable to develop significant cell-mediated
immune responses to that antigen[1,16]. The comparable delayed hypersensitivity responses to GPTG of strain 2 and strain 13 guinea pigs[3]

suggested that this may not be the case in our system. To investigate this point more thoroughly we have compared the ability of peritoneal exudate T cells from strain 2 and strain 13 guinea pigs immunized with 100 ug GPTG-CFA to proliferate in vitro in the presence of GPTG. As shown in Table 2 the T cell proliferative responses to GPTG were comparable in both strains even though the strain 2 animals all developed severe (3-4+) EAT and the strain 13 animals did not. Cells from all animals also responded well to the nonspecific T cell mitogen Con A and they did not respond to a non cross-reacting antigen keyhole limpet hemocyanin (KLH). These data would suggest that the defect in the 'low responder' strain 13 guinea pig is not at the level of development of sensitized effector T cells and that sensitized T cells per se are not sufficient for the development of EAT.

TABLE 2

COMPARISON OF IN VITRO PROLIFERATIVE RESPONSES TO GPTG IN STRAIN 2 AND STRAIN 13 GUINEA PIGS

|  |  | CPM $^3$HTdR Incorporation [a] after Incubation with: | | | |
|---|---|---|---|---|---|
|  | EAT | None | Con A | KLH | GPTG |
| Strain 2[b] | 4+ | 1032 | 36429 | 1027 | 18204 |
| Strain 2 | 3+ | 2178 | 164084 | 2559 | 15612 |
| Strain 2 | 4+ | 2059 | 137195 | 1478 | 23095 |
| Strain 13[b] | 0 | 898 | 107033 | 899 | 27996 |
| Strain 13 | 0 | 1672 | 237405 | ND | 24557 |
| Strain 13 | 1+ | 513 | 128137 | 1824 | 12092 |

[a]Mean CPM of triplicate cultures.
[b]Guinea pigs immunized with 100 ug GPTG + CFA 3-4 weeks previously.

Suppression of EAT by Antigen Pretreatment

Previous studies have shown that clinical signs of experimental allergic encephalomyelitis (EAE) can be prevented in rats if they are pretreated with encephalitogen in IFA prior to immunization with encephalitogen in CFA.[9] The suppression of EAE by antigen pretreatment was shown to be due to the activation of Ts by the encephalitogen - IFA injections.[9,10] We have used a similar antigen pretreatment regimen to determine 1) if EAT is also regulated by suppressor cells and 2) whether such suppressor cells might be preferentially activated in the EAT "low responder' strain 13 guinea pigs. As shown in Table 3, pretreatment with GPTG in IFA (100ug x 7 injections) did result in marked suppression of EAT when animals were subsequently challenged with GPTG in CFA. There was however, no evidence that the degree of suppression differed in strain 2 versus strain 13 guinea pigs. Pretreatment with GPTG in IFA also resulted in a significant reduction of anti-GPTG antibody in most, but not all,[14] experiments (Table 3) but there was no suppression of delayed hypersensitivity responses in pretreated animals[14] (data not shown).

TABLE 3

SUPPRESSION OF EAT IN STRAIN 2 AND STRAIN 13 GUINEA PIGS BY PRE-TREATMENT WITH GPTG IN IFA

| Pretreatment [b] | Severity of Lesions [a] | | | | | Mean [c] % Binding |
|---|---|---|---|---|---|---|
| | 0 | 1+ | 2+ | 3+ | 4+ | |
| A  None | 2 | 4 | 3 | 0 | 2 | 57.4 ± 7.5 |
| B  OVA/IFA | 0 | 1 | 0 | 3 | 1 | 61.7 ± 11.4 |
| C  GPTG/IFA | 16 | 8 | 1 | 1 | 0 | 18.1 ± 3.6 |
| D  None | 0 | 1 | 2 | 3 | 0 | 49.0 ± 6.7 |
| E  GPTG/IFA | 4 | 2 | 0 | 0 | 0 | 10.0 ± 3.1 |

[a] Number of animals with various degrees of severity of EAT 23 days after immunization with 2 mg GPTG-CFA.
[b] Animals in Group B were pretreated with ovalbumin (OVA) in IFA (100 ug x 7) and those in Groups C and E with GPTG in IFA (100 ug x 7). All animals were challenged with GPTG-CFA 2 weeks after the last pretreatment injection. Groups A-C, strain 13; Groups D and E, strain 2.
[c] Mean ± SEM binding of serum by $^{125}$I-GPTG.

We next determined whether cells and/or serum from GPTG-IFA pretreated animals could actively suppress EAT after transfer to normal syngeneic recipients. As shown in Table 4, 8/12 control strain 2 animals (Groups A and B) developed moderate to severe (2-4+) EAT and only 1 of 12 control animals failed to develop EAT. By contrast only 3/13 animals which received lymph node cells alone or in combination with serum from GPTG-IFA pretreated donors developed 2-4 + EAT (Groups D and E) and 10/13 had minimal (1+) or no disease. ($p < 0.05$ for Groups A and B vs. Groups D and E). Serum from GPTG/IFA pretreated donors was not suppressive (Group C). In this experiment recipients of lymph node cells also had reduced anti-GPTG titers as compared to controls ($p < 0.005$) but this was not observed in several other experiments in which a comparable degree of suppression of EAT was obtained (e.g. Table 5). Similar results were obtained when lymph node cells from GPTG-IFA pretreated strain 13 guinea pigs were transferred to

strain 13 recipients (Table 5). Lymph node cells from OVA-IFA pretreated animals were not suppressive. Whether the suppressor cells
induced by pretreatment with GPTG in IFA are T cells has not yet
been determined.

TABLE 4

SUPPRESSION OF EAT BY CELLS FROM GPTG/IFA PRETREATED STRAIN 2 GUINEA
PIGS

| | Cells Transferred[b] | Antiserum[c] | Mean % Binding[d] | Severity of EAT[a] | | | | |
|---|---|---|---|---|---|---|---|---|
| | | | | 0 | 1+ | 2+ | 3+ | 4+ |
| A | - | - | 67.3 $\pm$ 14.1 | 0 | 2 | 2 | 2 | 0 |
| B | - | + (OVA) | 67.3 $\pm$ 12.4 | 1 | 1 | 1 | 1 | 2 |
| C | - | + (GPTG) | 92.5 $\pm$ 5.1 | 0 | 0 | 0 | 4 | 2 |
| D | + | - | 30.0 $\pm$ 10.5 | 0 | 4 | 1 | 0 | 1 |
| E | + | + (GPTG) | 37.0 $\pm$ 8.6 | 3 | 3 | 1 | 0 | 0 |

[a]Number of animals with various degrees of severity of EAT 23 days
after immunization with 300 ug GPTG-CFA.
[b]Groups D and E received 6 x $10^8$ lymph node cells from strain 2
animals which were pretreated with GPTG-IFA (100 ug x 7).
[c]Serum from strain 2 animals pretreated as above (Groups C and E)
or from animals pretreated with OVA-IFA (Group B). 2 ml serum was injected i.p. on days 0,2,4,7 and 10.
[d]Mean $\pm$ SEM binding of serum by $^{125}$I-GPTG.

TABLE 5

SUPPRESSION OF EAT BY CELLS FROM GPTG/IFA PRETREATED STRAIN 13
GUINEA PIGS

| Cells Transferred[b] | Mean % Binding[c] | Severity of EAT[a] | | | | |
|---|---|---|---|---|---|---|
| | | 0 | 1+ | 2+ | 3+ | 4+ |
| OVA/IFA | 73.0 $\pm$ 5.6 | 0 | 3 | 4 | 0 | 0 |
| GPTG/IFA | 58.2 $\pm$ 6.2 | 4 | 2 | 1 | 0 | 0 |

[a]Number of animals with various degrees of severity of EAT 23 days
after immunization with 1 mg; GPTG/CFA.
[b]6 x $10^8$ lymph node cells from donors pretreated with OVA-IFA or
GPTG-IFA (100 ug x 7). Cells transferred i.p. to recipients
irradiated with 100 R 1 day earlier.
[c]Mean $\pm$ SEM binding of serum by $^{125}$I-GPTG.

Effect of Thymectomy and Irradiation on Induction of EAT

Previous studies have shown that the incidence and severity of
spontaneous autoimmune thyroiditis (SAT) in several species is augmented after neonatal thymectomy alone or in combination with low
doses of irradiation.[11,12,17,18]. The augmentation of SAT by thy-

mectomy and/or irradiation is thought to be due to the elimination of Ts which regulate SAT.[17,18]. The studies described below were done in order to determine if Ts might similarly regulate the production of EAT in the EAT resistant strain 13 guinea pigs. Contrary to our initial expectations, neonatal thymectomy (Tx) markedly reduced the incidence and severity of EAT (Table 6). The inhibition of EAT by Tx was accompanied by an approximately 50% reduction in anti-GPTG binding while delayed hypersensitivity skin tests were reduced only minimally. Low doses of irradiation (75-100R) had no effect on the incidence of EAT while higher doses of irradiation (150-350R) reduced the incidence of EAT (data not shown). Preliminary results suggest that Tx also results in a marked reduction of EAT and anti-GPTG antibody in strain 2 guinea pigs (Table 7) particularly when they are immunized with a low dose of GPTG. Delayed hypersensitivity responses were also significantly reduced by Tx in Strain 2 guinea pigs. To begin to assess the degree of T cell depletion in Tx animals we compared the ability of peripheral lymph node cells from sham Tx and Tx strain 13 guinea pigs to respond to T and B cell mitogens. As shown in Figure 1, there was no difference in the ability of lymph node cells from Tx and sham-operated animals to respond to the T cell mitogens, PHA and Con A or to the B cell mitogen LPS. All of the sham-operated animals developed EAT while none of the Tx animals developed EAT. Preliminary results also suggest that peritoneal exudate T cells from immunized Tx animals proliferate in vitro in the presence of GPTG nearly as well as T cells from control animals (not shown). Thus, it appears that a cell which is necessary for the induction of EAT and optimal antithyroglobulin antibody production is eliminated by neonatal Tx. The inhibition of these

TABLE 6

EFFECT OF THYMECTOMY ON INDUCTION OF EAT IN STRAIN 13 GUINEA PIGS

| | Severity of Lesions[a] | | | | | Mean % Binding[b] | Mean 24 hr Skin Reaction |
|---|---|---|---|---|---|---|---|
| | 0 | 1+ | 2+ | 3+ | 4+ | | |
| Sham-Operated | 1 | 3 | 7 | 8 | 1 | $69.0 \pm 4.1$ | $9.6 \pm 1.1$ |
| Thymectomized | 22 | 3 | 0 | 0 | 0 | $34.2 \pm 3.9$ | $6.3 \pm 1.2$ |

[a]Number of animals with various degrees of severity of EAT 24 days after immunization with 1 mg. GPTG-CFA.
[b]Mean $\pm$ SEM binding of serum by $^{125}$I-GPTG.
[c]Mean $\mp$ SEM diameter (mm) of skin reaction 24 hour after intradermal injection of 50 ug. GPTG.

parameters occurs even though Tx animals have considerable residual peripheral T cell function as measured by intact responses to T cell mitogens. The nature of the Tx-sensitive cell which is necessary for the induction of EAT is currently under investigation.

TABLE 7

EFFECT OF THYMECTOMY ON INDUCTION OF EAT IN STRAIN 2 GUINEA PIGS

|  | Severity of Lesions[a] | | | | | Mean % Binding[b] | Mean 24 hr. Skin reaction |
|  | 0 | 1+ | 2+ | 3+ | 4+ |  |  |
|---|---|---|---|---|---|---|---|
| A Sham-Operated | 1 | 2 | 2 | 1 | 1 | 32.1 ± 7.8 | 3.1 ± 1.4 |
| B Thymectomized | 6 | 1 | 0 | 0 | 0 | 12.8 ± 0.5 | 0.7 ± 0.7 |
| C Sham-Operated | 0 | 1 | 1 | 0 | 6 | 69.4 ± 8.2 | 9.7 ± 1.6 |
| D Thymectomized | 4 | 1 | 0 | 0 | 2 | 41.4 ± 6.0 | 2.3 ± 1.6 |

[a]Number of animals with various degrees of severity of EAT 24 days after immunization with 100 ug; (Groups A and B) or 1 mg; (Groups C and D) GPTG.
[b]See Footnote b, Table 6.
[c]See Footnote c, Table 6.

SUMMARY

Possible cellular mechanisms which might explain the genetic difference in susceptibility to EAT of strain 2 and strain 13 guinea pigs have been examined in this study. The results indicate that the basis for the genetic difference in susceptibility to EAT is apparently not due to a difference in the ability of low doses of GPTG to sensitize effector T cells in the two strains since they both develop comparable delayed hypersensitivity responses[3,4] and T cells of both strains proliferate comparably in the presence of GPTG in vitro (Table 2). The results to date have not, however, excluded the possibility that sensitization of helper T cells which are needed for optimal anti-GPTG antibody production might differ in the two strains. It has been shown that the EAT resistant strain 13 guinea pigs produce less antithyroglobulin antibody than do strain 2 guinea pigs after immunization with low doses of GPTG [3,4] which could indicate a defect at the level of helper T cells, B cells or macrophages.

We have also examined the possibility that the 'low responder' strain 13 guinea pigs might preferentially develop suppressor cells after immunization with GPTG. EAT could be significantly suppressed

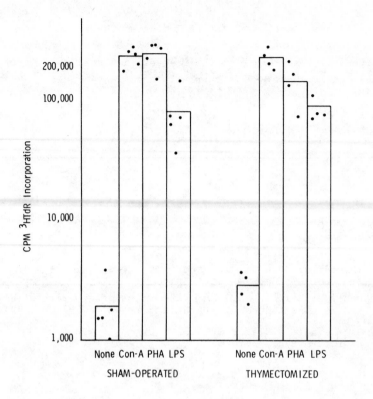

Fig. 1   In Vitro proliferative responses of peripheral lymph node cells of Tx and sham-operated Strain 13 guinea pigs to T and B cell mitogens.   Mean responses of 4 or 5 animals are shown by the bars. The dots represent mean CPM of triplicate cultures for each individual animal.

by pretreatment with GPTG in IFA (Table 3) and the suppressive effect could be transferred with lymphoid cells from pretreated animals (Tables 4 and 5).   These results suggest that suppressor cells do regulate EAT in the guinea pig.   However, the suppressor cells were readily induced in both strain 2 and strain 13 animals, i.e. there was no evidence that the ability to develop suppressor cells was greater in the 'low responder' strain 13 animals.

Finally, in an attempt to eliminate Ts by neonatal thymectomy 11,12,16,17 we found instead that thymectomy (Tx) resulted in a marked <u>inhibition</u> of EAT and anti-GPTG antibody production in both strains (Tables 6 and 7).   The nature of the Tx-sensitive cell which is necessary for the induction of EAT is currently under investigation.

## ACKNOWLEDGMENTS

This study was supported by research grant AM19409 from the U.S. Public Health Service. The authors thank Catherine Dunn, Nancy Hayes and Patra Mierzwa for excellent technical assistance and Debbie Harman for typing the manuscript.

REFERENCES:

1. Benacerraf, B. and McDevitt, H.O. (1972) Science, 175, 273.

2. Vladutiu, A.O. and Rose, N.R. (1971) Science, 174, 1137.

3. Braley-Mullen, H., Sharp, G.C. and Kyriakos, M. (1975) J. Immunol., 114, 371.

4. Braley-Mullen, H., Sharp, G.C. and Kyriakos, M., (1976) Immuno-genetics, 3, 205.

5. Gershon, R.K., Maurer, P.H. and Merryman, C.F. (1973) Proc. Nat. Acad. Sci. U.S.A. 70, 250.

6. Kapp, J.A., Pierce, C.W., Schlossman, S. and Benacerraf, B. (1974) J. Exp. Med., 140, 648.

7. Debré, P., Waltenbaugh, C., Dorf, M.E. and Benacerraf, B. (1976) J. Exp. Med., 144, 277.

8. Krakauer, R.S., Waldmann, T.A. and Strober, W. (1976) J. Exp. Med., 144, 662.

9. Swierkosz, J. and Swanborg, R.H. (1975) J. Immunol., 115, 631.

10. Welch, A.M. and Swanborg, R.H. (1976) Europ. J. Immunol. 6, 910.

11. Penhale, W.J., Irvine, W.J., Inglis, J.R. and Farmer, A. (1976) Clin. Exp. Immunol., 15, 6.

12. Wick, G., Kite, J.H. and Witebsky, E. (1970) J. Immunol., 104, 54.

13. Edelhoch, H. (1960) J. Biol. Chem., 235, 1326.

14. Braley-Mullen, H., Sharp, G.C., Kyriakos, M., Hayes, N., Dunn, C., Jepsen, P. and Sanders, R.D. Cell. Immunol. (in press).

15. Julius, M.H., Simpson, E. and Herzenberg, L.A. (1973) Europ. J. Immunol., 3, 645.

16. Schwartz, R.H. and Paul, W.E. (1976) J. Exp. Med., 143, 529.

17. Kojima, A., Tanaka-Kojima, B.S., Sakakura, T. and Nishizuka, Y. (1976) Lab. Invest. 550, 1976.

18. Whitmore, D.B. and Irvine, W.J. (1977) Clin. Exp. Immunol., 29, 474.

DISCUSSION

MACKAY: In speaking of thyroglobulin as "antigen", one must regard it more as an "antigen mosaic" with perhaps one determinant relevant to EAT. This could explain why there can be DTH reactions to non-disease relevant components of thyroglobulin.

Secondly, despite the early maturation of the guinea pig thymus, neonatal thymectomy have been expected to decrease, if anything, the severity of disease.

WARNER: Could you tell me whether the experiment on thymectomy in strain 13 was performed with 100 µg or 2 mg of GPTg for immunization. Since the aim of the experiment was to assess whether strain 13 is biased toward suppression, you would need to challenge with doses of GPTg that selectively induced EAT in strain 2, i.e., with 100 µg.

BRALEY-MULLEN: The thymectomy experiments with strain 13 were done using animals immunized with 1 mg GPTg. We wished to begin these experiments using a dose of antigen that would induce disease in the controls. If we had observed no effect or augmentation of disease in that situation, we then would have gone on to see if augmentation would occur in thymectomized animals immunized with 100 µg GPTg (where controls would not get disease). Since we observed such marked suppression of EAT after thymectomy, I think it is unlikely that disease would occur in low dose immunized thymectomized strain 13's.

VOLPE: Have you done experiments in which you have transferred lymphocytes from the thymectomized animals to either virginal recipients or those which have been already immunized, so as to possibly initiate or augment the degree of thyroiditis.

In those animals who had no morphological evidence of thyroiditis, but which manifested thyroid antibodies, how high did the titers go? In man, the presence of anti-thyroglobulin virtually always has a morphological correlate in terms of some lymphocytic infiltration within the thyroid gland. While I realize that thyroid antibodies might not have pathological significance (particularly in experimental thyroiditis), I wonder if you have performed immunofluorescent studies to determine whether the antibody ever localized in the thyroid gland?

BRALEY-MULLEN: In response to your first question, no, we have not done these experiments as yet. Secondly, we have not performed immuno-fluorescent studies on the thyroid.

ROSE: Are you justified in concluding that suppression of thyroiditis in guinea pigs immunized with GPTg plus IFA is due to T cells? Might it be due to antibody secreted by the lymphocytes transferred to a histocompatible host?

BRALEY-MULLEN: Our conclusion that suppression is due to suppressor cells is based on the finding that we can transfer suppression with cells but not with serum from pretreated animals. (We do not imply that these suppressor cells are necessarily T cells; this has yet to be investigated.) Identifying the cell type which is responsible for suppression should offer clues as to whether suppression could be due to antibody produced by the transferred cells. However in this regard, we have compared antibodies produced by pretreated versus control animals and we find no evidence that pretreated animals produce a different class of antibody that might be responsible for suppression (i.e., all detectable antibody in both groups is of the $IgG_2$ class).

KONG:   In regard to the comment of Dr. Mackay's that thyroglobulin is a very large molecule and a mosaic antigen, as a new person to autoimmunology, I had thought that the self-determinants are limited in number.   Perhaps Drs. Roitt and Rose could comment on this.

ROITT:   Human thyroglobulin is a large molecule of molecular weight 650,000 daltons.   However, patients with autoimmune thyroiditis only recognize between two and four epitopes.   The number of different epitopes could be even less if a degree of symmetry exists between them.

ROSE:   It is true that thyroglobulin is a large molecule and carries quite a large number of antigenic determinants that can be recognized by foreign species.   However, when it is used for immunization of members of the same species, thyroglobulin elicits a much narrower spectrum of antibodies; for example, only about five determinants are evident when rabbits are immunized with rabbit thyroglobulin.   I do agree, nevertheless, that it would be most desirable to break thyroglobulin into single-determinant fragments.   We have been attempting to do it for many years, but we have not yet had much success.

IN VITRO INDUCTION OF IMMUNOLOGICAL T LYMPHOCYTE MEMORY: STUDIES ON
THE REGULATION OF SELF TOLERANCE

HARTMUT WEKERLE, AVRAHAM BEN-NUN[+], URSULA HURTENBACH, and MARLOT
PRESTER
Max-Planck-Institut für Immunbiologie, Stübeweg 51, D-78 Freiburg
(Germany) and [+]The Weizmann Institute of Science, Rehovot, Israel

ABSTRACT

We describe the in vitro generation and isolation of memory T lymphocytes specific for unaltered and altered, cellular and soluble self
antigens. This method is based on the in vitro stimulation of self
reactive T lymphocytes and subsequent isolation of the responding
lymphoblasts in density gradients. In the absence of autoantigen, the
blasts are allowed to revert back to small, monospecific memory T
lymphocytes. Memory cells were generated against unaltered, TNP-modified self cellular antigens, as well as against the encephalitogenic
basic protein of myelin. They all are restricted by MHC determinants,
which are either present on the autoantigenic cells, or are provided
by helper cells presenting soluble autoantigens. The self reactive
T cells are probably clonally preformed in the normal immune system.
It appears that factors in the autologous serum prevent them from
being activated by self antigens in vivo.

INTRODUCTION

The question of the cellular basis of immunological self tolerance
is as old as immunology. It was Paul Ehrlich, who coined the term
"horror autotoxicus" to circumscribe the delicate situation of a protective system, which specifically and destructively reacts against
any foreign structure, but, at the same time, has to absolutely tole-
rate the body's own constituents. In original strict versions of the
Clonal Selection Theory, it has been postulated that self tolerance
is effected by elimination of self reactive lymphocyte clones during
ontogeny[1]. More recent investigations, however, leave little doubt
that, in contrast, such self reactive clones are normally present in
the intact immune system. This holds true for B lymphocytes[2], as well
as for T lymphocytes[3]. Hence, self tolerance must be based on suppressive, regulatory mechanisms which keep the potentially self reactive lymphocytes in a quiescent state.

The reasons, why the self reactive lymphocytes remained unrecognized for such a long time, were ideological as well as technical. Certainly, the prevailing dogma prevented many a researcher from considering autoreactivity when interpreting unexpected findings, particularly in vitro, as "background", "unspecific", or "polyclonal". Indeed, unequivocal determination of autoimmune specificity has been difficult, studying unseparated responder populations, as many conventional controls were impossible to carry out. We now report an approach which allows the isolation and characterization of virtually pure populations of secondary T lymphocytes with exclusive specificity for self antigens. This approach relies on the isolation of specifically in vitro sensitized autoreactive lymphoblasts and on their reversion back to small secondary lymphocytes in the absence of autoantigen. It is not in the scope of this contribution to review the present state of knowledge on autoimmunity. We rather wish to focus on discussing our own work concerning the in vitro induction of autoimmune memory T cells. We shall describe the generation and properties of secondary T cell memory populations responsive against autologous testis cells, against trinitrophenyl-substituted spleen cells, as well as against myelin basic protein.

MATERIALS AND METHODS

Animals. In these studies we used young adult rats from the strains Lewis, BN, AS2, their congenic counterparts L.BN and L.AS2, and F1 hybrids between Lewis x BN. Our animals were supplied from the breeding facilities of the Max-Planck-Institut für Immunbiologie.

Cell culture procedures. All procedures concerning the in vitro EAO model have been described previously[4-7].

Immunization of Lewis rats with myelin basic protein (BP). For immunization we extracted spinal cords from guinea pigs[8]. Recipient Lewis rats were s.c. injected with 200 ug BP suspended in complete Freund's adjuvant. Most of the injected rats can be expected to develop clinical EAE within 12 days. Draining lymph nodes were harvested after 9 days p.i. The lymphocytes were dissociated in glass tissue homogenizers, resuspended in Dulbecco's modified Eagle's medium supplemented with 1% fresh autologous Lewis rat serum and 0.05 mM 2-mercaptoethanol. After 5 days, the lymphoblasts were isolated from the cultures, as described[7], transferred into reversion cultures ($1 \times 10^6$ cells in 10 ml culture medium). After more than 3 days, most of the lymphoblasts have reverted back to small lymphocytes. They were tested for specific proliferation in microtiter plates.

TNP-modification of autologous stimulator cells. We followed the
method described by Shearer[9]. Rat spleen cells were dissociated and
freed of red blood cells by hyposmotic shock treatment. The cells were
resuspended in phosphate buffered saline containing 10 mM trinitroben-
zene phenolic acid at a pH 7.3. After incubation for 10 min. at room
temperature the cells were washed and resuspended in culture medium
(Dulbecco's modified Eagle's medium and 15% heat inactivated horse
serum (Gibco Ltd.)). We cocultured $30 \times 10^6$ fresh rat lymph node cells
along with $20 \times 10^6$ TNP-substituted irradiated (2,000R) autologous
spleen cells. At the peak of the stimulation, on day 5 in culture, the
responding lymphoblasts were isolated and processed as the other memory populations[7].

Elimination of autoreactive cells. Autoreactive cells were eliminated by two complementary procedures. Firstly, the reacting cultures
were fractionated in Ficoll density gradients. The dense fractions,
containing small lymphocytes were isolated. These were partly depleted
of self reactive cells, which mainly banded in the low density blast
fractions. Secondly, to remove the rest of self reactive dividing
cells present in these fractions, we applied the BUdR/light suicide
method described by Zoschke and Bach[10]. We cultured $5 \times 10^6$ lymphocytes in 5 ml medium in the presence of $3 \times 10^{-6}$ M BUdR in a dark in-
cubator for 36 hrs. Subsequently the cultures were exposed to neon
light for 90 min. The treated cells were washed and tested for their
reactivity in microtiter cultures.

RESULTS AND DISCUSSION

The in vitro Experimental Autoimmune Orchitis (EAO). The in vitro
induction of autoimmune memory T lymphocytes was originally established in an in vitro model of EAO. This system is based on culturing
freshly dissociated lymph node T cells together with autologous trypsin dissociated testis cells, and thus constitutes a primary autoimmune response. In autologous mixed testis/lymphocyte cultures, one
lymphocytic subpopulation starts to form rosette-like aggregates
around one particular subset of testicular cells. Morphological examination suggested that the testicular cells are related to Sertoli
cells. The lymphocytes rosetting around the testis cells belong to
the T lineage. Depletion experiments indicated that these rosetting
T cells are those cells which carry specific receptors for testicular
autoantigen[4,5]. During the subsequent days in culture, the rosetting
T cells start to transform to lymphoblasts and to proliferate. These
lymphoblasts are autosensitized, because, when reinjected into syn-

geneic recipients, they cause severe progressive orchitis. The specificity of the disease was established by the following controls. Fresh syngeneic lymphocytes administered with or without syngeneic testis cells cause only minimal lesions, and the same is true for ConA activated syngeneic lymphoblasts. In addition to in vivo autoreactivity, the EAO blasts were shown to specifically destroy syngeneic testis monolayer cells, which are the progeny of the rosette forming Sertoli-like cells. In criss-cross experiments, we found that EAO lymphoblasts strongly inhibit growth of syngeneic testis target cells, but only marginally affect syngeneic embryonic fibroblasts. Reversely, lymphoblasts autosensitized against syngeneic embryonic fibroblasts acted cytostatically against syngeneic fibroblasts, but barely inhibited the growth of syngeneic testis cells. This effector activity was found to be restricted by the self major histocompatibility gene complex (MHC), as suggested by experiments using targets from congenic rat combinations[5].

Induction of memory cells in EAO cultures. Our finding that in mixed autologous testis/lymphocyte cultures lymphoblasts are generated which specifically react against testicular self antigens, provided the basis for producing functionally pure memory cells. To achieve isolation of lymphoblasts from testis cells, debris, and small non-reactive small lymphocytes, we applied a discontinuous Ficoll density gradient, which permitted successful separation of various lymphoid and non-lymphoid cell subpopulations[6,7]. In these gradients, testis cells band in the topmost interphases, whereas small lymphocytes are found in the bottom fractions. From intermediate bands, we recover lymphoblast populations of a purity approaching 95%. Optimal cell separation is achieved at centrifugation speeds of 10,000 rpm for 1 hr.

We culture the purified lymphoblasts at low cell density in the absence of autoantigenic testis cells to allow them to revert back to small secondary lymphocytes. Embryonic fibroblasts may be added in order to condition the cultures. Within 3 days, most of the blasts vanish, and small or medium sized lymphocytes predominate. We harvest these populations and freeze them in medium containing 10% DMSO as cryoprotectant. When stored at $-80°$, biological activity of the secondary cells decreases after more than 2 months. Storage in liquid nitrogen prevents this loss of function.

Nature of the secondary EAO cells. Various lines of evidence suggest that the secondary EAO cells are the progeny of $Fc^-$ thymus dependent lymphocytes. We fractionated peripheral spleen and lymph node lymphocytes according to their capacity to bind antibody and complement loaden sheep erythrocytes. The non-rosetting T populations were maximally reactive against ConA, and contained less than 5% lymphocytes with surface Ig expression. B lymphocytes as well as phagocytozing macrophages, were highly enriched in the Fc/C'-rosetting populations. The lymphocytes that transform to lymphoblasts and give rise to memory cells in EAO cultures, were exclusively derived from $Fc^-$ T populations. Memory formation was not only found in cultures with peripheral T cells as responders, but also in autologous thymocyte/testis cultures. These cells, which probably were also members

TABLE 1

SECONDARY EAO CELLS CANNOT STIMULATE IN MLC

| Responders[a] | Stimulators[b] | | |
|---|---|---|---|
| | None | fresh Lewis LN | $2^{o}$EAO (Lewis) |
| Lewis LN | 789 ± 123 | 1,425 ± 112 | 1,705 ± 188 |
| AS2 LN | 437 ± 80 | 14,187 ± 633 | 1,779 ± 633 |

[a] fresh lymph node cells, 10 x $10^6$ cells/ml

[b] all stimulators irradiated (2,000 R); 10 x $10^6$ cells/ml; $^3$H-thymidine incorporation (cpm)

of a T lymphocyte subset, showed similar specificity properties as peripheral T memory cells. In contrast to the latter, they did not exert cytostasis against relevant target cells. Finally, it should be mentioned that secondary EAO cells are not able to elicit a mixed lymphocyte reaction, when irradiated and cultured together with immunocompetent allogeneic T cells. It may thus be speculated that they belong to a lymphocyte population that lacks Ia-like determinants which would be required for induction of MLC proliferation[11] (Table 1).

The stimulator cells in the EAO response. When *in vitro* generated secondary EAO cells are cultured in microtiter cells together with freshly dissociated syngeneic testis cells, a spectacular development can be observed. Within a few hours, the responder cells interconnect testis stimulator cells to form huge clumps. During the next 12 hours, most of the responders have retransformed to blasts, and an enormous proliferation begins. We are monitoring these proliferative events by adding $^3$H-thymidine to the cultures as early as after 24 hrs. in culture. The cultures are harvested after a further overnight incubation.

The [3]H-thymidine incorporation values faithfully reflect the morphologically observable secondary response. In a typical experiment Lewis secondary EAO cells incubated together with Lewis testis stimulator cells incorporated 16,133 cpm. The same cells, incubated with allogeneic AS2 testis cells are stimulated only to a value of 2,339 cpm.

As mentioned above, in primary cultures only one subpopulation among all the testis cells is able to specifically bind self reactive T lymphocytes and to trigger them. The same cell type was found to be responsible for cell triggering in the secondary cultures. We fractionated dissociated testis cells in discontinuous Ficoll density gradients of specially designed density profiles. We found that typical germinal epithelium cells did not trigger secondary EAO cells.

TABLE 2

SERTOLI-LIKE CELLS ARE THE EAO STIMULATORS

| Fraction[a] | Sertoli[b] Content | Stimulators[c] ($\times 10^5$/well) | | | |
|---|---|---|---|---|---|
| | | 0.5 | 1.0 | 2.0 | 4.0 |
| A' | None | 635 | 689 | 725 | 742 |
| A | (+) | 1,278 | 2,271 | 4,022 | 5,569 |
| B | + | 7,609 | 12,093 | 11,793 | 8,217 |
| C/D | +++ | 19,927 | 17,744 | 19,442 | 13,074 |

[a] freshly trypsinized Lewis rat testis cells were fractionated in Ficoll density gradients; A' is the least dense interface, C/D a pool of densest ones;

[b] determined in smears

[c] $1 \times 10^5$ $2^\circ$ Lewis EAO cells were cultured together with the stimulators for a total period of 60 hrs.; [3]H-thymidine was added 12 hrs before harvesting, incorporation expressed as cpm.

However, the stimulatory capacity was closely correlated to the content of Sertoli-like cells in the cell fractions and thus was greatly enriched in the densest bands (Table 2). The relevant autoantigen in the EAO is thus either expressed on the surface of Sertoli-like cells, or it is presented by these cells to the responder lymphocytes.

MHC and tissue restriction of the secondary EAO response. Recent evidence is accumulating suggesting that many, if not all, T lymphocyte responses involve the actual antigen as well as determinants specified within the major histocompatibility gene complex[12,13]. In our EAO system the importance of MHC determinants was already suggested in the cytostatic effector phase. In these studies we found that, in order to be recognized and to be cytostatically affected by effector T lymphoblasts, target cells had to be of testicular origin and had to share the MHC with the effectors[5]. Syngeneic MHC restriction was also found in the proliferative secondary EAO response, as shown by two independent approaches. Firstly, we generated secondary EAO cells of various genotypes. These memory cells were then cultured together with stimulator cells, which were syngeneic, allogeneic or congenic with respect to the MHC. We found that only in cases of MHC compatibility effective proliferative stimulation was obtained[7].

These results were complemented by serum blocking experiments. We pretreated stimulator cells with alloantisera in the absence of complement. This pretreatment was thought to block or to modulate defined membrane determinants without affecting stimulator cell viability. Furthermore, binding of antibodies to the responder cells, which could affect their recognizing capacity, was avoided by this approach. We found that only antibodies directed against MHC determinants were able to significantly depress the response in the secondary cultures. In contrast, antibodies which were directed against minor, non-MHC determinants, did not interfere with the response[7].

TABLE 3

STIMULATOR ACTIVITY IN LYMPHOID POPULATIONS

| Stimulator[a] Populations | Stimulators ($10^5$/well)[b] | | |
|---|---|---|---|
| | 2.5 | 5.0 | 10.0 |
| Lymph node unsepar. | 4,095 | 8,318 | 16,818 |
| Lymph node T | 1,564 | 3,105 | 5,420 |
| Lymph node B | 24,532 | 45,806 | 56,285 |
| Thymocytes | 4,632 | 11,179 | 29,642 |
| Testis unsepar. | 5,454 | 9,118 | ND |

[a] syngeneic lymph nodes were separated by the Fc/C'-SRBC technique in density gradients; responders: $1 \times 10^5$ Lewis secondary EAO cells

[b] stimulators and responders were cultured for 60 hrs.; $^3$H-thymidine incorporation (cpm)

As already indicated by cytostasis experiments[5], the EAO response contains a tissue specific determinant. This was corroborated in the memory proliferation system. In addition we discovered an unexpected cellular cross reactivity between testicular and lymphoid stimulator populations. When incubated with syngeneic testis, secondary EAO cells become strongly restimulated. Syngeneic embryonic fibroblasts, on the other hand, do not trigger a response. Surprisingly, we found that secondary EAO cells were also very strongly restimulated by lymphoid cells. As in the case of testis restimulation, this reaction showed MHC restriction. Several lines of evidence suggest that the stimulatory capacity of lymphoid cells is widely, if not exclusively, restricted to a macrophage-like cell. When we separated peripheral lymphocyte populations by means of the Fc/C'-sheep erythrocyte rosette assay, the stimulatory capacity was enriched in the $Fc^+C^+$ "non-T" fraction. $Fc^-C^-$ T cell populations, in contrast, did not trigger a significant proliferation. Absorption of non-T cells on glass beads reduced the stimulation. In addition to the peripheral non-T populations, unfractionated thymocytes showed a relatively high stimulatory capacity (Table 3). Since these populations contain extremely few $Ig^+$ B lymphocytes, but are rich in macrophages, it appears that macrophages are responsible for the secondary stimulation, rather than classical B cells. This was finally strengthened by our finding that in vitro cultured bone marrow macrophages, restimulated secondary EAO responses at extremely low cell numbers.

The cross reactivity between testicular and lymphoid populations thus reflects a cross reactivity between Sertoli cells and macrophages. This finding is interesting, as it is known that both cell types share a number of important properties. They are known to phagocytose cellular material, and they act as effective secretors of various hormonal factors.

Our results indicate that the secondary EAO response is restricted by the MHC, and that tissue specific antigens are involved in this reaction. Since the response is primarily directed against testicular cells, it was interesting to test, whether a response component was directed against the male specific H-Y antigen. There are claims that Sertoli cells are involved in production and expression of this remarkable antigen[14]. We restimulated secondary Lewis EAO cells with irradiated male and female Lewis lymphocytes. No consistent difference could be observed between both experimental groups (Table 4).

Are the self recognizing T cell clones preformed in the immune system? The clonal selection theory postulates that the clonal diversity is preformed and that the specificity repertoire is strictly conserved within the intact immune system. Recent experiments by Zinkernagel and his colleagues suggest that T lymphocytes are instructed within the thymic microenvironment to recognize self MHC determinants[15].

TABLE 4

MALE SPECIFIC ANTIGENS DO NOT DETERMINE SPECIFICITY IN EAO

| Responders[a] | Stimulators[b] | | | | |
|---|---|---|---|---|---|
| | | Lewis | | AS2 | |
| | None | female | male | female | male |
| 2° Lewis | 832 | 28,609 | 37,027 | 1,505 | 1,564 |
| 2° AS2 | 411 | 1,227 | 1,332 | 45,007 | 43,873 |

[a] $2x10^5$ secondary cells/well

[b] $10x10^5$ fresh irradiated (2,000R) lymph node cells; $^3$H-thymidine incorporation (cpm)

This may raise the possibility that, under certain circumstances, the capacity to recognize self may be acquired by interacting with self determinants. We tested this possibility by positively and negatively selecting self reactive lymphocytes from normal rat lymph node populations.

Normal $Fc^-/C^-$ lymph node T lymphocytes were primed in vitro against irradiated spleen cells. The lymphoblasts arising in these cultures were isolated and reverted back to secondary memory cells. Such memory cells are specific for macrophage-like non-T cells, which symmetrically cross-react with testicular Sertoli cells. In parallel, we sensitized rat T lymphocytes against TNP-substituted splenic cells in order to induce an anti-altered self reaction[9]. We isolated also these lymphoblasts to generate secondary memory cells against TNP-substituted self. In parallel we recovered those gradient fractions, which contained the quiescent small lymphocytes. These populations were depleted of lymphocytes responsive for unaltered or TNP-substituted self antigens. Negative selection was further completed by treatment with BUdR and light to eliminate all small lymphocytes involved in a mitotic cycle. We found, that the positively enriched secondary T cell populations were indeed specifically selected for reactivity to unaltered self antigens in the one population, and to TNP-altered self in the other. Between both populations there was, however, marked cross reactivity, which was probably due to incomplete chemical substitution of the self antigens. The negatively selected populations, which were depleted of cells responsive of unaltered or altered self antigens, responded well against third party allogeneic stimulator cells. In all cases, however, they were unreactive against self as well as against TNP-self stimulators (Table 5).

TABLE 5

POSITIVE AND NEGATIVE SELECTION OF ANTI-SELF AND OF ANTI-TNP-SELF T CLONES

| Responders[a] | Stimulators[b] | | | | |
|---|---|---|---|---|---|
| | None | Syn | TNP-Syn | Allo | TNP-Allo |
| 2° anti-self | 2,910 | 54,065 | 18,149 | ND | ND |
| 2° anti-TNP-self | 1,018 | 23,187 | 76,395 | ND | ND |
| (anti-self) depl. | 103 | 159 | 182 | 3,078 | 3,070 |
| (anti-TNP-self) depl. | 176 | 429 | 505 | 3,092 | 2,939 |

[a] $1\times10^5$ responders/well; either in vitro reverted secondary lymphocytes, or BUdR/light treated small lymphocytes depleted of self or TNP-self responsive cells

[b] $1\times10^6$ irradiated (2,000R) spleen cells, either untreated, or substituted with TNP, $^3$H-thymidine incorporation (cpm)

The finding that T cells once depleted for self reactive clones were unable to respond furthermore to similar self stimulators, may support the notion that self reactive clones are preformed in the immune system. Unexpected was the finding that in cultures depleted of anti-self reactive clones no response to TNP-modified self could be induced. This may suggest that modification of self antigens by TNP coupling does not create new determinants corresponding some foreign allogeneic MHC product, as widely assumed.

Self protective factors in autologous serum. So far, in all our experiments, the secondary EAO cells were generated in culture medium supplemented with heat inactivated horse serum and 2-mercaptoethanol. Several findings argued against the possibility that components from the heterologous serum could have bound to self antigens on the stimulator cells to trigger an anti-altered self response. Thus, secondary EaO cells generated in the presence of horse serum could be perfectly restimulated in fetal calf serum containing medium. It is known, that both sera display very little cross reactivity, at least as far as antibody formation is concerned. More important, we found that similar memory cells could be stimulated by syngeneic testis stimulators in the absence of any serum supplement. Although the total proliferation was somewhat reduced, the specificity of the response was fully maintained. We also found that the other non-self component in the culture medium, 2-mercaptoethanol, helped to improve culture conditions, but did not codetermine the specificity of the reaction[7].

In the presence of autologous serum no secondary activation of EAO cells was possible. This unreactivity was also observed, when rat serum was mixed together with horse serum suggesting that it was caused by some suppressive factor in rat serum, rather than by the lack of a horse serum component triggering the response. To investigate the role of autologous rat serum factors, we established primary EAO cultures in horse as well as in rat serum media. We found that in medium containing autologous rat serum as the only serum additive, some few self reactive lymphocytes started to undergo stimulation. This proliferation was, however, less than 10% as compared to EAO responses in

TABLE 6

SERUM EFFECT ON $2^\circ$ RESPONSE BY SECONDARY EAO CELLS PRIMED IN THE PRESENCE OF AUTOLOGOUS OR HETEROLOGOUS (HORSE) SERUM

| Serum additives[a] | | Stimulator cells[b] | | | | |
|---|---|---|---|---|---|---|
| $1^\circ$EAO | $2^\circ$EAO | | Lewis | | BN | |
| | | None | LN | Testis | LN | Testis |
| HS | HS | 1,665 | 19,390 | 6,873 | 4,023 | 1,728 |
| | RS | 819 | 1,959 | 1,454 | 1,037 | 714 |
| RS | HS | 2,324 | 19,243 | 9,550 | 6,813 | 5,014 |
| | RS | 685 | 8,675 | 3,967 | 1,740 | 1,367 |

[a] Lewis lymph node cells were primed either in medium with 10% horse, or 10% autologous serum; the secondary cultures contained each $1 \times 10^5$ responder cells in medium as indicated;

[b] fresh irradiated lymph node cells (2,000R) ($1 \times 10^6$ cells/well), or trypsin dissociated testis cells ($2 \times 10^5$ cells/well); $^3$H-thymidine incorporation (cpm)

horse serum medium. We isolated the lymphoblasts arising in rat serum EAO cultures and let them revert back to memory cells ($2^\circ$EAO-RS). In secondary microwell cultures, $2^\circ$ EAO-RS cells were tested for their specificity and compared to secondary EAO cells generated in horse serum ($2^\circ$EAO-HS). Table 6 shows that $2^\circ$ EAO-H cells responded strongly and specifically in horse serum supplemented secondary cultures. As expected, the $^3$H-thymidine uptake by the same populations was reduced to 10% in the presence of autologous serum. $2^O$ EAO-RS cells, in contrast, could be restimulated well even in the presence of autologous serum. Surprisingly, their specific reactivity was further increased, when restimulated in horse serum medium. Since these responder cells were never in contact with horse serum heterologous determinants before, the increase of responsiveness could not be due to introduction of foreign antigens. It is more plausible that rat serum contains suppressive factors that normally depress T cell anti-self responses. In our culture conditions, some few self reactive clones can escape this suppression, and their secondary progeny consequently will be more resistant to their suppressive action, when restimulated in vitro. In the absence of these suppressive factors, the same responder cells can mount a maximal self response, when confronted with the relevant autoantigens. We therefore believe that the enhanced EAO proliferation in horse serum medium is due to the lack of self suppressive, or self protective factors.

So far our data do not provide information on the specificity of these self protective factors. It should, however, be mentioned, that in an earlier series of experiments, we found that the cytostatic effector activity of sensitized EAO lymphoblasts against autologous testis monolayer forming cells was completely neutralized by addition of as little as 0.5% autologous serum. The serum factors preventing EAO cytostasis appeared to display a certain degree of specificity, as syngeneic serum had a higher protective titer than allogeneic serum, and among syngeneic sera, male sera were more active than female ones[16]. If indeed these self protective serum factors would prove to be immuno-specific for self determinants, it could be envisaged that they are soluble self antigen in a tolerogenic form, as discussed before[17]. Alternatively, these factors could be anti-idiotypic antibodies specifically affecting potentially self reactive clones[18], or antigen specific suppressor factors[19].

Apart from specific self protective factors, unspecific immunosuppressive serum factors could also be involved in the maintenance of self tolerance. Among the growing list of physiologically immunosuppressive serum factors[20], $\alpha$-fetoprotein is of particular interest, as it has been shown to particularly affect T cell mediated immune responses[21].

Memory T lymphocytes against encephalitogenic basic protein (BP).
Our studies using the in vitro EAO model indicate that T cell memory
can be generated against cell bound tissue specific autoantigens. In
another set of experiments we attempted to generate monospecific memory T lymphocytes against soluble myelin basic protein (BP) determinants, the autoantigens relevant in induction of Experimental Allergic Encephalomyelitis (EAE). This model differs from our others in
two respects. Firstly, here we are using in vivo primed cells for memory selection. Secondly, the encephalitogenic antigen is myelin material from guinea pigs. Although the antigens involved are not
strictly autologous, they are, nevertheless able to induce an autoimmune disease, and it is probable that this disease is caused by
lymphocytes included in the populations which we select by our procedures.

Lymph node cells from Lewis rats primed in vivo with 200 ug crude
guinea pig spinal cord Basic Protein (BP) respond by strong proliferation, when cultured together with either BP or with tuberculin PPD
(Table 7).

TABLE 7

SPECIFICITY AND SYNGENEIC HELPER REQUIREMENT IN THE IN VITRO EAE
MEMORY RESPONSE

| Responders[a] | Helper[b] | Stimulant[c] | | |
|---|---|---|---|---|
| | | BP | PPD | ConA |
| primed Lewis | -- | 15,826 | 18,251 | ND |
| naive Lewis | -- | 1,224 | 7,397 | 103,569 |
| 2°anti BP | -- | 4,735 | 812 | 16,564 |
| " | BN irrad. | 4,820 | -835 | 31,650 |
| " | Lewis irr. | 26,924 | 1,469 | 22,789 |

[a] fresh primed or naive Lewis lymph node cells ($1 \times 10^6$ cells/well)
cultured for 84 hrs., or Lewis 2°anti-BP cells ($1 \times 10^5$ /ml) cultured for 60 hrs.

[b] naive lymph node cells, irradiated (2,000R)

[c] BP: Basic Protein crude extract (10 μg/well); PPD (10μg/well);
ConA (5μg/well), $^3$H-thymidine incorporation (cpm): net values
(experimental groups -unstimulated background); ND: not done.

We isolated lymphoblasts from BP restimulated cultures and converted them to memory T lymphocytes in the absence of antigen. These memory populations selected for responsiveness to BP were tested for specificity against BP, against PPD, and against the polyclonal T cell activator ConA, as a control for their maximal proliferation capacity. Table 7 shows that indeed these populations were highly enriched for BP specific T cells. Whereas no significant response against PPD was detectable, the BP response equalled the maximum response to ConA. In contrast to freshly isolated in vivo primed lymphocytes, the in vitro generated memory populations require the help of syngeneic accessory cells to fully respond against the autoantigen. This help is provided by irradiated syngeneic fresh Lewis lymph node cells. Allogeneic BN cells do not provide this help. In contrast to the specific BP response, activation by ConA does not require syngeneic help, as previously reported[22].

CONCLUSIONS

The experiments described provide evidence that the intact immune system contains T lymphocyte clones which can be activated against determinants of the host organism. These determinants can be recognized in a genuine, unaltered state. Autologous serum contains factors which effectively, but not absolutely, prevent activation of the self reactive lymphocytes.

These findings indicate that elimination of self recognizing lymphocytes during embryogenesis may not be the only basis for physiological self tolerance, as suggested by the original clonal selection theory[1]. In the study of autoimmune diseases and their therapy, it therefore may be important to focus the attention on possible defects in the regulatory mechanisms that could lead to pathological triggering of self recognizing T cell clones. Moreover, application of our techniques may facilitate studies on possible physiological roles of the self reactive lymphocyte clones in the organisation and development of the intact immune system.

REFERENCES

1. Burnet, F.M. (1959) The Clonal Selection Theory of Acquired Immunity, Vanderbilt University Press, Nashville, TN
2. Clagett, J.A. and Weigle, W.O. (1974) J.Exp.Med., 139, 643-660.
3. Cohen, I.R. and Wekerle, H. (1973) J.Exp.Med., 137, 224-238.
4. Wekerle, H. and Begemann, M. (1976) J. Immunol., 116, 159-161.
5. Wekerle, H. and Begemann, M. (1978) Eur. J. Immunol., 8,294-3o2.

6. Wekerle, H. (1977) Nature, 267, 357-358.

7. Wekerle, H. (1978) J.Exp.Med., 147, 233-250.

8. Hirshfeld, H., Teitelbaum, D., Arnon, R. and Sela, M. (1970) FEBS Letters 7, 317-320.

9. Shearer, G.M. (1974) Eur. J. Immunol. 4, 527-533.

10. Zoschke, D.C. and Bach, F.H. (1970) Science, 170, 1404-1406.

11. Niederhuber, J.E. and Frelinger, J.A. (1976) Transpl. Rev. 30, 101-121.

12. Rosenthal, A.S., Lipsky, P.E. and Shevach, E.M. (1975) Fed. Proc. 34, 1743-1748.

13. Miller,J.F.A.P., Vadas, M.A., Whitelaw, A. and Gamble, J. (1976) Proc. Nat. Acad. Scie. 73, 2486-2490.

14. Ciccarese,S. and Ohno, S. (1978) Cell, 13, 643-650.

15. Zinkernagel,R.M., Callahan, G.N., Althage, A., Cooper, S., Streilein, J.W. and Klein, J. (1978) J.Exp.Med., 147, 897-911.

16. Wekerle, H. and Begemann, M. (1977) Transpl. Proc. 9, 775-778.

17. Cohen, I.R. and Wekerle, H. (1977) in Autoimmunity, Talal, N. ed., Academic Press, New York pp. 231-265.

18. Eichmann, K. (1978) Adv. Immunol., in press.

19. Tada, T., Taniguchi, M. and Takemori, T. (1975) Transpl. Rev. 26, 106-129.

20. Cooperband, S.M., Nimberg, R., Schmid, K. and Mannick, J.A. (1976) Transpl. Proc. 8, 225-242.

21. Murgita, R.A. and Tomasi, T.B. (1975) J.Exp.Med. 141, 440-452.

22. Pilarski, L.M., Bretscher, P.A. and Baum, L.L. (1977) J.Exp.Med. 145, 1237-1249.

DISCUSSION

VOLPE: Why did you choose the model of Sertoli cells for these studies? Was there anything special about these cells that made them particularly useful?
What proportion of the blood lymphocyte counts did the Sertoli-lymphocyte rosettes comprise?
Have you used any other models aside from Sertoli cells?

WEKERLE: In response to your last question, we are now generating memory T populations specific for myelin basic protein, which may serve in in vitro studies on EAE.
As to the proportion of normal rat lymphocytes binding to Sertoli cells, this is very hard to assess exactly. Judging from depletion experiments, I would estimate the proportion of rosetting lymphocytes to be less than 1%. Finally, we chose the EAO model by a chance observation. We looked for an easily dissociable cell population suitable as a cytotoxicity target and thought of testis cells. This primary goal was not reached because of excessive spontaneous isotope release, and probably because of special antigenic properties of these cells. In the same cultures we observed, however, the rosetting phenomenon and subsequent lymphocyte autosensitization.

ENGLEMAN: The reaction you describe is reminiscent of the autologous mixed lymphocyte reaction, in which autologous T cells proliferate in response to autologous non-T cells. If you "prime" with non-T lympho-cytes and perform the secondary culture with Sertoli cells, is there a good secondary response? Are Sertoli cells Ia positive, and might this antigen be the sensitizing antigen? Lastly, wandering macrophage-like cells, such as Langerhans' cells are known to be Ia positive. Is there any chance that such cells might be a contaminant in your testes preparation?

WEKERLE: The relationship between our EAO model and "spontaneous" T cell responses against autochthonous non-T cells is indeed very close. We primed rat T cells against their non-T counterparts and generated secondary cells. These cells can also be secondarily triggered by syngeneic Sertoli cells. In this context it should be mentioned that macrophages and Sertoli cells share some remarkable properties. Both are phagocytes and both are known to secrete a wide number of factors, among them plasminogen activator and prosta-glandins. Whether Sertoli cells express integral membrane Ia deter-minates is not known to me. They may, however, present engulfed sperm Ia antigens.
Finally, Langerhans' cells, to my knowledge, are defined as macrophage-like cells within the epidermis. We have no indication for similar cells in testes.

WARNER: Have you determined whether the proliferating secondary responders are capable of generating cytotoxic T cells? The use of this assay would then permit a more complete analysis of the nature and cross reactivities of antigen on various cell types, since the cold cell target competitive inhibition assay could be used.

WEKERLE: We have not tested our secondary EAO cells in short-term cytolysis assays. The target cell proportion is too low, and spon-taneous isotope release is too high. We assayed these cells, however, in a cytostasis system, lasting about three days, and found strong and specific cytostatic activity. This activity may be due to other cell functions than to typical CML action, such as factor production by DH type of responding lymphocytes. I should like, however, to point out that we find it quite relevant not to limit our attention to CNL cells, which as Cantor told us constitute less than 5% of all T cells and whose biological significance is still not established. Using proliferation assays, we may, however, cover a wider spectrum of responsive T cells.

SVEJGAARD: Pursuing Dr. Engleman's question: could the suppression seen with rat serum be due to Ia antigens in the serum, and have you tried to inhibit the stimulation by treating the stimulating cells with anti-Ia serum?

WEKERLE: Anti-Ia-sera in the rat are still very difficult to obtain. We shall do the experiment as soon as we can get some. Second, so far we do not know whether the autologous serum effect on the secondary EAO proliferation model is immunospecific. I should, however, mention that in a primary EAO cytostasis assay, we found that addition of autologous serum diluted to 1:200 can completely inhibit the reaction. In this model, we found that (a) autologous serum had a higher inhibitory titer than allogeneic serum, and (b) syngeneic male serum was more effective than syngeneic female serum. This looks, therefore, like a specific factor. In this case, soluble Ia antigens would be a serious candidate.

# FINE STRUCTURE OF GENETIC CONTROL OF AUTOIMMUNE RESPONSE TO MOUSE THYROGLOBULIN

YI-CHI M. KONG, CHELLA S. DAVID, ALVARO A. GIRALDO, MOSTAFA ELREHEWY*, AND NOEL R. ROSE
Department of Immunology and Microbiology, Wayne State University School of Medicine, 540 E. Canfield, Detroit, Michigan  48201 and Department of Immunology, Mayo Medical School, Rochester, Minnesota 55901

ABSTRACT
   Using lipopolysaccharide as adjuvant, we have defined the regions of the *H-2* complex that govern the response of mice to mouse thyroglobulin. An *Ir-Tg* gene at the *K* or *I-A* region interacts with a secondary gene at the *D* end which modified inflammation of the thyroid, but has little effect on antibody levels.

   Experimental autoimmune thyroiditis can be induced in mice by injection of mouse thyroglobulin (MTg) as an emulsion with complete Freund's adjuvant (CFA)[1]. As in the case of human chronic thyroiditis, the experimentally induced disease usually involves both antibody production to MTg and infiltration of the thyroid by mononuclear cells, including lymphocytes and macrophages.  The severity of thyroiditis is dependent upon the genetic constitution of the animal and is *H-2*-linked[2].  A few selected *H-2* types and their responses to MTg in CFA are summarized in Table 1.

TABLE 1
*H-2*-LINKED RESPONSE TO MTg[a]

| *H-2* Haplotype | Response to MTg in CFA | | Classification |
| | Antibody | Thyroiditis | |
|---|---|---|---|
| *k* | High | Severe | Good |
| *s* | High | Severe | Good |
| *q* | High | Severe | Good |
| *a* | Moderate | Moderate | Intermediate |
| *b* | Moderate | Mild | Poor |
| *d* | Moderate | Mild | Poor |
| *f* [b] | Moderate | Mild | Poor |

[a]Modified from Vladutiu and Rose[2].
[b]Recent finding.

*Present address:  Microbiology and Immunology Department, Faculty of Medicine, Assiut University, Assiut, Egypt.

Good responder strains, having an $H$-$2^k$, $H$-$2^s$, or $H$-$2^q$ haplotype, respond with high levels of antibodies to MTg, measured by passive hemagglutination, and severe thyroiditis, characterized by mononuclear cell infiltration involving about 60% or more of the thyroid gland. Poor responder strains have an $H$-$2^b$ or $H$-$2^d$ haplotype and show only moderate levels of antibody to MTg and mild inflammation of the thyroid, involving no more than 10% of the gland. $H$-$2^a$ mice produce both moderate levels of antibody and cellular infiltration and are intermediate responders. Interestingly, the $H$-$2^a$ strain is a recombinant of good responder $k$ and poor responder $d$ genes. More recently, we have found $H$-$2^f$ mice to belong to the poor responder category.

Adoptive transfer studies of Vladutiu and Rose[3] demonstrated that it was the good responder thymus-derived (T) cells that were responsible for disease production. Utilizing the observation that responsiveness to MTg is an autosomal dominant trait manifested in $F_1$ hybrids between good and poor responders, they showed that severe thyroiditis could be produced in thymectomized, lethally irradiated recipient mice only if the cells used for reconstitution contained T cells from good responder mice. Whether the bone marrow-derived (B) cells were from good or poor responder strains was much less important. These studies established that the $H$-$2$-linked susceptibility to thyroiditis is T-cell based.

Although the thyroglobulin used in these studies was prepared from the same species, the mouse, it was injected as a water-in-oil emulsion. The possibility existed that the animals were responding to altered or denatured self-determinants produced during emulsification or in the resultant granuloma. The autoimmune response might result from T cells reactive to some altered determinants similar to those present on hapten-substituted[4] or heterologous[5] thyroglobulin which can induce autoimmune thyroiditis without the benefit of CFA. These observations lend support to the hypothesis that self-tolerance exists only at the T-cell level and that autoimmune thyroiditis results from the cooperation between T cells responding to cross-reactive or altered self-determinants and B cells which can bind MTg[5,6].

However, it seemed unlikely that the $H$-$2$-linked genetic control of susceptibility to thyroiditis governed T-cell responsiveness only to foreign determinants on cross-reactive molecules without regard to the autoantigenic potential of self-determinants. In order to study the genetic regulation of autoantigenic potential, we sought adjuvants with the capacity to induce immune response to an antigen when given separately so that no alteration of the antigen could occur. One such adjuvant is bacterial lipopolysaccharide (LPS), a polyclonal B cell activator which enhances the immune response to T-dependent antigens[7]. By comparing congenic good (B10.BR) and poor (B10.D2) responder mice, the autoimmune response after LPS administration was found to correspond to the *H-2*-linked responsiveness to MTg in CFA[8]. Furthermore, by using nude mice and T-cell-depleted, B-cell-replenished mice, a requirement for T cells was demonstrated for the adjuvant effect of LPS for MTg.

It is clear from these and other studies that neither the presence of antibody nor its level directly determine thyroid lesions. We have begun studies to see if antibody class plays a decisive role in the induction of disease. In previous experiments using either CFA[1,8] or LPS adjuvant[8], a dose of 250 µg of MTg was used for immunization on days 0 and 7. Antibodies were not detectable before day 14 in good responder mice and appeared even later in poor responder mice[8]. These antibodies were mercaptoethanol resistant, probably of the IgG class. Once the antibodies appeared, they could reach high levels with only mild inflammation in the thyroid, as in the CFA-treated poor responder mice[8]. In good responder mice also, high antibody levels could be attained when polyadenylic-polyuridylic acid complex (poly A:U) was used as adjuvant without concommitant thyroid infiltration[9]. In these studies, some mercaptoethanol-sensitive antibodies, probably IgM, were detectable on day 14. One possible explanation for not detecting early IgM antibody might be the presence of circulating antigen after injecting 250 µg of MTg with or without adjuvant. To test this possibility, three groups of 4-6 female mice each of strains B10.A(2R) and B10.A(5R) were injected with 250 µg alone, with LPS or in CFA on days 0 and 7. Serum samples were obtained on days 6 and 14 for antibody determination using hemagglutination as well as antigen determination by means of hemagglutination-inhibition. On day 6, all 30 serum samples were negative for antibody to MTg. However, they inhibited the hemagglutination reaction between MTg-coated erythrocytes and 1:500 and 1:1000 dilutions of rabbit hyperimmune antiserum, indicating an antigen concentration in the serum of approximately 12 ng/ml. By day 14, 7 days after the second immunizing dose, two samples from the group given MTg plus LPS showed the presence of IgG antibody but the remaining 28 samples contained antigen which inhibited the reaction of a 1:500 dilution of antiserum. IgG antibodies were then detected on day 21 and subsequently.

To determine if reducing the MTg dose would result in early detection of IgM antibody as well as in thyroiditis, B10.BR and B10.D2 mice

were immunized with 20 µg of MTg plus LPS or MTg in CFA. Antibody
levels were determined on day 7 just before the second MTg and adjuvant
injection and at weekly intervals thereafter. As seen in Figure 1,
thyroiditis still occurred in almost 100% of good responder B10.BR
mice treated with MTg plus LPS or MTg in CFA. In antibody production,
mercaptoethanol-sensitive, presumably IgM antibodies, were observed
on day 7 only in mice given LPS as adjuvant. The antibody switched
to the IgG class by day 14, at which time mercaptoethanol-resistant
antibody also appeared in mice given MTg in CFA. In contrast, B10.D2
mice showed a delayed onset of antibody production, appearing only on
day 14 in the LPS-treated group and on day 21 in the CFA-treated group.
No lesions were evident in either adjuvant-treated group.

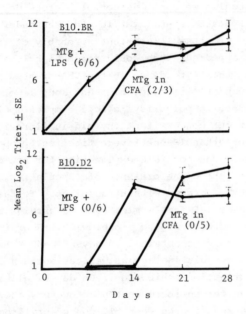

Fig. 1. Comparison of the adjuvant effect of lipopolysaccharide
(LPS) and complete Freund's adjuvant (CFA) on the kinetics of antibody response to mouse thyroglobulin (MTg). LPS (20 µg) was given
i.v. 3 hr after MTg (20 µg) i.v. on days 0 and 7; MTg in CFA was
given s.c. The ratio of mice with thyroiditis is presented in
parenthesis.

Since, in B10.BR mice, lesions were induced with both LPS and CFA and IgM antibody was observed only in the presence of LPS, the IgM antibody class appears to play little direct role in producing thyroid pathology. However, the early appearance of IgM antibody is a useful indicator of the responsiveness of a recombinant strain to MTg. Since the T-cell-based control of responsiveness to MTg leading to thyroid infiltration can be brought forth by LPS with unmodified MTg, the termination of self-tolerance apparently does not involve a bypass of T cells or require T cells cross-reactive with foreign or modified thyroglobulin.

That T cells can be activated by native MTg in good responder mice was further confirmed by using poly A:U as an adjuvant[9]. After the administration of MTg and poly A:U, antibody production was observed only in B10.BR mice, but not in the congenic B10.D2 mice. These studies confirm the autoantigenicity of native MTg and provide evidence for the presence of self-reactive T cells controlled by an *Ir-Tg* gene(s) located within the *H-2* region.

Earlier work by Tomazic *et al.*[10] using MTg in CFA has indicated that the genetic control is associated with the centromeric (left) side of the *H-2* complex, probably the *K* and *I-A* regions. Their data also suggest that the *D* end may also exert an influence on the response to MTg, reducing thyroid infiltration. Using LPS as adjuvant and available intra-*H-2* recombinants, we have confirmed and extended their study and observed a suppressive effect of the *D* end on thyroid lesion in three sets of experiments. In the first set of experiments, the good responder *k* alleles were kept constant to the left of the *I-C* subregion and the remainder of the *H-2* gene complex was varied as shown in Table 2. The responses will be described briefly. As far as antibody titers were concerned, all six strains showed early appearance of antibody on day 7; in all five tested, the antibodies were mercaptoethanol-sensitive. Antibody levels increased in all strains by day 28 and were not very different from one another. Since antibody levels were not reduced, there appeared to be no interference with helper activity.

However, the severity of thyroiditis varied from strain to strain. Compared to B10.BR mice with *k* across the entire *H-2* region, substitution with *d* alleles from the *I-C* subregion to the *D* end in C3H.A mice led to a lower degree of infiltration. C3H.A mice shared the same *d* alleles at the *I-C* subregion and the *S* and *G* regions with B10.A(2R) and B10.M(17R) mice. Yet, the pathologic response of each strain varied from high [in B10.M(17R) mice] to medium [in C3H.A and B10.A(2R)

TABLE 2

SUMMARY OF IMMUNE RESPONSES TO MTg IN RECOMBINANT STRAINS EXPRESSING
*k* ALLELE AT THE *K* END

| Strain | *H-2* Gene Complex | | | | | | | | | | Response to MTg[a] |
|--------|-----|-----|-----|-----|-----|-----|-----|-----|-----|-----|-----|
|        | *K* | *A* | *B* | *J* | *E* | *C* | *S* | *G* | *D* | *TL* | |
| B10.BR | *k* | *k* | *k* | *k* | *k* | *k* | *k* | *k* | *k* | *a* | High |
| B10.AKM | *k* | *k* | *k* | *k* | *k* | *k* | *k* | *k* \| | *q* | *a* | Medium |
| B10.AM | *k* | *k* | *k* | *k* | *k* | *k* | *k* | *k* \| | *b* | *b* | Medium |
| C3H.A | *k* | *k* | *k* | *k* | *k* \| | *d* | *d* | *d* | *d* | *a* | Medium |
| B10.A(2R) | *k* | *k* | *k* | *k* | *k* \| | *d* | *d* | *?* | *b* | *b* | Medium |
| B10.M(17R) | *k* | *k* | *k* | *k* | *k* \| | *d* | *d* | *d* \| | *f* | *a* | High |

[a]Antibody titers were comparable in all strains; extent of thyroid
infiltration varied from high to medium.

mice] depending upon the genes at the *D* region. Thus, there was a
suppressive effect to the right of the *G* region, somewhere between the
*G* and *D* regions, and was seen with the *d* and *b* alleles at the *D* end
but not with the *f* allele. Similarly, when the *I-C* subregion and the
*S* and *G* regions contain *k* alleles as seen in B10.BR, B10.AKM, and
B10.AM mice, the *q* genes (B10.AKM) and the *b* genes (B10.AM) at the
*D* end reduced the pathologic changes.

The *TL* region did not seem to play a role, since B10.BR and B10.AKM
mice share the *a* allele and differ only at the *D* region. Similarly,
C3H.A and B10.M(17R) mice differed in severity of thyroiditis but
share the same $TL^a$ genes. The difference between these two strains
is also at or near the *D* region. We are currently testing *TL* and *Q*
locus congenic mice to verify this point.

Table 3 summarizes experiments using strains with good responder *s*
alleles at the *K* end including *I-A* and *I-B* subregions. Early mercaptoethanol-sensitive antibodies were observed on day 7; later antibodies
were of the IgG class but did not reach high levels. In the thyroid,
cellular infiltration was again reduced apparently by genes located
at the *D* end, between the *G* and *D* regions. On the other hand, the *TL*
region had little influence on the response to MTg. Of interest is
the very low antibody response of B10.S(9R) and B10.HTT mice. Since
they also share the *k* allele at the *I-E* subregion, the helper T cell
activity may also be suppressed by genes located between the *I-J*, *I-E*
and *I-C* subregions.

TABLE 3

SUMMARY OF IMMUNE RESPONSES TO MTg IN RECOMBINANT STRAINS EXPRESSING
$s$ ALLELE AT THE $K$ END

| Strain | $H$-$2$ Gene Complex | | | | | | | | | | Response to MTg[a] |
|--------|---|---|---|---|---|---|---|---|---|---|-----|
|  | $K$ | $A$ | $B$ | $J$ | $E$ | $C$ | $S$ | $G$ | $D$ | $TL$ | |
| B10.S | $s$ | $s$ | $s$ | $s$ | $s$ | $s$ | $s$ | $s$ | $s$ | $b$ | Medium |
| B10.S(7R) | $s$ | $s$ | $s$ | $s$ | $s$ | $s$ | $s$ | $s$ | $\mid d$ | $a$ | Low |
| B10.S(9R) | $s$ | $s$ | $\underline{?}$ | $k$ | $k$ | $\mid d$ | $d$ | $d$ | $d$ | $a$ | Low |
| B10.HTT | $s$ | $s$ | $s$ | $s$ | $\mid k$ | $k$ | $k$ | $k$ | $\mid d$ | $c$ | Low |

[a]Both antibody titers and extent of infiltration were low in
B10.S(7R), B10.S(9R) and B10.HTT, compared to B10.S.

In the third set of experiments, we tested poor responder $d$ alleles
at the $K$ and $I$-$A$ regions and recombinants varying at the $D$ end (Table
4). Another poor responder strain, B6, was included for comparison.
Generally, day-7 antibodies were undetectable and infiltrations were
mild and found only in some of the animals. As seen in D2.GD mice,
the poor responder $d$ alleles at the $K$ and $I$-$A$ regions determined the
response with little influence to the right of the $I$-$A$ subregion. A
low response was seen whether the $D$ region had the $d$ or $b$ allele.
Again, the $TL$ region exerted little effect on this response. But, when
the $k$ allele was present at the $D$ region, the response was very low.

TABLE 4

SUMMARY OF IMMUNE RESPONSES TO MTg IN RECOMBINANT STRAINS EXPRESSING
$d$ ALLELE AT THE $K$ END

| Strain | $H$-$2$ Gene Complex | | | | | | | | | | Response to MTg[a] |
|--------|---|---|---|---|---|---|---|---|---|---|-----|
|  | $K$ | $A$ | $B$ | $J$ | $E$ | $C$ | $S$ | $G$ | $D$ | $TL$ | |
| B10.D2 | $d$ | $d$ | $d$ | $d$ | $d$ | $d$ | $d$ | $d$ | $d$ | $c$ | Low |
| D2.GD | $d$ | $d$ | $\mid b$ | $b$ | $b$ | $b$ | $b$ | $b$ | $b$ | $b$ | Low |
| B10.HTG | $d$ | $d$ | $d$ | $d$ | $d$ | $d$ | $d$ | $\underline{?}$ | $b$ | $b$ | Low |
| B10.BDR2 | $d$ | $d$ | $d$ | $d$ | $d$ | $d$ | $d$ | $\underline{?}$ | $b$ | $b$ | Low |
| B6 | $b$ | $b$ | $b$ | $b$ | $b$ | $b$ | $b$ | $b$ | $b$ | $b$ | Low |
| C3H.OH | $d$ | $d$ | $d$ | $d$ | $d$ | $d$ | $d$ | $d$ | $\mid k$ | $b$ | Very Low |

[a]Both antibody titers and extent of thyroid infiltration were low in
all strains. In addition, C3H.OH had very low, transient antibody
titer on day 14 only.

440

In summary, we have used LPS as an adjuvant to define regions of
gene interactions in the response to self-determinants of MTg.  Compared to CFA, which has high concentrations of mycobacteria specially
designed to induce a vigorous autoimmune response, LPS is a weaker
immunopotentiator.  Yet, it brings out antibody of the IgM class which
appears to be an early indicator of responsiveness to MTg, and it am-
plifies genetic differences in autoimmune response.  Gene complementation with *D*-end influences has been reported in recovery to Friend
leukemia virus[11], resistance to radiation leukemia virus[12], and suppression in contact hypersensitivity[13].  Our findings demonstrate
gene complementation between *I* region and the *D* end for a self-antigen.

The cell type that mediates the suppression and its target are unknown.  The extent of suppression apparently depends upon the deriva-
tion of the *Ir-Tg* gene at the *K* end and the *D*-end gene(s).  If we
postulate that the *D*-end influence is expressed as an immune suppression (*Is*) gene(s)[14] acting through suppressor T cells ($T_S$), its possible site of action may be illustrated by a simple diagram.

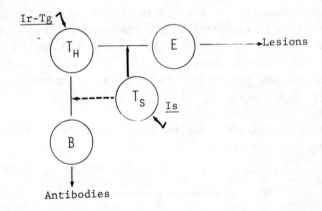

When the *Ir-Tg* gene is derived from a good responder $H\text{-}2^k$ strain,
and the *Is* gene from *b*, *d* or *q* allele at the *D* end, antibody levels
remain high and only infiltration is reduced.  The action of $T_S$ is
then to prevent the interaction between the helper T cell ($T_H$) and
the precursor of the undefined effector cell (E) directly or indirectly
responsible for thyroid damage.  This scheme does not exclude the possibility that the *Is* gene may be expressed on the same cell as the *Ir*
gene[15].  In the case of *Ir-Tg* gene from good responder $H\text{-}2^s$ strain,

antibody levels may sometimes be suppressed.  This effect may require additional gene interactions from the $I$-$J$, $I$-$E$ or $I$-$C$ subregion and $T_S$ acts to interfere with $T_H$-B cell cooperation as seen in responses to certain antigens[16,17].  When the $Ir$-$Tg$ gene is derived from a poor responder $H$-$2^d$ strain, the $D$-end effect is difficult to discern, since cellular infiltration is already mild or nonexistent.  One striking observation is that $D^k$ genes seem to mediate further suppression of $d$ genes at the $K$ end.  These cellular interactions have been verified by *in vitro* T-cell proliferation assays of recombinant strains described in the next paper[18].

## ACKNOWLEDGMENTS

This study is supported in part by NIH grants CA 18900 and AM/AI 20023.  We appreciate greatly the technical help of Margaret Clark and Ruey Chen and the assistance of Teresa Boczar in preparing the manuscript.

## REFERENCES

1.  Rose, N.R., Twarog, F.J. and Crowle, A.J. (1971) J. Immunol., 106, 698-704.

2.  Vladutiu, A.O. and Rose, N.R. (1971) Science, 174, 1137-1139.

3.  Vladutiu, A.O. and Rose, N.R. (1975) Cell. Immunol., 17, 106-113.

4.  Weigle, W.O. (1965) J. Exp. Med., 121, 289-307.

5.  Clagett, J.A. and Weigle, W.O. (1974) J. Exp. Med., 139, 643-660.

6.  Allison, A.C. (1977) in Immunological Tolerance, Katz, D.H. and Benacerraf, B. eds., Academic Press, New York, pp. 25-49.

7.  Johnson, A.G., Gaines, S. and Landy, M. (1956) J. Exp. Med., 103, 225-246.

8.  Esquivel, P.S., Rose, N.R. and Kong, Y.M. (1977) J. Exp. Med., 145, 1250-1263.

9.  Esquivel, P.S., Kong, Y.M. and Rose, N.R. (1978) Cell. Immunol., 37, 14-19.

10. Tomazic, V., Rose, N.R. and Shreffler, D.C. (1974) J. Immunol., 112, 965-969.

11. Chesebro, B., Wehrly, K. and Stimpfling, J. (1974) J. Exp. Med., 140, 1457-1467.

12. Meruelo, D., Lieberman, M., Ginzton, N., Deak, B. and McDevitt, H.O. (1977) J. Exp. Med., 146, 1079-1087.

13. Moorhead, J.W. (1977) J. Immunol., 119, 1773-1777.

14. Debre, P., Kapp, J.A., Dorf, M.E. and Benacerraf, B. (1975) J. Exp. Med., 142, 1447-1454.

15. Schwartz, R.H., Yano, A. and Paul, W.E. (1977) Regulatory Genetics of the Immune System, Academic Press, New York, pp. 479-488.

442

16. Pierres, M., Germain, R.N., Dorf, M.E. and Benacerraf, B. (1977) Proc. Natl. Acad. Sci., 74, 3975-3979.

17. Tada, T., Taniguchi, M. and David, C.S. (1976) J. Exp. Med., 144, 713-725.

18. Christadoss, P., Kong, Y.M., ElRehewy, M., Rose, N.R. and David, C.S. (1978) this volume.

DISCUSSION

VLADUTIU: In general when speaking on complementation of Ir genes, one understands two synergistic genes. Could the postulated D-end gene be in fact an immune response (Ir) gene instead of an immune suppressor (Is) gene? There is no direct evidence of suppression and I wonder if the D gene is not synergistic with the K/A gene.

KONG: By gene complementation, we are only using the terminology of Dorf et al., who showed the requirement of two Ir genes for the immune response to a terpolymer. However, their Ir genes are located within the I region and ours are at the K and I-A regions and the D-end.

VLADUTIU: Could it be a synergistic effect?

KONG: Yes, they could be synergistic rather than one suppressing the other.

MULLEN: Your summary slide implied that lesions of thyroiditis are mediated by effector T cells. Do you have any direct evidence for that?

KONG: The diagram of $T_E$ (effector T cells) is just a way of representing mononuclear cells infiltrating the thyroid glands. They may be killer cells, cells involved in antibody-dependent cell-mediated cytotoxicity, etc. There is no conclusive evidence that autoantibody mediates murine thyroiditis. We have seen many animals given different adjuvants with high antibody levels but no lesion. This cannot be explained unless one postulates that the antibodies observed in the presence of lesions and those seen without lesion belong to different subclasses (IgG). We are presently investigating this possibility. From studies presented here, in which mice with comparably high antibody levels showed reduced thyroiditis because of having a D-end allele of d rather than k (Table 2), one would have to postulate that the D-end genes somehow altered the antibody class and infiltration is lessened.

ROITT: I agree with Dr. Kong that the processes operating to cause damage within the gland are still poorly identified. My colleague, Mr. Lain de Carvalho has looked at the possibility that differences in the class of antibody produced are responsible for the high and low responders. He has measured IgM, IgA, $IgG_1$, $IgG_{2a}$, and $IgG_{2b}$ but found no difference in the distribution of antibody between these various isotypes.

MULLEN: Has anyone been able to transfer thyroiditis in the mouse using sensitized cells that have been separated into T and B cell fractions?

ROSE: The roles of T cells, B cells and antibody in producing the histological picture of experimental thyroiditis is still undefined. It is possible under very special circumstances to produce thyroiditis in mice by injection of antibody-containing serum. However, we have seen significant damage only in responder strains of mice, and we strongly suspect that the lymphocytic infiltration results from active sensitization following upon an initial Arthus or immune complex insult. In the rat, we have succeeded in transferring the lesions of thyroiditis to syngeneic recipients with lymph node suspension but not with antiserum. But of course transferred cells might still secrete antibody.

ROITT: In addition to the many modes of immunological tissue
destruction which have been so far recognized, Dr. I. Cohen's data
suggests a further possibility. Thyroglobulin (normally leaving
the thyroid acinar cells) coats mononuclear myeloid cells which
could then be killed by cytotoxic T cells and release injurious
components.

GENETIC CONTROL OF T-LYMPHOCYTE PROLIFERATIVE AUTOIMMUNE RESPONSE TO
THYROGLOBULIN IN MICE

PREMKUMAR CHRISTADOSS,[1] YI-CHI M. KONG,[2] MOSTAFA ELREHEWY,[2] NOEL R.
ROSE,[2] AND CHELLA S. DAVID[1]

[1]Department of Immunology, Mayo Clinic and Medical School, Rochester,
Minnesota; [2]Department of Microbiology and Immunology, Wayne State
University, School of Medicine, Detroit, Michigan

INTRODUCTION

Injection into mice of crude mouse thyroid extract or purified mouse
thyroglobulin (MTg) mixed with Complete Freund's Adjuvant (CFA) induces
lymphocytic thyroiditis.[1] In these mice autoantibodies to mouse thyroglobulin can be demonstrated.[2] The animals have thyroid lesions with
marked infiltration of lymphocytes and macrophages, sometimes virtually
obliterating the normal thyroid structure. Mouse strains can be classified as good responders to MTg with severe thyroid lesions or poor
responders with little or no disease. The level of antibodies produced
can also be classified as high or low. Early studies showed that
autoimmune murine thyroiditis is closely associated with histocompati-
bility-2 (H-2) type of the strain.[3] Further studies with the recombinant H-2 haplotypes suggested mapping of an immune response gene controlling the antibody production and cellular infiltration in the K
region or *I-A* subregion of the H-2 gene complex.[4] It was obvious that
other genes mapping in other parts of the H-2 gene complex or closely
linked to it, also regulated the level of antibody produced and the
severity of the disease. The genetic basis of good or poor response to
thyroglobulin is associated with the thymus-derived (T) lymphocytes.[5]

An *in vitro* assay would lend itself to more intricate genetic and
cellular probes in mapping and functional studies of the gene(s) controlling response to thyroglobulin. Hence, we tried to adapt Alkan's
recently developed T cell proliferation assay[6] to measure autoimmune
response to thyroglobulin. Lymphocytes obtained from draining lymph
nodes of mice primed with the antigen are allowed to proliferate after
exposure *in vitro* to the homologous antigen and the response measured
by the uptake of [$^3$H] thymidine. This response has been shown to be
specific and mediated by T cells.

As a first step we wanted to confirm the *H-2*-linked *Ir* gene control
to thyroglobulin shown previously by antibody assay and by cellular
infiltration. The results show that the T lymphocyte proliferative

response to mouse thyroglobulins is under *H-2*-linked *Ir* gene control (probably at *I-A*) similar to the humoral and pathologic responses. Studies reported in this paper also suggest that two other genes mapping to the right of *I-E* subregion regulate the magnitude of the response to thyroglobulin. In Part II of this study we found that most mouse strains responded to heterologous (bovine) thyroglobulin. Priming *in vivo* with bovine thyroglobulin could generate cross-reactive T-cell proliferation *in vitro* to porcine thyroglobulin in only certain H-2 haplotypes. Primary stimulation with bovine thyroglobulin did not result in autoimmune T cell proliferation upon secondary challenge with mouse thyroglobulin.

MATERIALS AND METHODS

  Animals. The congenic and recombinant strains of mice used in this study were produced and maintained in the immunogenetic mouse colony at Mayo Clinic.

  The antigens. The murine thyroglobulin (MTg) was prepared as follows. Mouse thyroids were homogenized and then clarified by centrifugation at 69,000 g for 30 minutes. The resultant extracts had a protein concentration of 1.84 to 3.01 mg/ml. The saline extract was further purified by column chromatography on Sephadex G-200. The purity of MTg was verified by immunoelectrophoresis. Bovine thyroglobulin (BTg - Type 1) and porcine thyroglobulin (PTg - Type 2) were purchased from Sigma Chemicals, St. Louis, Missouri.

  Immunizations. Eighty μg of MTg or 100 μg of BTg mixed 1:1 in Complete Freund's Adjuvant containing 1 mg/ml of *Mycobacterium tuberculosis* strain H37Ra (Difco Labs., Detroit, Michigan) was injected subcutaneously at the base of the tail. Seven days after immunization, the animals were sacrificed and the inguinal, lumbar and caudal lymph nodes were collected and placed in cold Hanks' medium (Ca-free, Mg-free) and washed three times. After washing the pellet was resuspended in 10 ml of culture medium [RPMI 1640 (GIBCO) supplemented with penicillin and streptomycin, 100 units/ml and 100 μg/ml, respectively].

  Cell cultures. Two hundred μl of culture medium containing $4 \times 10^5$ viable lymph node cells was cultured in flat-bottom microculture plates with 50 μg final concentration of MTg or 100 μg of BTg or PTg in RPMI 1640 medium containing glutamine and 2.5% horse serum for 3 to 5 days. Each antigen was run in three replicate wells. Positive PPD controls (40 μg/ml) were added to three wells and three negative control wells without antigens were also included. The cells were pulsed with [³H] thymidine 16 hours before harvest. Thymidine uptake is expressed as

mean CPM $\pm$ standard error of the mean and as stimulation indices (SI).
Experiments with each strain were repeated three times. The experiment which gave the median SI is shown in the tables. The low responders never gave an index higher than 1.5. The high responders
showed some variation in the degree of response.

RESULTS

Dose response studies. Experiments were carried out to determine the
most optimal priming dose and the challenging dose. Doses of 20, 40,
80 and 120 µg of thyroglobulin were used for immunization. A dose
between 40 and 80 µg gave good proliferative responses. But even at
the optimum dose of MTg the stimulation indices are only a fraction of
what we get with PPD, suggesting a weak immunigenicity of MTg. In most
of the experiments in this study a standard dose of 50 µg of MTg was
injected. For the challenge, 100 µg was found to be the least amount
of MTg that would give a significant response and was employed throughout the studies.

Response to MTg-independent haplotypes. Response to MTg in mice of
independent H-2 haplotypes was studied by using mostly mice of a C57BL
background. The use of congenics would rule out any influence of background genes. As shown in Table 1, strains B10.RIII ($H\text{-}2^r$), B10.G
($H\text{-}2^q$), and B10.S ($H\text{-}2^s$) gave stimulation indices of 5.42, 4.04 and
3.01, respectively and were classified as high responders. Strain
B10.K ($H\text{-}2^k$) gave a stimulation index of 2.48, not as high as the other
high responders, but it was classified as a high responder strain since
other $H\text{-}2^k$ strains (C3H, CBA) gave good responses. Strains B10 ($H\text{-}2^b$),
B10.D2 ($H\text{-}2^d$), B10.M ($H\text{-}2^f$) and B10.WB ($H\text{-}2^j$) gave stimulation indices
between 0.30 and 1.11 and were all classified as low responders. To
compare the H-2 haplotypes in non-B10 background several other strains
were also tested. For example, A.BY ($H\text{-}2^b$), BALB/c ($H\text{-}2^d$), A.CA ($H\text{-}2^f$),
C3H ($H\text{-}2^k$), DBA/1 ($H\text{-}2^q$) and A.SW ($H\text{-}2^s$) were tested and found to respond similarly to the strain with the same H-2 haplotypes on a B10
background, even though the strength of the stimulation sometimes
varied. In conclusion, H-2 haplotypes $k$, $q$, $r$ and $s$ were high responders to mouse thyroglobulin, whereas $b$, $d$, $f$ and $j$ were low responders. The results with the T cell proliferation assay correspond
very closely with the results which have already been obtained on the
basis of antibody production and cellular infiltration.

TABLE 1

T-LYMPHOCYTE PROLIFERATIVE RESPONSE TO MOUSE THYROGLOBULIN IN INDEPEN-
DENT H-2 HAPLOTYPES

| Mouse strains | H-2 haplotype | Cpm (mean + SD) | | Stimulation index |
|---|---|---|---|---|
| | | Medium | Thyroglobulin | |
| B10 | b | 800+75 | 600+50 | 0.7 |
| B10.D2 | d | 3600+100 | 1100+200 | 0.3 |
| B10.M | f | 4000+650 | 4500+300 | 1.1 |
| B10.WB | j | 6500+600 | 5500+400 | 0.8 |
| B10.K | k | 1600+100 | 4000+50 | 2.5 |
| B10.G | q | 3600+400 | 14400+500 | 4.0 |
| B10.RIII | r | 4600+700 | 24800+2000 | 5.4 |
| B10.S | s | 1200+100 | 3700+300 | 3.0 |

Response to MTg in recombinant haplotypes. In an attempt to map the
genes controlling the T lymphocyte proliferative autoimmune response to
mouse thyroglobulin, several recombinant H-2 haplotypes were tested.
In Table 2, we show results of selected recombinants which are informa-
tive on the genetic mapping. Strains B6 and B6 (Tl+) gave similar results, suggesting no direct role for TL region in the immune response.
Recombinant strain B10.A(4R), a high responder, carries the $H$-$2^k$
alleles in the $K$ and $I$-$A$ regions and the low responder $H$-$2^b$ alleles in
the other regions of the H-2 gene complex. This finding suggests that
a gene controlling immune response to thyroglobulin maps in the $K$ end,
most probably in the $I$-$A$ subregion where most of the $Ir$ genes have been
mapped. This again confirms previous data obtained in studies of the
humoral response and cellular infiltration. B10.A(2R), which differs
from B10.A(4R) in regions $I$-$C$ and $S(G?)$, gave a stimulation index of
1.74, approximately 50% of the SI seen in B10.A(4R). This suggests that
genes mapping in either $I$-$C^d$ and/or $S^d$ $(G^d?)$ regions might have a negative effect on the response to MTg.

B10.S $(H$-$2^s)$ was shown before to be a high responder. Recombinant
B10.S(7R) which differs from B10.S only in the $H$-$2D$ and $TL$ regions
showed a lower stimulation index (1.73). This suggests that genes
either in $H$-$2D^d$ or $TL^a$ regions might also have a negative influence on
the response to thyroglobulin. Recombinant strain B10.S(9R) which
carries the high responder $H$-$2^s$ allele in the $K$ and the $I$-$A$ regions,
but the low responder $(H$-$2^d)$ alleles in the $I$-$C$, $S$, $G$ and $D$ regions,
gave a very low stimulation index (0.76). Recombinant B10.HTT which

carries $I-J^s$, $I-C^k$, $S^k$, $G^k$ and $TL^c$ gave an intermediate response. These results suggest that there may be two 'suppressive' genes mapping to the right of the $I-E$ subregion; B10.S(9R) has both of them while B10.S(7R) and B10.HTT have only the one associated with the $H-2D$ end.

TABLE 2

GENETIC MAPPING OF T-LYMPHOCYTE PROLIFERATIVE RESPONSE TO MOUSE THYRO-GLOBULIN

| Mouse strain | MHC alleles | | | | | | | | | | Stimulation index (range) |
|---|---|---|---|---|---|---|---|---|---|---|---|
| | K | A | B | J | E | C | S | G | D | TL | |
| B6 | b | b | b | b | b | b | b | b | b | b | 0.8 (0.5-1.0) |
| B6(T1+) | b | b | b | b | b | b | b | b | b | a | 0.9 (0.4-1.1) |
| B10.A(4R) | k | k | b | b | b | b | b | b | b | b | 3.9 (3.1-5.2) |
| B10.A(2R) | k | k | k | k | k | d | d | ? | b | b | 1.7 (1.3-2.1) |
| B10.S | s | s | s | s | s | s | s | s | s | b | 3.0 (2.4-3.7) |
| B10.S(7R) | s | s | s | s | s | s | s | s | d | a | 1.7 (1.6-1.9) |
| B10.S(9R) | s | s | ? | k | k | d | d | d | d | a | 0.8 (0.7-1.2) |
| B10.HTT | s | s | s | s | k | k | k | k | d | c | 1.5 (1.3-2.0) |
| B10.G | q | q | q | q | q | q | q | q | q | ? | 4.0 (3.2-5.6) |
| B10.T(6R) | q | q | q | q | q | q | q | ? | d | a | 1.5 (1.3-1.9) |
| B10.AQR | q | k | k | k | k | d | d | d | d | a | 1.0 (0.7-1.4) |
| B10.BDR | d | d | d | d | d | d | d | ? | b | b | 0.6 (0.4-0.8) |
| B10.A(5R) | b | b | b | k | k | d | d | d | d | a | 0.8 (0.7-1.3) |

_____ - High responder allele(s)

_ _ _ _ _ _ _ - 'Suppressive' allele(s)

Strain B10.G ($H-2^q$) as shown before is a high responder. Recombinants B10.T(6R) which differ from B10.G only in the $H-2D$ and the $TL$ regions gave an intermediate response. Recombinant B10.AQR, which expresses $K^q A^k B^k J^k E^k C^d S^d G^d D^d$, gave a much lower stimulation index. These results show that the 'suppressive' genes derived from $H-2^d$ haplotype is functional also in strains expressing $A^k$ and $A^q$ as well as $A^s$. Two recombinants, B10.BDR and B10.A(5R) which are reciprocal combinations of $H-2^d$ and $H-2^b$, are both low responders and gave low stimulation indices suggesting no complementation in this combination.

T lymphocyte proliferation to bovine thyroglobulin. Mice were primed with bovine thyroglobulin *in vivo* and challenged *in vitro* with bovine (BTg), porcine (PTg) and mouse thyroglobulin (MTg). In this experiment we were assaying a homologous response to BTg and cross reactive responses to PTg and MTg. The response to MTg was considered autoimmune.

The results with the independent haplotypes are shown in Table 3. The homologous response with BTg varied from a low of 4.10 to a high of 32. So, to evaluate the three different responses objectively we had to look within each haplotype. For example, $H-2^b$ haplotype gave similar response to BTg and PTg, but the response to MTg was very low. The same results were seen in B10.K ($H-2^k$) and B10.S ($H-2^s$). But with B10.D2 ($H-2^d$) the immune response to BTg is high, 9.60, but the cross-reactive responses to both PTg and MTg were very low. Lack of cross-reaction to PTg was also seen in B10.G ($H-2^q$) and B10.RIII ($H-2^r$). All these haplotypes gave a comparatively low response to MTg. Since we did not measure the cellular infiltration to the thyroid gland in these animals we cannot conclude whether immunization with BTg results in autoimmune thyroiditis. The cross-reactive response to PTg was high in $H-2^b$, $H-2^k$ and $H-2^s$ haplotypes and low in $H-2^d$, $H-2^f$, $H-2^q$ and $H-2^r$ haplotypes.

TABLE 3

T LYMPHOCYTE PROLIFERATION OF MICE PRIMED WITH BOVINE THYROGLOBULIN AND CHALLENGED WITH BOVINE, PORCINE AND MOUSE THYROGLOBULIN

| Mouse strains | H-2 haplotype | Stimulation indices | | |
|---|---|---|---|---|
| | | Bovine Tg | Porcine Tg | Mouse Tg |
| B10 | b | 6.5 | 7.3 | 1.5 |
| B10.D2 | d | 9.6 | 1.2 | 2.6 |
| B10.M | f | 4.0 | 2.5 | 2.6 |
| B10.K | k | 6.5 | 5.5 | 2.3 |
| B10.G | q | 9.7 | 3.3 | 2.1 |
| B10.RIII | r | 4.1 | 2.2 | 1.3 |
| B10.S | s | 32.0 | 26.0 | 3.8 |

To map the gene(s) involved in cross-reactive response, we tested re-combinant strains involving the $H-2^d$ haplotype. B10.BDR-2, which differs from the B10.D2 only in the $H-2D$ region, also gave low response to PTg. C3H.OL, which carried $K^d I^d S^k D^k$ gave a poor response to PTg while recombinant D2.GD which expresses $K^d$ and $I-A^d$, but $H-2^b$ alleles at the other regions gave a good response. This result suggests that gene(s) mapping in the $I-B^d$, $I-J^d$, $I-E^d$ and/or $I-C^d$ have a suppressive effect on cross-reactive immune response to PTg. Since DBA/2 ($H-2^d$) also gave a low response, this effect is probably not due to a non-H-2-linked background gene. Recombinants B10.A and B10.S(9R) gave good responses suggesting that either the $I\text{-}C^d$ allele is not involved or they are unable to suppress responses associated with $H\text{-}2^k$ and $H\text{-}2^s$. This possibly could be another example of coupled complementation or inter-action, proposed by Dorf and Benacerraf.[7]

TABLE 4

GENETIC MAPPING OF CROSS REACTIVE T-LYMPHOCYTE PROLIFERATION TO PORCINE THYROGLOBULIN

| Mouse strain | MHC alleles K A B J E C S G D TL | Stimulation index | |
| --- | --- | --- | --- |
| | | Bovine thyroglobulin | Porcine thyroglobulin |
| B10.D2 | d d d d d d d d d c | 9.6 | 1.2 |
| B10.BDR-2 | d d d d d d d ? b b | 12.1 | 1.9 |
| D2.GD | d d b b b b b b b b | 10.3 | 12.7 |
| C3H.OL | d d d d d d k k k b | 17.6 | 1.8 |
| B10.A | k k k k k d d d d a | 4.7 | 4.4 |
| B10.S(9R) | s s ? k k d d d d a | 16.4 | 12.7 |

DISCUSSION

   In these studies we have confirmed by the T cell proliferation assay previous results obtained by antibody production and cellular infiltra-tion on the genetic control of autoimmune response to thyroglobulin in mice. As shown previously there is good evidence that an $Ir$ gene con-trols response to thyroglobulin and maps in the $I\text{-}A$ (or $K$) subregion. In the studies reported here we have identified other gene(s) which regulate the magnitude of immune response to thyroglobulin. This again confirms studies on humoral response and cellular infiltration (Kong et al., this volume). These genes which lower the response to Tg could be inducing suppression, but so far we have not done studies to show presence of suppressor cells. Results in this study suggest at least two genes involved in inducing 'suppression', one presumably mapping in the $I\text{-}C$ subregion, and the other in the vicinity of the $H\text{-}2D$ region.

   Studies by Rich et al.[8] have shown that the $I\text{-}C$ subregion controls alloantigen mediated MLR suppressive factors. Presumably, the same gene could be involved in inducing suppression to determinants on thyroglobulin molecule, since the thyroglobulin used in these studies are not from a particular haplotype. Several others have shown that gene(s) mapping in the vicinity of the $H\text{-}2D$ region influences resis-tance or suppression to certain tumors.[9,10] These genes presumably suppress response to self-antigens, and could regulate the response to thyroglobulin which should be classified as a self-antigen. When both the $I\text{-}C^d$ and $H\text{-}2D^d$ alleles are expressed there is a cumulative effect of suppression suggesting that these two genes produce an additive effect.

In the cross-reactive immune response to mice primed with bovine thyroglobulin and tested against porcine thyroglobulin, gene(s) to the right of the $I\text{-}A$ subregion and left of $S$ region derived from $H\text{-}2^d$ haplotype also induces a negative effect. Presumably, this could be the same gene inducing the autoimmune suppression.

The T cell proliferation assay adapted in this study enables one to do several manipulative cellular assays which are difficult to do *in vivo*. For example, we plan to treat the responding cells with antisera against specific subregions of the H-2 complex, for example, anti-K, anti-A, anti-J, anti-C, anti-E, anti-D, etc., with complement and see whether the suppressive effect can be altered. Such studies are underway.

SUMMARY

Genetic control of autoimmune response to mouse thyroglobulin and cross reactive immune response to porcine thyroglobulin and mouse thyroglobulin was studied by an *in vitro* T cell proliferation assay. Results suggest that:

A. T-lymphocyte autoimmune proliferation to thyroglobulin in mice is under $H\text{-}2$-linked *Ir* gene control.
   (1) An *Ir* gene mapping in $I\text{-}A$ (or $H\text{-}2K$) subregion controls autoimmune response to thyroglobulin.
   (2) Two genes with regulatory effect (*Is* genes?) map between $I\text{-}E$ subregion and $TL$ region ($I\text{-}C$ and $H\text{-}2D$).

B. Cross-reactive immune response to porcine thyroglobulin in mice primed with bovine thyroglobulin is also mediated by genes mapping within the H-2 gene complex with either responsive ($I\text{-}A$) or suppressive ($I\text{-}B$, $I\text{-}J$, $I\text{-}E$ or $I\text{-}C$) effects.

ACKNOWLEDGEMENTS

The authors thank Drs. Kenji Okuda and Christopher J. Krco for valuable advice, Suresh Savarirayan for superb breeding and husbandry of our mouse colony and Sharon Ames for skillful preparation of the manuscript. This study was supported by grants from the National Institute of Arthritis, Metabolism and Digestive Diseases (AM 20023), National Cancer Institute (CA 24473, CA 18900), the National Institute of Allergy and Infectious Diseases (AI 14764) and the Mayo Foundation.

REFERENCES

1.  Rose, N.R., Twarog, F.J. and Crowle, A.J. (1971) J. Immunol., 106, 698-704.

2.  Twarog, F.J. and Rose, N.R. (1968) J. Immunol., 101, 242-250.

3.  Vladutiu, A.O. and Rose, N.R. (1971) Science, 174, 1137-1138.

4.  Tomazic, V., Rose, N.R. and Shreffler, D.C. (1974) J. Immunol., 112, 965-969.

5.  Vladutiu, A.O. and Rose, N.R. (1975) Cell. Immunol., 17, 106-113.

6.  Alkan, S.S. (1978) Eur. J. Immunol., 8, 112-118.

7.  Dorf, M.E. and Benacerraf, B. (1977) in Ir Genes and Ia Antigens, McDevitt, H.O. ed., Academic Press, New York.

8.  Rich, S.S., David, C.S. and Rich, R.R. (1978) J. Exp. Med., (in press).

9.  Cheseboro, B., Wehrty, K. and Stimpfling, J. (1974) J. Exp. Med., 140, 1457-1461.

10. Meruelo, D., Lieberman, M., Ginzton, N., Deak, B. and McDevitt, H.O. (1977) J. Exp. Med., 146, 1079-1087.

DISCUSSION

WARNER:  Could you give the audience some feeling for the repro-
ducibility of the rather modest differences that are being interpreted
as, for example, one versus two negative effect genes.  Stimulation
indices of 1.5 versus 30, for example, may be subject to many variables
even within a strain, depending upon control values which in turn can
be affected by environmental flora changes and other unknown factors.

CHRISTADOSS:  I have tested all the strains more than four times.
The S.I. were always consistant for a given strain.  For example,
BIO.A(4R) gave a S.I. of 3.93 while BIO.A(2R) gave a S.I. of 1.74,
the latter having the suppressive gene in $c^d s^d$ allele.

|  | 1 exp | 2 exp | 3 exp | 4 exp |
|---|---|---|---|---|
| B10.A(4R) | 4.3 | 3.8 | 3.93 | 4.9 |
| B10.A(2R) | 1.8 | 1.5 | 1.74 | 1.43 |

Similarly, B10.S gave a S.I. of 3.01 and B10.S(7R) gave a S.I. of 1.73,
the latter having the low responder d allele at the D region.

|  |  |  |  |  |
|---|---|---|---|---|
| B10.S | 4.51 | 3.92 | 3.01 | 3.5 |
| B10.S(7R) | 1.22 | 1.52 | 1.73 | 1.53 |
| B10.S(9R) | .82 | .74 | 0.76 | .92 |

B10.S(9R) has the low responder d allele from C to D.
I took all the results obtained from the third experiment.  Since
these results correlate with the results obtained by Dr. Kong in the
antibody production and thyroid infiltration, we concluded that there
is (are) suppressive gene(s) to the right of the IC subregion (presumably one in the IC and the other in the vicinity of the D region).

MORSE:  Have $F_1$ and backcross mice been studied to determine if
these differences segregate in association with the H-2 complex?

CHRISTADOSS:  These studies are in progress.

MULLEN:  Dr. Kong found D end suppression in mice immunized with
MTg + LPS.  Did you use MTg + CFA for immunization of your mice where
you found D end suppression in the T cell proliferative assay?

CHRISTADOSS:  I used MTG + CFA for immunization.  Even Dr. Kong has
found D end suppression by immunizing MTG + CFA.

AUTOIMMUNE RESPONSES TO SPERMATOZOA IN VASECTOMIZED RATS AND MICE
OF DIFFERENT INBRED STRAINS

Pierluigi E. Bigazzi, M.D., Department of Pathology, University of
Connecticut Health Center, Farmington, Conn., 06032

Recent studies of the genetic control of organ-specific autoimmune
responses have given contrasting results. Vladutiu and Rose[1] initially showed that there was a striking correlation between response
to thyroid antigens and H-2 type, with mice of $H-2^k$, $H-2^s$ and $H-2^q$
genotypes behaving as good responders to thyroglobulin and mice of
$H-2^b$ and $H-2^d$ genotypes behaving as poor responders. Later studies
from the same group demonstrated that good and poor responder strains
differed more in the intensity of pathological changes in the thyroid
than in the level of antibodies to thyroglobulin[2]. Similar investigations in rats with experimentally induced thyroiditis showed
that immune responsiveness in terms of antibody formation to thyroglobulin was not linked to AgB (H-1) antigens, whereas Ag-B4 and
Ag-B2 might be linked to the allele favoring susceptibility to
experimentally induced thyroid damage[3]. Susceptibility to experimental allergic encephalomyelitis in rats has been found to be
closely linked to the Ag-B (H-1) histocompatibility locus, but not
identical to it[4,5]. Finally, an increased susceptibility to Heymann
nephritis is linked to the Ag-B1 locus[6], while the production of
antibodies to the glomerular basement membrane after administration
of mercuric chloride is under the control of two or three genes,
one of which is linked to the Ag-B (H-1) histocompatibility locus[7].

In our studies, we have used as an experimental model vasectomized
rats and mice. It is well known that vasectomy is followed by an
autoimmune response to antigens of spermatozoa and, in rabbits, by
an immune complex-mediated membranous orchitis[8]. Thus, we have

investigated differences in immune responses to spermatozoa observed after vasectomy in different inbred strains of rats and mice.

A total of 500 rats from 11 strain have been vasectomized and 250 rats have been sham-vasectomized as controls. Bilateral vasectomies were performed by sterile techniques and avoiding trauma to the testes and their blood vessels. Each vas deferens was isolated, ligated and a segment approximately 1 cm long was cut out. Sham vasectomies were performed in a similar fashion, with the exception that the vas deferens was not cut out. Circulating antibodies to spermatozoa were detected by indirect immunofluorescence on rat sperm smears. For further details on the techniques used we refer to our previous publications[8-10].

As shown in Table I, vasectomized rats of the Lewis and Brown Norway strains had a high incidence of antibodies to spermatozoa, with mean titers that were also high. Vasectomized rats of the

TABLE I

Production of antibodies to spermatozoa in vasectomized rats of different strains.

| STRAIN | HISTOCOMPATIBILITY | | CIRCULATING ANTIBODIES TO SPERMATOZOA (%) | TITERS |
|--------|-------|-----|------|--------|
| | Ag-B | H-1 | | |
| Lewis | 1 | 1 | 80 | $137 \pm 39$ |
| Brown-Norway | 3 | n | 47 | $48 \pm 30$ |
| Buffalo | 6 | b | 13 | 10 or less |
| Wistar-Furth | 2 | w | 12 | 10 or less |
| ACI | 4 | a | 11 | 10 or less |
| Fischer | 1 | 1 | 0 | 0 |
| Dark Agouti | 4 | a | 0 | 0 |
| August 28807 | 5 | c | 0 | 0 |
| BDII | 2 | w | 0 | 0 |
| BDIX | - | d | 0 | 0 |
| Sprague-Dawley | 1&2 | 1&w | 0 | 0 |

Buffalo, Wistar-Furth and ACI strains had a low incidence of antibodies to spermatozoa and titers were also low. Finally, vasectomized rats of the Fischer, Dark Agouti, August, BDII, BDIX and Sprague-Dawley strains did not have circulating antibodies to spermatozoa. Sera from sham-vasectomized and untreated rats, as well as
sera from prevasectomy bleedings obtained from Lewis, ACI, Buffalo
and Wistar-Furth rats did not contain antibodies to rat spermatozoa.
The only exception was the prevasectomy serum from one BN rat, that
gave a positive reaction with a titer of 80. The serum of this
animal was found to be positive until 7 months after vasectomy, with
titers lower than in the prevasectomy serum.

The time course of the production of antibodies to spermatozoa is
shown in Figure 1. The maximum incidence was reached at 3 months
after vasectomy in Lewis, Brown Norway and ACI and at 4 months in
Buffalo and Wistar-Furth.

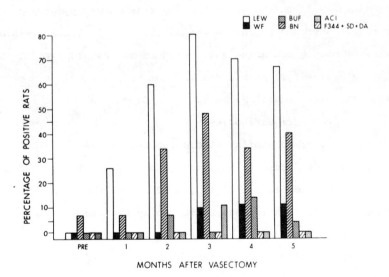

Fig. 1. Incidence and time sequence of sperm antibody response in
bilaterally vasectomized rats of different strains.
Abbreviation: LEW, Lewis; WF, Wistar-Furth; BUF, Buffalo;
BN, Brown Norway; F344, Fisher; SD, Sprague-Dawley; DA, Dark Agouti.
(Reproduced from Bigazzi et al.[9], with permission of Science).

Positive reactions obtained with sera from Lewis, Brown Norway,
ACI, Buffalo and Wistar-Furth sera were localized in the acrosomal
region of rat spermatozoa. No other major patterns of staining were
observed. Positive reactions were noted using sperm smears from
Lewis, Brown Norway, ACI, Buffalo and Wistar-Furth as well as Fischer,
Dark Agouti and Sprague-Dawley rats. This indicates that the acrosomal antigens involved in this immune response are present even in
the spermatozoa of those rat strains like the Fischer, Dark Agouti
and Sprague-Dawley that did not produce antibodies to spermatozoa.
Thus, the lack of an immune response was not due to lack of sperm
antigens, but to other reasons.

The antibodies to spermatozoa in the sera of positive animals were
capable of reacting with spermatozoa from animals of the same inbred
strain and also with spermatozoa from the positive animals themselves; thus, they can be defined as autoantibodies.

Preliminary investigations on Fl hybrids between good and poor
responders have shown that approximately 11% of (Lewis x Wistar-
Furth) Fl have circulating antibodies to spermatozoa, while no
(Lewis x Fischer) Fl show a similar immune response. Approximately
15% of (August x Lewis) Fl and 22% of (Lewis x August) Fl have been
found to have sperm antibodies.

We have performed similar studies on a total of 200 vasectomized
mice of 11 inbred strains and, as shown in Table II, have found
that vasectomized mice of C57L/J, BALB/cJ and DBA/lJ had a high
incidence of circulating antibodies to spermatozoa, while vasectomized mice from 5 other strains had a much lower incidence and
animals from 3 other strains had no sperm antibodies. Positive
results by indirect immunofluorescence were observed with sera
obtained 7 and 10 months after vasectomy and none of the sera obtained from the same animals prior to vasectomy or from sham-vasectomized control mice contained similar antibodies when tested at the

TABLE II

Production of antibodies to spermatozoa in vasectomized mice of different strains.

| STRAIN | H-2 | CIRCULATING ANTIBODIES TO SPERMATOZOA (%) |
|--------|-----|-------------------------------------------|
| C57L/J | b | 78 |
| BALB/cJ | d | 75 |
| DBA/1J | q | 67 |
| C57Bl/6J | b | 30 |
| DBA/2J | d | 30 |
| SWR/J | q | 10 |
| C57Bl/KsJ | d | 10 |
| CBA/J | k | 10 |
| C57Bl/10J | b | 0 |
| A/WySn | a | 0 |
| BUB/BnJ | q | 0 |

same dilutions. Positive reactions were localized in the acrosomal region of mouse spermatozoa and no other major patterns of staining were observed.

Positive reactions were obtained using sera from DBA/1J, C57L/J and C57Bl/6 strains on mouse sperm smears from the same strains and also on smears from C57Bl/10J, A/WySn, BUB/BnJ, BRVR and outbred Swiss White mice. These findings show that the antigens involved are present even in the spermatozoa of those strains that do not produce antibodies to acrosomal antigens. Since the antibodies to spermatozoa observed in the serum of positive mice were able to react with spermatozoa from the positive animals themselves, they can be defined as autoantibodies.

In conclusion, we have demonstrated that different strains of rats and mice differ in their response to sperm antigens following vasectomy. A large percentage of vasectomized Lewis and Brown Norway rats produce antibodies to acrosomal antigens of spermatozoa, as compared to very small numbers of rats from other inbred strains. As a matter of fact, vasectomized rats from some strains such as the Fischer, Dark Agouti, August, BDII and BDIX were consistently negative for sperm antibodies, even though their spermatozoa possess acrosomal antigens. Thus our studies show that Lewis and BN behave as high responders to sperm antigens, whereas the other strains may be considered poor responders. These differences are not easily explained on the basis of a simple correlation between major histocompatibility loci and immune response genes controlling autoimmune responses to sperm antigens. Rats that may be considered good responders such as the Lewis and the Brown Norway belong to different genotypes, Ag-B1 and Ag-B3. On the other hand, two strains of rats belonging to the same Ag-B1 genotype differ in their response, i.e. Lewis rats produce sperm antibodies and Fischer do not.

Similarly, mice of C57L/J, BALB/cJ and DBA/1J behave as good responders to sperm antigens, while mice of other strains behave as mediocre or poor responders. Again, these differences in autoimmune responses do not seem to be linked to a major histocompatibility locus, since mice of the $H-2^b$ genotype may be good, mediocre or poor responders and the same situation occurs with mice of $H-2^d$ and $H-2^q$ genotypes.

Thus, the locus controlling the production of autoantibodies to spermatozoa in vasectomized rats and mice may not be contained within the Ag-B (H-1) and H-2 complex. Immune response genes not linked to H-2 have been previously reported[11] and the production of thymocytotoxic autoantibodies and autoantibodies to single stranded DNA does not seem to be linked to the H-2 complex [12,13]. Similarly, formation of autoantibodies to thyroglobulin in rats is not linked to the major histocompatibility complex[3].

Previous studies on the genetic control of experimentally induced organ-specific autoimmune reactions have relied on the injection of heterologous antigens or autoantigens in adjuvants. On the other hand, our observations of differences in the immune response of different rat and mouse strains to autoantigens of spermatozoa are based on a simple surgical procedure (vasectomy) and do not involve the administration of adjuvants. This may prove a definite advantage for further studies of the role of immune response genes in auto-immune reactions.

ACKNOWLEDGEMENT

This research was supported by Contract 72-775 from the National Institute of Child Health and Human Development of the National Institutes of Health.

REFERENCES

1. Vladutiu A.O. and Rose N.R. (1971) Science 174:1137-1139.

2. Tomazic V. and Rose N.R. (1977) Eur.J.Immunol.7:40-43.

3. Rose,N.R. (1975) Cell. Immunol. 18: 360-364.

4. Gasser,D.L. et al. (1973) Science 181:872-873.

5. Williams,R.M. and Moore,M.J. (1973) J. Exper. Med. 138: 775-783.

6. Stenglein,B. et al.(1975) J. Immunol. 115: 895-897.

7. Druet,E. et al. (1977) Eur. J. Immunol. 7: 348-351.

8. Bigazzi,P.E. et al. (1976) J. Exper. Med. 143: 382-404.

9. Bigazzi,P.E. et al. (1977) Science 197: 1282-1283.

10. Kosuda, L.L., Bigazzi, P.E. (1978) Invest. Urol. (in press).

11. Gasser, D.L. (1969) J. Immunol. 103: 66-70.

12. Steinberg, A.D. et al. (1971) Immunology 20: 523-531.

13. Raveche, E.S. et al. (1978) J. Exper. Med. 147: 1487-1502.

DISCUSSION

LENNON:  If the autoantibody specificities are confined to products of mature sperm, this perhaps may not be a cause for concern in vasectomized humans.  Is there any evidence in vasectomized animals for widespread testicular damage (e.g., to the Sertoli cells) or for immune complex damage to tissues outside the testes?

BIGAZZI:  At present, testicular damage has been observed in vasectomized rabbits.  Studies on the incidence of orchitis in rats and mice with circulating antibodies to spermatozoa are still in progress.

WARNER:  I would just raise the point for the record that your data with mice do not necessarily preclude the role of MHC linked genes.  Formal studies would require the use of congenic strains, since it is possible that other background genes could be obscuring the detection of MHC linked genes.

BIGAZZI:  This is correct.  Such studies are now in progress and that is actually the reason why we extended our investigations from rats to mice, since congenic strains are more easily available in the latter species.

KONG:  In vasectomized men, there is usually more than one type of antibody or antibodies to different antigens.  Which antibody are you measuring?  And if you look at kinetics of the antibody response, would you see differences in the different strains?

BIGAZZI:  As far as classes of antibodies are concerned, we have shown that sperm antibodies in rabbits may belong to IgG, IgA, and IgM classes, but with a higher incidence of IgG antibodies.  The kinetics of antibody response have also been investigated in rats, and we have observed that "good responders" such as Lewis and BN reach a maximum incidence of antibody production three months after vasectomy.